Primate Communication
A Multimodal Approach

Primates communicate with each other using a wide range of signals, such as olfactory signals to mark territories, screams to recruit help while fighting, gestures to request food and facial expressions to initiate play.

Primate Communication brings together research on different forms of interaction and discusses what we know about primate communication via vocal, gestural, facial, olfactory and integrated multimodal signals in relation to a number of central topics. It explores the morphological, neural and cognitive foundations of primate communication through discussion of cutting-edge research.

By considering signals from multiple modalities and taking a unified multimodal approach, the authors offer a uniquely holistic overview of primate communication, discussing what we know, what we don't know and what we may currently misunderstand about communication across these different forms. It is essential reading for researchers interested in primate behaviour, communication and cognition, as well as students of primatology, psychology, anthropology and cognitive sciences.

Katja Liebal is an Assistant Professor for Evolutionary Psychology at the Freie Universität Berlin. Current research interests include the gestural communication of primates, factors that motivate prosocial behaviour in apes and socialization practices in apes and humans of different cultural backgrounds. She co-edited the books *Gestural Communication in Nonhuman and Human Primates* (2007) and *Current Developments in Gesture Research* (2012).

Bridget M. Waller is a Reader in Evolutionary Psychology in the Department of Psychology at the University of Portsmouth, and is the Director of the Centre for Comparative and Evolutionary Psychology. She is interested in human and nonhuman primate facial expression and emotion, and how they contribute to sociality and social bonding. She has developed the Facial Action Coding Schemes for several primate species and established the Macaque Study Centre at Marwell Zoo.

Anne M. Burrows is an Associate Professor at the Duquesne University in Pittsburgh. Her research centres on the evolution of mammalian facial expression musculature and primate sensory ecology, especially olfactory perception. She significantly contributed to the development of the Facial Action Coding Schemes for several primate species by conducting the corresponding anatomical studies of their facial musculature and is the co-editor of *The Evolution of Exudativory in Primates*.

Katie E. Slocombe is a Senior Lecturer at the Psychology Department at the University of York and the Scientific Director of the Budongo Trail Chimpanzee Exhibit at Edinburgh Zoo. Her research interests centre around a comparative approach to language evolution. Current research focuses on chimpanzee vocal communication, the extent they can use calls to refer to objects and events in the external environment and the psychological mechanisms underlying their call production.

Primate Communication

A Multimodal Approach

KATJA LIEBAL
Freie Universität Berlin, Germany

BRIDGET M. WALLER
University of Portsmouth, UK

ANNE M. BURROWS
Duquesne University, Pittsburgh, USA

KATIE E. SLOCOMBE
University of York, UK

CAMBRIDGE
UNIVERSITY PRESS

University Printing House, Cambridge CB2 8BS, United Kingdom

Published in the United States of America by Cambridge University Press, New York

Cambridge University Press is part of the University of Cambridge.

It furthers the University's mission by disseminating knowledge in the pursuit of
education, learning and research at the highest international levels of excellence.

www.cambridge.org
Information on this title: www.cambridge.org/9780521195041

First published 2014

Printing in the United Kingdom by TJ International Ltd. Padstow Cornwall

A catalogue record for this publication is available from the British Library

Library of Congress Cataloguing in Publication data
Liebal, Katja.
Primate communication: A multimodal approach / Katja Liebal, Bridget M. Waller, Anne M. Burrows,
Katie E. Slocombe.
    pages   cm
Includes bibliographical references and index.
ISBN 978-0-521-19504-1 (Hardback) – ISBN 978-0-521-17835-8 (Paperback)
1. Primates–Behavior.   2. Animal communication.   I. Title.
QL737.P9L56 2013
599.815–dc23   2013014372

ISBN 978-0-521-19504-1 Hardback
ISBN 978-0-521-17835-8 Paperback

Additional resources for this publication at www.cambridge.org/9780521195041

Contents

Preface   *page* ix
Acknowledgements   xiii

Part I Introduction to primate communication

1   What is primate communication?   3

   1.1   What is communication?   3
          1.1.1   Understanding communication at different levels   6
   1.2   What are primates?   8
          1.2.1   Primate phylogeny   9
          1.2.2   Primate social systems   13
          1.2.3   Evolution of primate social systems   18
   1.3   What do primates need communication for?   21
          1.3.1   Communication across different primate social systems   24

2   The morphology of primate communication   31

   2.1   Olfactory communication   31
          2.1.1   The perception of olfactory signals   34
          2.1.2   The production of olfactory signals   35
   2.2   Auditory communication   36
          2.2.1   The production of vocalizations   37
          2.2.2   The perception of vocalizations   40
   2.3   Visual communication   40
          2.3.1   The perception of visual signals   43

3   The neural substrates of primate communication   46

   3.1   Anatomy of the primate brain   46
   3.2   Evolution of the primate brain   48
   3.3   The processing of communicative signals   50
          3.3.1   Olfactory communication: perception   50
          3.3.2   Olfactory communication: production   52

	3.3.3	Visual communication	52
	3.3.4	Vocal communication: perception	61
	3.3.5	Vocal communication: production	63
3.4	Multimodal integration on a neural level		66
	3.4.1	Perception: integration of multiple sensory modalities	66
	3.4.2	Production of multiple communicative signals	68

Part II Approaches to primate communication

4   The methods used in primate communication   73

4.1	Olfactory communication		73
	4.1.1	When do primates produce olfactory signals?	74
	4.1.2	Structure and form of olfactory signals	75
	4.1.3	What mechanisms underlie the production of olfactory signals?	76
	4.1.4	What does a sniffer understand from olfactory signals?	77
	4.1.5	Brain mechanisms underlying olfactory signal production and perception	77
4.2	Gestural communication		77
	4.2.1	When do primates produce gestures?	79
	4.2.2	Structure and form of gestures	81
	4.2.3	What mechanisms underlie the production of gestures?	82
	4.2.4	What does a receiver understand from gestures?	83
	4.2.5	Brain mechanisms underlying gesture production and perception	84
4.3	Facial expression		85
	4.3.1	When do primates produce facial expressions?	86
	4.3.2	Structure and form of facial expressions	87
	4.3.3	What mechanisms underlie the production of facial expressions?	89
	4.3.4	What does a receiver understand from facial expressions?	89
	4.3.5	Brain mechanisms underlying facial expression production and perception	91
4.4	Vocal communication		92
	4.4.1	When do primates produce calls?	93
	4.4.2	Structure and form of vocalizations	94
	4.4.3	What mechanisms underlie call production?	97
	4.4.4	What does a listener understand from the calls?	97
	4.4.5	Brain mechanisms involved in vocal production and perception	99
4.5	General data collection and coding methods		100
4.6	Pseudoreplication		101

5   A multimodal approach to primate communication   104

5.1	What kind of research has been done on primate communication?	104
5.2	What are the findings of the systematic review?	105

		5.2.1	Why do vocal, gestural and facial research differ systematically in their approaches?	107
		5.2.2	What are the consequences of differences in approaches to vocal, gestural and facial work?	108
		5.2.3	How does evidence from primate communication inform our understanding of language evolution?	109
	5.3	The way forward		112
		5.3.1	What is multimodal communication?	115
		5.3.2	Why do animals produce multimodal signals?	115
		5.3.3	How can multimodal signals be classified?	117
		5.3.4	Multimodal signals in non-primate species	118
		5.3.5	Multimodal signals in primates	121
	5.4	Looking to the future		126

Part III Cognitive characteristics of primate communication

6	**Acquisition**			**131**
	6.1	Ontogeny of communicative signals		132
		6.1.1	When does communication emerge?	132
		6.1.2	What can affect the emergence of communication in an individual?	133
		6.1.3	What does change in relation to life stage mean?	136
	6.2	Mechanisms underlying ontogeny of communicative signals		139
		6.2.1	Ontogenetic ritualization	139
		6.2.2	Imitation	140
		6.2.3	Genetic transmission	142
		6.2.4	Evidence for the three mechanisms	142
	6.3	Can primates generate and modify communicative signals as adults?		143
		6.3.1	Generation of novel signals	144
		6.3.2	Modification of signals in response to the social environment	146
		6.3.3	Modification of signals in response to the social context	147
		6.3.4	Modification of signals in response to the group	149
		6.3.5	Modification of signals in response to physical factors	151
	6.4	What could a multimodal approach bring to this topic?		152

7	**Flexibility**			**154**
	7.1	Why is it interesting to investigate flexibility?		154
	7.2	What do we mean by flexibility?		155
		7.2.1	Flexibility in usage	155
		7.2.2	Flexibility in the receiver	160
		7.2.3	Flexibility in combining signals into sequences	162
	7.3	What could a multimodal approach bring to this topic?		167

8   Intentionality   169

         8.1   What is intentional communication?   169
         8.2   Why is the intentional nature of communicative acts of interest? 170
         8.3   How can intentional communication be identified?   172
               8.3.1   Social use   173
               8.3.2   Visual-orienting behaviour and gaze alternation   177
               8.3.3   Influence of the recipient's attentional state   179
               8.3.4   Attention-getting behaviours   180
               8.3.5   Persistence and elaboration   183
               8.3.6   Response-waiting   187
               8.3.7   Flexibility   187
         8.4   How have intentionality criteria been applied in empirical studies? 188
         8.5   Validity of criteria to measure intentionality   188
         8.6   What could a multimodal approach bring to this topic?   192

9   Referentiality   194

         9.1   What is reference?   194
         9.2   Why is reference interesting to investigate in primates?   195
         9.3   Gestural reference: pointing   195
               9.3.1   Production of pointing gestures   195
               9.3.2   Comprehension of human pointing gestures   198
         9.4   Functionally referential signals   199
               9.4.1   Production criteria   200
               9.4.2   Perception criteria   200
               9.4.3   Predator avoidance contexts   202
               9.4.4   Food discovery contexts   206
               9.4.5   Social contexts   209
         9.5   What could a multimodal approach bring to this topic?   212

Part IV Approaches to the evolution of primate communication

10   A multimodal approach to the evolution of primate communication   217

         10.1   Why adopt a multimodal approach?   217
         10.2   What is the difference between phylogenetic and functional questions? 218
         10.3   How will a multimodal approach help us understand the evolution of
                communication?   223

         Glossary   230
         References   238
         Species index   281
         Subject index   284

Preface

Why this book? An anecdote

In 2006, some of the authors attended a conference on the Evolution of Language. There were a considerable number of talks that reported results from comparative research on nonhuman primates and how this might inform research into the origins of human language. However, although each of us was working on primate communication, we rarely met at this conference. Each of us attended the sessions according to her (modality specific) expertise, while largely ignoring presentations related to other modalities. Since, it has become more and more apparent to us that isolating communicative modalities is a common practice across the field of primate communication, regardless of whether the overarching research question is the same.

This conference gave the crucial impetus for this book. Writing it was a fascinating, though challenging process: each of us with a background of research in one particular modality, had her own perspective on primate communication, and a corresponding bias regarding the other modalities. After writing this book, this has fundamentally changed. Of course, each of us is still an expert in 'her' modality and there are still many open questions and some debates remain unsolved. Writing this book has, however, changed our perspectives not only on other modalities, but also raised our awareness in regard to open questions or pitfalls in our own respective areas of expertise.

Although this book advocates the necessity of a multimodal approach, realizing and applying it to the investigation of primate communication very much depends on the corresponding research questions. With this book, we hope to spark the reader's interest in the importance of a multimodal perspective and fuel discussions about advantages and limitations of a multimodal approach to primate communication. Taking this new perspective on primate communication and making the effort to work on shared definitions and methods will enable researchers to ask new and fascinating questions. Answering them will provide new insights and perspectives on primate communication.

What can readers expect from this book?

This book is aimed at an audience that includes researchers in primate communication and related fields of research, as well as students of primatology, psychology, biology, ethology, anthropology and cognitive sciences, but also general readers, who are interested in:

- an integrated multimodal approach to important cognitive aspects of communication
- different communicative modalities, including olfaction, vocalizations, facial expressions and gestures
- an overview of cutting-edge research and the current state of the art in primate communication
- a critical discussion of current problems, pitfalls and misunderstandings in primate communication
- the biological foundations and cognitive mechanisms underlying primate communication
- the causes and consequences of studying modalities separately
- current approaches and the most common methods used in research into primate communication
- the relevance of comparative research on primate communication for understanding the evolution of human language.

The book consists of four major parts, comprising an introduction to communication and the features that characterize primates (Part I), methods and approaches used to study primate communication (Part II), cognitive characteristics underlying primate communication (Part III), and approaches to its evolution (Part IV). These parts and the corresponding chapters differ in regard to several dimensions:

First, while some chapters of the book (Chapters 1–3) focus on the morphology and physiology of primate communication, others (Chapters 6–9) specifically address the psychological characteristics and the cognitive skills underlying primate communication.

Second, Chapters 1–4 will mostly use a unimodal perspective to introduce definitions and methods separately for each modality, while the remaining book pursues a multimodal approach to primate communication by discussing evidence from different modalities in an integrated way.

Third, Chapters 1–4 put the emphasis on the biological foundations of primate communication and include olfaction as a separate modality, while the remaining chapters mostly focus on facial expressions, gestures and vocalizations. This is not to neglect the importance of olfaction, but to emphasize that virtually nothing is known in regard to the psychological mechanisms underlying olfactory communication.

Chapter 1 introduces important terms, definitions and approaches to primate communication. It gives an overview of the diversity of primate species together with the features characterizing them and describes our approach to communication with special reference to its importance for maintaining complex social groups. **Chapter 2** explains the morphology of those structures involved in the production and perception of the different modalities, with special reference to their evolutionary pathways, while **Chapter 3** provides an overview of the neural substrates involved in the production and perception of olfactory signals, facial expressions, vocalizations and gestures separately. **Chapter 4** summarizes the variety of methods used to investigate the production and comprehension of the different modalities as well as their structural and functional properties. **Chapter 5** is central to the purpose of this book since it provides an overview of the current state of the art of research into primate communication and highlights the importance but apparent absence of a multimodal approach, particularly in regard to understanding the evolution of human language. **Chapter 6** is dedicated to the developmental aspects of primate communication and addresses the issue of how primates acquire their communicative repertoire in regard to the underlying mechanisms. **Chapter 7** concerns the flexibility of primate communication in regard to both the production and comprehension of signals, with special attention to their use across different functional contexts and their combination into signal sequences. **Chapter 8** represents a critical evaluation of the criteria used to identify intentional acts of communication, summarizes the empirical evidence for intentional signalling and suggests a set of criteria that are applicable across the different modalities. **Chapter 9** summarizes evidence for the referential use of primate signals. Finally, **Chapter 10** elaborates on different approaches to the evolution of primate communication by highlighting the importance of a multimodal approach to fully understand the complexity of primate communication and its implications for the evolution of human language.

What is not covered in this book?

To avoid any misplaced expectations, this section will summarize those aspects that will not be part of this book.

First, despite also being primates, humans only play a minor role in this book. Covering the full extent of human communication is beyond its scope. However, we often use human communication as the point of reference, as the focus of research is often explicitly comparative between humans and other primates. For example, we use this perspective when describing potential precursors to human language in other primates and reviewing the research investigating the language skills of nonhuman great apes. The focus of this book, however, is on nonhuman primates and the fascinating variety and skills that characterize their communication.

Second, despite some chapters referring to possible scenarios of the evolution of human language and evaluating potential problems related to these current approaches, this book does not aim to offer a new alternative theory of language evolution. Instead, the focus is on evaluating the empirical evidence from comparative research that current theories build on. We aim to highlight invalid or premature conclusions, and suggest some future directions for research into primate communication, which are prerequisites for the development of an informed multimodal theory of language evolution.

Third, although this book reports some of the major and most influential findings, it does not aim at providing a detailed comprehensive overview of the existing literature on primate communication such as presenting communicative repertoires for different species. Instead, the emphasis is on integrating different perspectives on primate communication that often differ as a function of the modality and a critical discussion in regard to open questions and problems related to these different approaches, illustrated by empirical evidence from a variety of studies. This is the prerequisite for pointing out the way forward, to rethink and redefine terms and criteria (and how they are operationalized) and to develop new methods to study primate communication in a multimodal, integrated way.

Notes

Primates: Although primates include humans, in this book we use the term *primates* to refer exclusively to nonhuman primates unless otherwise stated.

Signals: If specific signals are described, they are highlighted by inverted commas, e.g. 'playface' or 'ground slap'.

Glossary: Some key terms or phrases are defined and explained in some more detail at the end of this book.

Species: Common names for species are used throughout this book. However, in the appendix, each common name is listed together with the Latin name for the species or genus referred to in this book.

Acknowledgements

We couldn't have finished the book without the help of a lot of people who contributed in many ways. We thank Kate Arnold, Ed Donnellan, Pawel Fedurek, Julia Keil, Kathrin Kopp, Manuela Lembeck, Jérôme Micheletta, Nadja Miosga, Richard Moore, Bruce Rawlings, Anne Marijke Schel, Thom Scott-Phillips, Mary Silcox, Tim Smith, Jared Taglialatela, Sebastian Tempelmann, Sam Thornton, Michael Tomasello and Simon Townsend for their enormously helpful comments on varying stages of the manuscript. We are grateful to Rhett Butler, Cathy Crockford, Katie Cronin, Tobias Deschner, Seth Dobson, Julia Fischer, Thomas Geissmann, Giyarto (Macaca Project), Daniel Haun, Catherine Hobaiter, Paul Kuchenbuch, Mark Laidre, Manuela Lembeck, Jérôme Micheletta, Michal Olszanowski, Lisa Parr, Simone Pika and Edwin van Leeuwen for contributing photos to this book. We thank Alina Feinholdt and Tim Smith for drawing the figures and Daniel Haun and Pascal Hecker for advice and editing of the photos and figures throughout the book. We owe special thanks to Kathrin Kopp for her help with the index, literature search and for organizing the references. Many thanks also to Nicole Stein and Linda Scheider for their help with the glossary and the index. Special thanks go to the Max Planck Institute for Evolutionary Anthropology in Leipzig for the access to the library. Last but not least we are indebted to Martin Griffith, Chloe Harries, Judith Shaw (freelancer), Ilaria Tassistro and their colleagues from Cambridge University Press for their continued support. Finally, we owe very special thanks to our families, friends and colleagues for their encouragement and inspiring discussions.

Part I

Introduction to primate communication

1   What is primate communication?

This chapter starts with a general introduction to the two main components of this book: primates and communication. After describing how the term communication is used in this book and how communication can be approached on different levels, primates are introduced with a focus on the variety of their social systems and potential factors that influence their formation. Finally, this diversity of primate social systems is discussed with special reference to the role of communication in maintaining such complex social systems. However, while the aim of this chapter is to provide an overview of primates and their communication in different social systems, Chapter 10 will focus in more detail on the evolution of primate communication with special reference to the role of social organization in the emergence of particular communicative systems.

1.1   What is communication?

Scientists are often particularly interested in the communication of primates, as opposed to other animals, in the hope that it will shed light on the evolution of human communication. The assumption underlying this comparative approach is that as monkeys and apes are phylogenetically closely related to humans, both human and nonhuman primates must share characteristics in the way they communicate. One specific hope, of course, is that an understanding of primate communication will lead to a better understanding of the evolution of human language.

Human communication (especially language) involves various complex (and perhaps evolutionarily recent) cognitive processes. A comparative approach searching for the precursors of human communication in other primates, therefore, often targets these processes for comparison. As such, the study of primate communication can be driven in a slightly different direction to the study of other animal communication, where such cognitive processes are not looked for. It is important that this theoretical and methodological bias is acknowledged in order to make true comparisons between primate species and other animals. We can attempt to untangle this problem by going back to basics and considering precisely what communication *means*, and how we can study it in different ways. There is much debate over how to conceptualize animal communication, however, and scientists are often criticized for overusing language metaphors (**Box 1.1**). In the following section, different approaches to communication will be discussed, and then communication will be defined for the purpose of this book.

Box 1.1 Do researchers rely too heavily on language metaphors to conceptualize animal communication?

Many researchers study animals with the goal of trying to understand the phylogenetic history of human linguistic abilities (see Chapter 5 and 10), and consequently animal communication research has been heavily shaped by terms and concepts from the study of human language. Rendall, Owren, and Ryan (2009) recently reignited the debate about the utility of such an approach and heavily criticized the use of linguistic terms in animal communication research. They argue that casual and inconsistent use of terms such as 'information' and 'meaning' lead to unnecessary confusion and facilitate inappropriate conceptualization of communication. In particular, the use of such terms encourages researchers to rely (consciously or not) on Shannon's (1948) intuitively appealing *Theory of Information*. This theory, originally conceived in an engineering context to describe the effective reproduction of a symbol across noisy physical channels, involves the signaller encoding and transmitting information for subsequent decoding by the receiver. Rendall, Owren and Ryan (2009) argue, however, that such a system can only apply when there is parity between the signaller and receiver in terms of common representational processes to allow accurate encoding and decoding of the information in the signal. Thus the signal has to be represented by the sender and the receiver in the same way. It is clear, however, that many communicating animals may not be capable of sharing such representations and that the goals and processes of signallers and receivers may not be the same (Rendall, Owren and Ryan, 2009). As such this conceptual framework is argued to be inappropriate for animal communication and some would also suggest human language (e.g. Scott-Philips, 2010).

   As an alternative framework, Rendall and colleagues (2009) and Scott-Philips (2010) advocate thinking of communication in terms of influencing others, not informing them. This builds on seminal arguments proposed by Dawkins and Krebs (1978), who suggest that signallers primarily produce signals to manipulate receivers. Other researchers see more utility in preserving the concept of 'information' in animal signals (Seyfarth *et al.*, 2010) and deny that the use of terminology such as 'information transfer' necessitates reliance on language metaphors and Shannon's *Theory of Information*. Seyfarth and colleagues clarify that information can be usefully defined as reducing uncertainty in the receiver, allowing them to predict current or forthcoming events. Font and Carazo (2010) attempt to unify these 'information' and 'influence' camps with a view of communication where signallers influence receiver behaviour in ways that increase their fitness and receivers 'eavesdrop', extracting information from signals that the signaller may not intend to have provided them with. Interestingly, this is not dissimilar to traditional views of animal communication within ethology.

   Whilst it is undoubtedly the case that the pursuit of understanding the information available to and extracted by receivers from animal signals has positively advanced our understanding of animal communication in a number of ways (Seyfarth *et al.*, 2010), the search for consensus on a more appropriate conceptual framework in which to embed animal communication research is ongoing.

In **behavioural ecology**, communication has been defined as 'the process in which actors use specially designed signals or displays to modify the behaviour of reactors' (Krebs and Davies, 1993, p. 349). The phrase *specially designed* is crucial, as it qualifies that the signals or displays must have been selected via evolutionary processes, specifically for the purpose of communication. Importantly, such signals should be decoded by the receiver successfully (Burling, 2005), otherwise they would not evolve. There could be many *cues* that indicate something to a conspecific, and which can be used by the conspecific to alter their behaviour, but unless this has been shaped by evolution to form a specific adaptation, it would not count under this definition. For example, two monkeys sit facing each other, grooming. One monkey sees a conspecific join the group, and shifts her eyes to watch this new individual. The monkey grooming her observes this eye movement, and follows her gaze. Information about the incoming individual is shared through this process, but does this count as communication? A behavioural ecology perspective would argue that this is not communication, but instead that the eye movement functioned as a cue to information (for the receiver). Aspects of this process could, however, be acted on through evolution and become fixed as communicative signals or displays. Eye movements could, for example, become more salient through adaptive change in the morphology of the eyeball and the coloration of the sclera, and thus gain signal function (Kobayashi and Kohshima, 2001). Interestingly, white **sclera** is only typical for humans, possibly having evolved as a specific adaptation for cooperative communication (Emery, 2000) (see also Chapter 2, section 2.3.1).

Another distinction divides communication into that which involves behaviour, and that which does not. In his classic work *The Behaviour of Communicating*, the ethologist W. J. Smith defines communication as 'the behaviour that enables the sharing of information between interacting individuals as they respond to each other' (Smith, 1977, p. 2). Importantly, this definition excludes static and invariant displays, such as the brightly coloured wing of a butterfly or the peacock's tail. Of course, the line might be blurred in many cases, as the salience of a static signal can be increased through behaviour. The dance of the male stickleback, for example, is paired with his red colouration in the mating season. The distinction between behavioural and non-behavioural communication, however, is a useful one in order to narrow the focus of study, and one that is adopted throughout this book. As such, although there is a vast range of fascinating and important primate signals, they will not be considered here unless explicitly related to behaviour. Thus, the facial colouration of the mandrill, the sexual swellings of female crested macaques and the striped tails of ring-tailed lemurs, for example, are not considered further within this book.

One aspect of communication that has not featured in the previous definitions, and yet is a feature that is central to the interest to many scientists, is that of intention. For the aforementioned Krebs and Davies (1993), Smith (1977) and many other scientists with a more biological leaning, the issue of proximate *motivation* on the part of the sender is largely irrelevant. The role that the communication plays in the social interaction, and how this has been shaped by evolution, is their primary concern. For scientists with a more psychological perspective, in contrast, whether or not the communicative events are **intentional** is a key question. Primarily, this stems from the goal to shed light on the evolution of human communication (and language in particular) through studying primates. As much of human communication is intentional, many psychologists are interested in whether primate communication is also intentional, and thus may have been an important precursor to language. Thus, Tomasello (2008, p. 14) uses a narrower definition, and defines communicative signals as 'signals that are chosen and produced by individual organisms flexibly and strategically for particular social goals, adjusted in various ways for particular circumstances. These signals are *intentional* in the sense that the individual controls their use flexibly towards the goal of influencing others.' Note that Tomasello (2008, p. 14) distinguishes *signals* from *displays*, and refers to displays as 'prototypically physical characteristics that in some way affect the behaviour of others', which are 'created and controlled by evolutionary processes', which seems to indicate that the sender has no voluntary control over their production. Determining whether a signal is intentional and voluntary is not straightforward, however, and so the issue of intentionality and how this relates to communication will be discussed in more depth in Chapter 8.

For the purpose of this book, communication is defined as *social behaviours that transmit information from one individual to another*. How these behaviours can be identified is, of course, part of the challenge and process of study. Throughout this book, we refer to four different modes of primate communication that have (historically) been studied separately: gesture, facial expression, vocalization and olfaction (see Chapter 4 for a more detailed discussion of this categorization). Our aim is to argue that these modes should be studied simultaneously and (where possible) using similar methods in a multimodal approach (see Chapter 5 for a discussion on how to define multimodality).

1.1.1   Understanding communication at different levels

The distinction between psychologists and biologists is always hazy within the study of animal behaviour, as the topic necessarily sits on the cusp of these two disciplines. There are times, however, when it is helpful to understand the difference between these two parent disciplines in order to unpack the nuances between differing approaches. Historically, psychologists have been more interested in the mechanistic aspects of behaviour (the immediate, causal processes), whereas biologists have focused more on the functional aspects (why/how that behaviour has evolved). One reason that scientists use different definitions for communication is that they are looking at communication from these two different perspectives, which can boil down to a focus on the **proximate** level, or a focus on the **ultimate** level.

In 1963, Niko Tinbergen was well aware of the divisions that existed between scientists interested in behaviour, and following on from Mayr (1961) endeavoured to unify the field in his landmark paper 'On aims and methods in ethology' (Tinbergen, 1963). His main aim was to highlight that there are four different but related levels at which behaviour operates and thus can be studied, but that a focus on each level makes a valuable and necessary contribution to the study of behaviour. Importantly, he emphasized that all levels share a common goal – to understand behaviour. Often called Tinbergen's four questions, or *four whys*, he described these four levels and in doing so, he constructed a framework for the study of behaviour which has offered clarity ever since. **Box 1.2** shows a unit of communication (a facial expression), and outlines how this could

---

Box 1.2 Analysing signals at Tinbergen's four different levels: 'why' is that primate showing a 'playface'? (Photo: Jérôme Micheletta)

Level of explanation		Example: 'Playface' of crested macaques
Proximate: *How* an organism works (focus on individuals)	Causal/ Mechanistic	Psychological (e.g. cognitive, emotional) and physiological processes underlying the behaviour: *He is feeling playful, the facial muscles are activated by the corresponding neural substrates and the facial nerve.*
	Developmental/ Ontogenetic	Developmental processes that have led to the production of the behaviour: *He started producing the 'playface' during playful encounters from birth, possibly due to innate processes, and its use may have been refined through experience.*
Ultimate: *Why* an organism works like that (focus on species)	Phylogenetic/ Historical	The history of the behaviour over evolutionary time: *Crested macaques inherited the behaviour from ancestral primate species, which used a similar facial expression during play.*
	Functional/ Evolutionary	The adaptive function of the behaviour for the species: *Communicating motivation to play is adaptive as it helps coordinate and maintain play bouts, which is a useful social behaviour.*

be explained at the four different levels. First, Tinbergen described *causation*. This relates to the 'preceding events, which can be shown to contribute to the occurrence of the behaviour' (Tinbergen, 1963, p. 418). Specifically, this relates to the mechanism associated with the behaviour, the physiology and the cognitive and emotional experience. Second, Tinbergen described *survival value*. Here, the function of the behaviour is considered in

terms of why it has been selected by evolution, and how it is adaptive. What survival value does the behaviour offer? What is it good for? Third, Tinbergen described *ontogeny*. Ontogeny relates to the development of an individual during its lifetime. Of particular interest are how different variables influence the manifestation of certain behaviours, and the extent to which behaviours can be modified or are present from birth. Fourth, Tinbergen described *evolution*. Often confused with survival value, this level does not directly consider adaptation, but instead the phylogenetic history and relationships that exist between **extant** and extinct species that reflect evolutionary ancestry. Here, it is the evolutionary path of an organism that is of interest, not necessarily the selection pressures that directed that path, or the solutions to the problems posed. These issues are related, of course, as are all of the four levels. Historically, psychologists have tended to focus on causation and ontogeny (termed *proximate* levels), and biologists have tended to focus on survival value and evolution (termed *ultimate* levels).

The resounding lesson from Tinbergen (1963) is that it is vital to understand at which level you are asking questions, in order to get the right answers. Also important is an acknowledgement that communication operates at all these different levels, and understanding that ultimate aspects might aid understanding of proximate aspects, and vice versa. However, ultimate and proximate explanations are often confused or are used as contradictory alternatives, which they are not (see Chapter 10, for a more detailed discussion).

1.2   What are primates?

The order Primates[1] is a monophyletic group of mammals united by a single common ancestry (Fleagle, 1999; Schultz, 1969). A taxonomic group must be characterized by a set of shared morphological traits possessed by no other group (Mayr, 1963, 1969; Simpson, 1961). These shared specializations, or shared derived features, are called **apomorphic** characters and they help define one taxonomic group relative to others. While behavioural and ecological characteristics are important in helping to create a definition of animal taxa, particularly close attention is paid to morphological characteristics because these will be preserved in the fossil record and will allow the evolution of a taxon to be traced over time. However, one of the problems with primates as an order is that it is relatively difficult to identify a clear fossil record of early primates.

Broadly speaking, primates are defined as an order based upon a general set of morphological and behavioural characteristics summarized in **Box 1.3**. Living primates possess a suite of morphological characteristics that help us unite them into a single order. These characteristics include a grasping big toe (hallux) and thumb (pollex) and low-crowned molars. However, unlike other mammalian orders, primates lack any one shared specialized characteristic that is clear and unambiguous as, for example, the blowhole shared by all cetaceans. The defining morphological features of

---

[1] Clearly, humans belong to the Primate order and are part of the great ape family Hominidae, along with orangutans, gorillas, chimpanzees, and bonobos (Groves, 2001). However, in this book we use the term 'primates' to refer exclusively to nonhuman primates unless otherwise stated.

Box 1.3 How are primates defined relative to other mammals?

Morphology

- Limbs adapted for prehensile (grasping) lifestyle rather than clinging or clawing
- Grasping pollex (thumb) and hallux (big toe) to enhance prehension
- Nails, not claws, associated with development of tactile pads on hands and feet
- Shortened snout or face
- Binocular vision
- Low-crowned molars, reflecting diet centred more on fruits than grass
- Cerebral cortex expansion with relatively larger head

Behaviour

- Highly social throughout life
- Increased reliance on vision
- Unique reproductive behaviours (see **Box 1.4**)

---

primates do not come in all at once in the fossil record, leaving us with a mass of fossils that are controversial as to whether or not they are indeed primates.

In addition to the morphological traits summarized in **Box 1.3**, reproductive features including the cyclical nature of female ovulation, pregnancy and the birth process, parental care of offspring and adaptations of offspring for growth and development are also of central importance (**Box 1.4**). An understanding of these life-history patterns in any mammal group is central to an appreciation of the evolution of primate societies and communication (Martin, 1990).

One of the major behavioural traits of primates relative to other mammals is their tendency to be highly social throughout all life stages,[2] an attribute that is reflected in the morphology, behaviour and reproductive features of primates. While many mammals temporarily group together under various circumstances, the associations are typically short term and the individuals involved in the aggregation change over time. In primates, group membership tends to be highly regular (Clutton-Brock and Harvey, 1977; Fleagle, 1999; Sussman, 1999). Because this book is highly influenced by this particular distinguishing characteristic of primates, we focus on how their highly social behaviour influences their life histories and the development and evolution of their communication.

1.2.1 Primate phylogeny

Today, primates naturally occur almost exclusively in the tropical and sub-tropical zones of South America, Africa and Asia. While many aspects of primate taxonomy and phylogeny are hotly debated, living primates are generally grouped into suborders

---

[2] Exceptions to this include the highly social naked mole rats (Sherman, Jarvis and Alexander, 1991), meerkats (Clutton-Brock *et al.*, 2000), hyenas (Kruuk, 1972), wolves (Mech *et al.*, 1998), lions (Schaller, 1972) and some cetaceans (Mann, 2000).

Box 1.4 Distinctive features of primate reproductive behaviour

Reproductive organs

Prosimians retain the primitive mammalian condition of a **bicornuate uterus**, while **anthropoids** have developed a **simplex** uterus. This may be related to the reduction in litter size seen in anthropoids, which typically only have one offspring at a time.

Female ovulation

Most mammals are induced ovulators; that is, an ovum is released by the female in response to the act of mating. Primate females are spontaneous ovulators. Females release an ovum at regular intervals regardless of the act of mating. Spontaneous ovulation implies a close coordination between males and females so that the male can detect when the female is in oestrus. This is typically done through olfactory signals (pheromones) but also through visual signals (e.g. sexual swellings in female chimpanzees).

Pregnancy and birth

Female primates have a very long gestation (pregnancy) period relative to maternal weight. This allows the developing foetus to spend a greater period of time developing the large brain that primates are born with, relative to other mammals. Unlike most other mammals, primates typically only gestate a single, high-quality offspring with a high brain weight relative to body weight. Callitrichids, such as marmosets and tamarins, have evolved an interesting reproductive strategy of usually giving birth to twins. This is accompanied by their unusual social system of polyandry, with only one breeding female having access to several males.

Parental care of offspring

Unlike most other mammals, female primates spend a long period of time in post-partum anoestrus while they lactate and suckle the newborn. This investment in a (typically) single offspring at a time requires the mother to stay in close contact with the infant, usually by carrying it whilst the infant clings to the mother's fur until it can begin to feed more independently. This period of intensive investment by the mother often requires provisioning or **allocare** from other social group members in species that live in groups. The callitrichids again have a unique parental care strategy among primates, since they form cooperative polyandrous groups. In these species, the newborns are typically carried by the males or by the nonreproductive females so that the single breeding female can return to oestrus.

Adaptations in the offspring for growth and development

Primate infants are characterized as having a markedly slower postnatal development relative to other mammals, a period where they are learning social behaviours and other skills. This slow period of development has been linked to the relatively late onset of sexual maturity in primates.

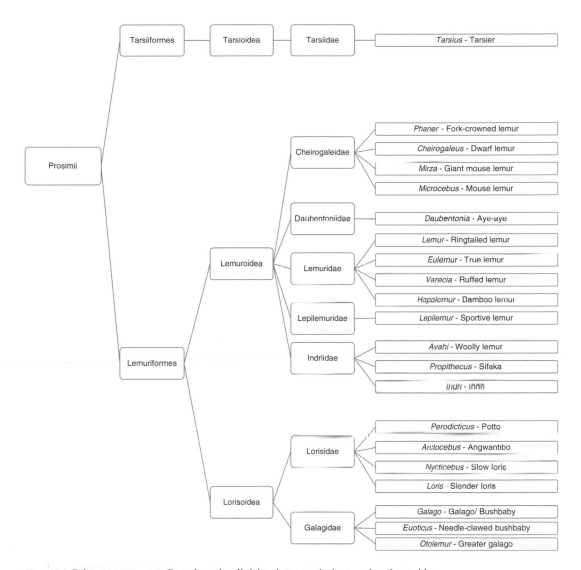

Figure 1.1 Primate taxonomy. Based on the division into prosimians and anthropoids.

in two ways (e.g. Fleagle, 1999). In one scheme, primates are grouped into two suborders known as the prosimians and the anthropoids. Prosimians include lemurs, lorises and tarsiers. Anthropoids include Old World monkeys, New World monkeys, apes and humans (see Figure 1.1).

Alternatively, some taxonomists, such as Luckett (1976) and Groves (2001) prefer to group primates into suborders **strepsirrhines** and **haplorrhines**. Under this scheme all prosimians except for tarsiers would be placed into strepsirrhines and all anthropoids plus tarsiers would be placed into haplorrhines (see **Box 1.5**). This scheme of classification is preferred by some because it provides an indication of which group of living prosimians

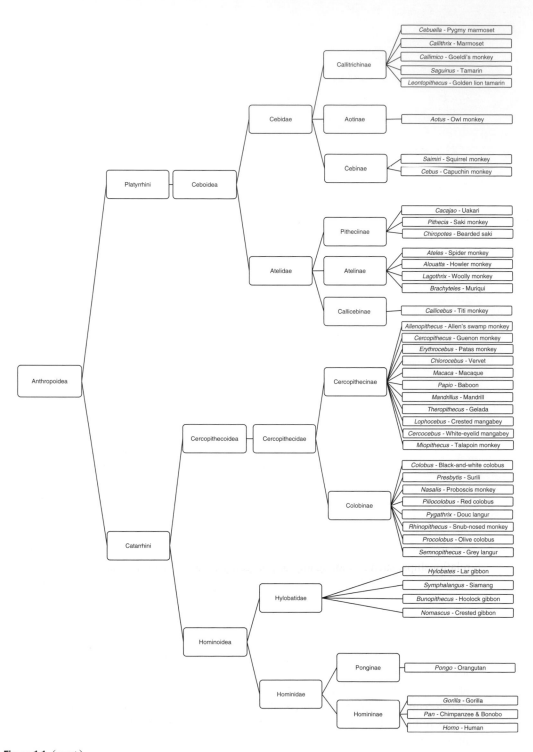

Figure 1.1 (*cont.*)

(tarsiers) may be more closely related to anthropoids and what derived morphological traits are shared between tarsiers and anthropoids. For the sake of simplicity, the scheme dividing primates into prosimians and anthropoids is used throughout this book.

Prosimians are considered to be the more primitive suborder because they retain many general morphological characteristics possessed by the first primates (Conroy, 1990; Fleagle, 1999). The majority of prosimians are small-bodied, nocturnal species with a relatively low brain size to body size ratio compared to anthropoids. They also tend to be far less **gregarious** than anthropoids with smaller group sizes.

The first appearance of primate-like mammals in the fossil record likely represented an adaptive shift by the animals to occupy a previously unfilled niche. It is reasonable to expect that this shift would be seen in the morphology of the animals and thus be recognizable in the fossil record as a new order. However, it can be difficult to recognize these early primates due to the relative lack of morphological signatures. There is general agreement, though, that primate-like mammals first appeared in the fossil record around 65 million years ago in the Paleocene epoch. These animals, the plesiadapiforms, went on to give rise to the diverse array of both extant and extinct primates known today (Bloch and Silcox, 2006; Bloch et al., 2007).

1.2.2   Primate social systems

One of the most conspicuous characteristics of primate social systems is their variety of forms. Some members of primate species only come together with **conspecifics** for short periods of time or limited activities (such as sleeping) while others live in very large groups most of the time. Overall though, most primates live in groups where membership is relatively stable compared to most mammals.

To characterize the complexity of primate social systems, it is important to differentiate between three distinct facets of such societies (Kappeler and van Schaik, 2002). *Social organization* refers to the size, sexual composition and the spatiotemporal cohesion of a social system. Three different types of primate social organization are differentiated: **solitary**, pair-living and group-living species. The *mating system* is defined by the numbers of mating males and females in a group and thus their sexual relationships, such as monogamy or polygamy (Dixson, 1998). Finally, *social structure* refers to social relationships between individuals such as male–male, female–female or female–male interactions, with each of them being characterized by a distinct patterning like differences in the nature, frequency or intensity of **affiliative** and agonistic interactions (de Waal, 1989; Hinde, 1976).

Although these facets are considered distinct elements of primate social systems, they are clearly interrelated. For example, the social organization like spatial cohesion and the ratio of males and females constrains which and how many individuals are available for social relationships, which influences both the social structure and mating system. This is further complicated by the fact that there is a considerable degree of variability within each species, so that it is difficult to assign one of each of these facets to each species (Kappeler and van Schaik, 2002). In the following section, the variety of primate social systems is demonstrated by referring to some typical examples with

Box 1.5 Primate taxonomy

The division of the primate order into the suborders Prosimii and Anthropoidea is a gradistic scheme that gives no information on the derived traits used to group lemurs and lorises. In the scheme, tarsiers are grouped with lorises and lemurs. Generally, this scheme is not favoured by people interested in evolutionary relationships among primates. Most investigators instead prefer the suborder grouping of Strepsirrhini and Haplorrhini, a phyletic scheme that takes into account the derived morphological characteristics of lemurs and lorises. Here, tarsiers would be grouped with monkeys and **apes**. However, many paleontologists are less sure that a tarsier/anthropoid grouping represents an accurate evolutionary relationship, especially when fossil primates are included.

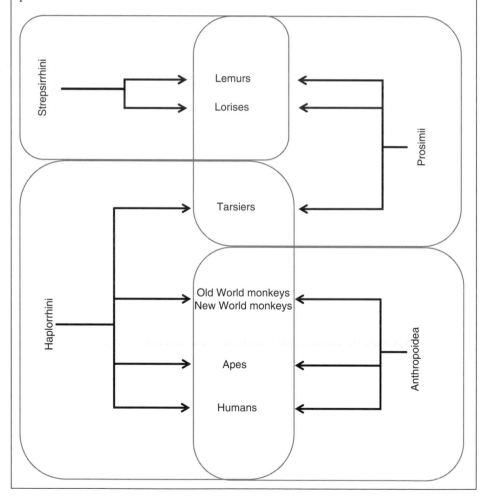

focus on the mating systems of these species, supplemented by some further information on their social organization or structure (see Figure 1.2).

A solitary lifestyle is rather rare among primates, since individuals may at least temporarily aggregate into bigger groups. For example, although adult orangutan males usually live alone and spend less than 2% of their time with other conspecifics (Galdikas, 1985), adolescent individuals or adult females with offspring can aggregate to form temporary social groups during feeding in the same fruit tree or travelling, particularly if food is abundant (Galdikas and Vasey 1992; MacKinnon 1974; Rijksen 1978; van Schaik, 1999). Furthermore, although many nocturnal prosimians are considered solitary species, individuals of both sexes may come together in sleeping sites during the day. Individuals establish ranges that might overlap with ranges of other individuals, which is referred to as noyau (Figure 1.2a). Depending on the sex of the owner of these ranges, different mating systems may emerge. If a dominant male has a range that overlaps those of several females, this is referred to as a polygynous mating system (Figure 1.2d), which is found, for example, in mouse lemurs and bushbabies (Bearder, 1987; Charles-Dominique, 1978; Martin, 1973; Pullen, Bearder, and Dixson, 2000). In a promiscuous mating system (Figure 1.2e), several males have ranges that overlap with those of several females (Kappeler, 1997; Sterling and Richard, 1995).

The monogamous mating system (Figure 1.2b) is relatively rare among primates (as is the case in mammals in general) compared to other animals such as birds (Kleiman, 1977; Reichard and Boesch, 2003). It has been described for prosimian species (e.g. bamboo lemurs and indris), for several New World monkeys (e.g. owl monkeys, titi monkeys and callitrichids) and particularly for gibbons (e.g. Kinzey, 1981; Palombit, 1996; Robinson, Wright, and Kinzey, 1987). In these species, territorial and mate defence are typically intense and are often **intrasexual** in nature. Groups usually consist of a nuclear family group comprising one heterosexual pair and its offspring, with maturing individuals leaving their natal group. Only callitrichids form extended family groups and thus larger groups consisting of one breeding pair with their offspring of all age classes including mature individuals.

Polyandry (Figure 1.2c), a form of polygamy, is exceedingly rare and is only practised by callitrichids (marmosets and tamarins) (Digby, 1995; Digby, Ferrari, and Saltzman, 2007; Rylands, 1993). Group size is highly variable and can comprise two up to 20 individuals, with only one reproductive female, one or several adult males and the reproductive female's offspring, including already mature individuals. Thus, breeding is monopolized by a single dominant female (French, 1997; Tardif et al., 2003; Terborgh and Goldizen, 1985). This is accomplished through the female suppressing the ovulation of other females by releasing olfactory cues that suppress the secretion of reproductive hormones (Abbott et al., 1997). The nonreproductive females then act as alloparents or 'helpers' in the cooperative care of the young, to whom they are related (see also **Box 1.4**) (Tardif, Harrison, and Simek, 1993; Tardif et al., 2002). Males in polyandrous groups also often provide direct parental care by transporting infants (Digby, 1995; Rylands, 1986). Both the presence of 'helper females' and multiple males are most likely responsible for the evolution of the callitrichids' unique reproductive

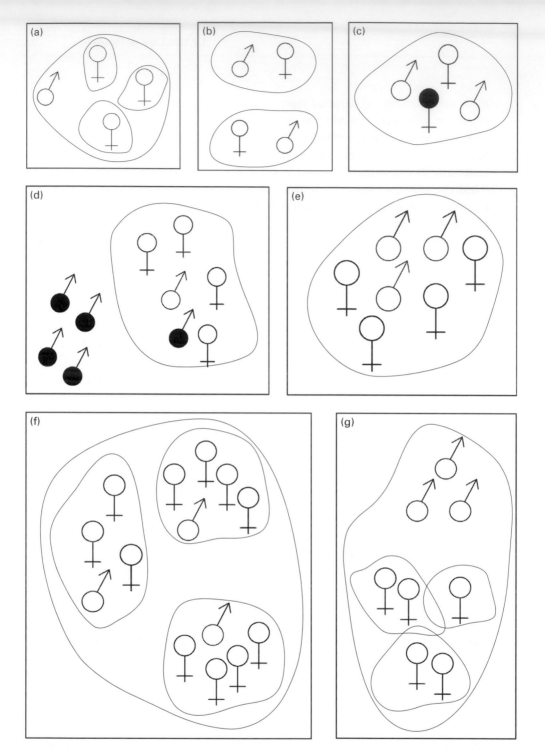

Figure 1.2 Types of primate social systems with focus on mating systems. (a) Noyau (polygynous): range of dominant male overlaps with those of several females; (b) Monogamy: one male and one female (with offspring) defend their territory against other pairs; (c) Polyandry: only the dominant female is reproductive and mates with several males; (d) Polygyny: only one male mates with several females, maturing nonreproductive males form all-male groups; (e) Promiscuity: multi-male, multi-female groups, several males mate with several females. (f) Multi-level and (g) fission–fusion societies each comprise several subgroups. Only adult individuals are shown: ♂ = male, ♀ = female, solid circles = nonreproductive male or female (modified from figure 1 "Basic types of primate social organization (after J. G. Fleagle, Primate Adaptation and Evolution, 2nd ed., Academic Press, San Diego, 1999)", in the article Primate social organization by Anthony Di Fiore published on AccessScience. Copyright © McGraw-Hill Global Education Holdings, LLC. All rights reserved.)

features of twinning and early post-partum ovulation in females (Digby, Ferrari, and Saltzman, 2007) (see also **Box 1.4**).

Much more commonly, polygamous primate species practise polygyny (Figure 1.2d), where one male mates with more than one female. Two different types of polygynous groups are differentiated as a function of the number of males in such groups. Some groups have only one male, while others may have multiple males in a group. One-male polygynous species include many cercopithecine species (e.g. patas monkeys and guenons) and colobine species (e.g. grey langurs and colobus monkeys). Group size varies greatly but seems to most commonly range from three to 20 individuals (Cords, 1987; Jolly, 2007) although group sizes of up to 50 individuals have been reported for some species of guenons (Struhsaker and Leland, 1988). This considerable variation in group sizes may be due to the relatively frequent immigration attempts of non-resident males. Generally, the resident male aggressively resists immigration attempts by other males. However, it may tolerate multiple males in the group if several females are in oestrus at the same time and as a consequence, the resident male may be unable to defend all of the reproductive females in the group (Cords, 2002; Enstam and Isbell, 2007). If other males are present, there is usually a rigid, linear dominance hierarchy between them, while females usually have a very weak – if any – dominance hierarchy, with only few aggressive encounters between them (Enstam and Isbell, 2007). While females stay with their natal group, young males emigrate out of the group once they reach sexual maturity and either form all-male groups or live alone (Enstam and Isbell, 2007; Struhsaker, 1975). These young males try to immigrate into a different one-male group to displace the resident male. As a consequence, the new male usually kills the infants sired by the resident male to gain himself reproductive opportunities (Harris and Monfort, 2003; Struhsaker and Leland, 1987).

The second type of polygynous groups are multi-male groups characterizing some prosimians like ring-tailed lemurs and some sifakas, some Old World monkeys, including baboons and macaques, several species of New World monkeys, such as capuchins, squirrel monkeys and howler monkeys, and is also frequently found in one subspecies of gorillas, the mountain gorilla. Group size in multi-male groups varies considerably, ranging between seven and 50 individuals in New World monkeys or from eight to 90 individuals in some macaques. There is usually an overt, rigid dominance hierarchy in multi-male groups, which can be present in both sexes. However, males are always dominant to females, except for ring-tailed lemurs, who have a female-dominated hierarchy. As in one-male groups, females usually stay with their natal group, while males disperse around the time of sexual maturity to establish their own group.

Both one-male groups and multi-male groups can form multi-level societies (Figure 1.2f), with several of those smaller units forming larger groups such as bands or even troops comprising several hundreds of individuals, like in baboons and geladas (Kawai *et al.*, 1983; Stammbach, 1987).

Groups that consist of several males and females can also be characterized by a social organization referred to as **fission–fusion** society (Figure 1.2g). Thus, individuals from a large, main community associate temporarily in smaller subgroups that vary in size and composition, as found in spider monkeys, chimpanzees and bonobos (Chapman, 1990;

Nishida, 1968). Community group size in chimpanzees and bonobos varies greatly and ranges from 16 up to 150 individuals (Stumpf, 2007). Spider monkeys tend to have smaller community group sizes comprising 16–42 individuals (Di Fiore and Campbell, 2007), most likely because of their strictly **arboreal** environment compared to the more terrestrial chimpanzees and bonobos. In both spider monkeys and chimpanzees, females are the dispersing sex, emigrating at around the time of sexual maturity, while males tend to be philopatric and develop strong social bonds with other males (Nishida, 1979; Strier, 1992). In both species, males are dominant as evident in a rigid, linear male-centred dominance hierarchy (Di Fiore and Campbell, 2007; Nishida, 1979), while the society of bonobos centres around the females (Kano 1992; Susman, 1984). As in chimpanzees, male bonobos establish dominance relationships, but aggression between males and between the sexes is less intense than in chimpanzees and conflicts are often settled in a non-agonistic manner (de Waal 1995; Furuichi and Ihobe, 1994). Spider monkeys and chimpanzees are territorial with males patrolling the boundaries (Di Fiore and Campbell, 2007; Wilson and Wrangham, 2003). Interactions between different chimpanzee groups are usually aggressive and may result in violent and sometimes fatal attacks (Goodall *et al.*, 1979; Nishida, 1979; Nishida *et al.*, 1985).

1.2.3   Evolution of primate social systems

There are many excellent reviews available that attempt to identify the factors that influence the evolution of the diversity of primate social systems (e.g. Clutton-Brock and Harvey, 1977; Dunbar, 1987; Kappeler, 1997; Janson, 2000; Kappeler and van Schaik, 2002; Thierry, 2008; van Schaik and van Hooff, 1983). Fundamentally, the variation seen among primate social systems is due to the same factors driving variation among all mammal species (see also Figure 1.5). While social patterns and organizations are *group*-level characteristics, these factors are best envisioned as the strategies that *individual* animals undertake to find food, avoid predators, and reproduce (Krebs and Davies, 1993; van Schaik and van Hooff, 1983). Where these three fundamental individual level requirements are best met in the company of conspecifics, groups will form. In the following section we review an influential model that offers explanations as to the effect important factors such as predation pressure and food availability have on the evolution of primate societies.

1.2.3.1   The Socioecological Model

One theory that inspired a large body of research is the *Socioecological Model*, which links the variation of primate social systems to ecological factors including the distribution and density of food as well as predation pressure (see **Box 1.6**). In his influential paper, Wrangham (1980) initially suggested that the formation of dominance hierarchies and the defence of resources by stable groups of related females is more likely in the presence of distributed, often high-value food that usually results in intense or regular competition over this resource. If food is more evenly distributed, the formation of stable kin groups is less likely and the dominance relationships between females are less consistent. Since then, this model experienced a variety of modifications and extensions, including the consideration of the influence of predation risk on group size, and the differentiation between scramble and contest competition, both within and between groups (Isbell, 1991; Koenig, 2002; Sterck, Watts and van Schaik, 1997). In general, the Socioecological Model rests on three main propositions that group living: (1) reduces the predation risk for the individual and improves the success of defending resources against other groups; (2) has an effect on the intensity of competition over resources between group members, particularly if the resource is unevenly distributed in space and/or time; and (3) in cases where resources are not evenly distributed, the increasing competition over food in larger groups offsets the benefits of group living and thus constrains group size (Clutton-Brock and Janson, 2012). In the following, a few examples are summarized that demonstrate the role of food distribution and density as well as predation for the formation of particular social systems in primates.

Food distribution and food density
While primates in general feed on a broad array of items, in most species usually one type of food, such as leaves, fruits or insects, predominates in their diet. The group size and home range of a species is influenced by the spatiotemporal distribution of these resources, the size of the patches in which they occur and the density of the foods within these patches that vary according to the food type, the environment and season. In general, species that mostly feed on widely distributed food resources like leaves, which are available throughout the year, usually have smaller ranges and are mostly non-territorial, for example, leaf monkeys (surilis). Species that mostly feed on spatially and/ or temporally clumped food resources like fruits that are concentrated on certain trees at particular times of a year require larger ranges to find enough food and are often territorial to defend those resources against other groups, like ring-tailed lemurs, chimpanzees and macaques (Janson, 2000; Janson and Chapman, 1999; van Schaik and van Hooff, 1983; Wrangham, 1980). Furthermore, because fruits are also temporally distributed, this leads to seasonal changes in ranging behaviour and, in some cases, group cohesion. If fruits are scarce, large groups may need to split into smaller temporary groups in order to reduce feeding competition, as seen in the fission–fusion dynamics in groups of chimpanzees, bonobos and spider monkeys.

Predation pressure
Intensity and frequency of predation have been discussed as being major factors influencing the evolution of social group size, group composition and group cohesion in primates (Isbell, 1994; Miller and Treves, 2007; Sussman and Kinzey, 1984; van Schaik and van Hooff, 1983; Wrangham, 1987). The major influence on group living in **diurnal** primates has been related to predation pressure (Dunbar, 1988; van Schaik, 1983). These species, especially their offspring, are subject to predation pressure, regardless of their environment. Gathering together in large groups may serve several functions such as increased vigilance ability and it may support the development of display behaviours, alarm calls, and 'mobbing' strategies against the predator (Cheney and Wrangham, 1987; Kappeler, 1997). For example, while most primate species typically flee under the presence of a potential predator, some primate species, such as baboons and red colobus

Box 1.6 The Socioecological Model

Social relationship	Ecological condition		Competitive regime				Behavioural expression			
	Predation risk	Food distribution/ population density	Within-group scramble	Within-group contest	Between-group contest	Infanti-cide	Dominance asymmetry	Coalition formation	Nepotism	Dispersal
Dispersal-egalitarian	High	Food dispersed Low density	High	Low	Low	Yes	Weak	Rare	No	Common
Resident-egalitarian	High	Food dispersed High density	High	Low	High		Weak	Rare	No	Rare
Resident-nepotistic	High	Food clumped Low density		High	Low		Strong	Common	Yes	Rare
Resident-nepotistic-tolerant	Low	High density		Potentially high	High		Variable	Common	Yes	Rare

This table illustrates the main predictions of the Socioecological Model (missing information means no assumption). The model assumes that variation in predation risk and the abundance and distribution of food resources shape the relationships of females by determining the competitive regimes to which they are submitted. These relationships have different components, but agonistic relationships are of particular importance. They vary along three interrelated dimensions: (1) from **egalitarian** (no or poorly defined dyadic dominance relations) to **despotic** (clearly established, often linear dominance hierarchies); (2) from individualistic (ranks of female relatives are independent of each other) to nepotistic (female relatives rank close together); and (3) tolerance (with increasing tolerance, the severity of aggression decreases). In addition to these social factors, demographic factors are important. Females are either philopatric and stay in their group (residents; mostly in despotic systems) or they leave their group (dispersal). When predation risk is high, animals form cohesive groups to reduce it. According to variation in food distribution and population density, group living may generate three different social relationships: dispersal-egalitarian, resident-egalitarian, resident-nepotistic. They correspond to different levels of contest and scramble competition, within or between groups that influence female reproductive success. Within-group competition increases with group size. If some individuals can exclude others from food patches, within-group contest competition prevails. If food is dispersed, scramble competition predominates within groups, but contest competition between groups may be more or less elevated, depending on population density; when habitat is saturated, between-group competition is elevated. A low predation risk produces a fourth social relationship: resident-nepotistic-tolerant. This results from relatively high within- and between-group contest. Competitive regimes induce different patterns of social relationships. Contests for clumped food lead females to make coalitions with kin. High within-group contest is conducive to strong dominance asymmetry, which translates into stable hierarchies and the use of formal signals of submission. Dispersal does not occur when females need to cluster with relatives to win aggressive contests. In some dispersal-egalitarian species, where scramble competition does not entail significant costs, females gather in groups to reduce

Continued

Box 1.6 *(cont.)*

rates of infanticide by males aiming to improve their access to reproductive females by shortening the time to fertility (modified after Thierry, 2008; see also Sterck, Watts and van Schaik, 1997; van Schaik, 1989)

---

monkeys, have developed aggressive defence behaviours executed by high-ranking males against certain predators. Species that attempt to drive off a predator typically contain large groups of individuals that include multiple adult males.

The nocturnal prosimians and other small-bodied primates (such as the callitrichids), tend to live in very small social groups and therefore they do not have the ability to develop anti-predator strategies used by the diurnal species living in large groups. As a consequence, their anti-predator strategies focus on concealment, vigilance and fleeing in addition to being active during the night and hiding away in a sleeping site during the day (Charles-Dominique, 1977; Miller and Treves, 2007).

1.2.3.2   Critiques and alternative approaches

Although the Socioecological Model dominated much of the research into primate social systems, there is an ongoing debate whether it is suitable to describe sufficiently the complexity and variability of primate sociality, which more recently culminated in the proposal to abandon this theory altogether (Thierry, 2008). The major objection against this theory is the frequent mismatch between the predictions of the Socio-ecological Model and the observed variation in the social systems and behaviour of primates (for an overview, see Table 2 in Clutton-Brock and Janson, 2012).

These inconsistencies between the predictions of the model and the actual variety of primate social systems can be explained by the fact that many other factors contribute to the formation and variability of groups, including variation in body size and reproductive strategies (Clutton-Brock and Harvey, 1977; Clutton-Brock and Janson, 2012). Further-more, there is evidence that closely related species, like macaques, often share similar patterns of social organization and structure, despite differences in their feeding ecology (Thierry, Iwaniuk and Pellis, 2000). This suggests that the variety of primate social systems can at least partly be attributed to phylogenetic differences and that some aspects of primate sociality are highly conservative (Matsumura, 1999; Thierry, 2008). As a consequence, both ecological and phylogenetical factors need to be considered to understand the diversity and complexity of primate social systems (Clutton-Brock and Janson, 2012; Thierry, 2008).

1.3   What do primates need communication for?

In order to understand how a communicative system works, it is essential to consider the social function of communication. As discussed in more detail in section 1.1.1, this could be at an ultimate level (i.e. considering survival value and how it has evolved) or

Figure 1.3 Contexts of primate interactions. (a) Reassurance behaviour in male–male interactions in chimpanzees; (b) greeting behaviour of male crested macaques; (c) mother–infant interactions in chimpanzees; (d) aggressive interactions between chimpanzees; next page: (e) sexual behaviour in chacma baboons; play behaviour in (f) gorillas and (g) Barbary macaques; and (h) grooming context in chacma baboons. (Photos: Cathy Crockford (e, h); Julia Fischer (g); Katja Liebal (d); Edwin van Leeuwen (a, c); Jérôme Micheletta (b); Simone Pika (f))

at a proximate level (i.e. how does it relate to the immediate motivation of the sender). It is often very hard to identify the specific survival and reproductive value of individual behaviours (e.g. Tinbergen's survival value), but at the very least it is important to nest behaviour within a wider, functional context for that organism. In a sense, it is through considering the problems that a primate has to solve that we can hypothesize what specific communicative events are for. Broadly, primates have to solve the problem of survival, reproduction and (often) living in groups (see section 1.2.2). The latter is, of course, connected to survival and reproduction, but is such an important factor in all social behaviours, that it seems to merit its own category. Communication as an aid to survival could manifest as response to predators or response to and defence of food sources. Sexual communication and mother–infant interaction are examples of communication as an aid to reproduction. Communication as an aid to group living could

Figure 1.3 (*cont.*)

manifest as play signalling, alliance formation, conflict resolution and dominance displays (see Figure 1.3 for some examples).

Given a blank slate, there may be many useful ways to solve specific problems with communication, but design is constrained by a number of factors. In addition, evolution tends to operate on what it already has, rather than evolving something from scratch (Dunbar, 1993). Specifically, constraints are imposed by the environment, and by characteristics of the receiver. In terms of the environment, the **modality** of communication may be constrained by how and when information can

Table 1.1 Comparison between the different sensory modalities of primate signals (modified from Alcock, 1984).

Feature	Sensory modality of signal		
	Olfactory (chemical)	Auditory (sound)	Visual
Range	Medium	Long	Short
Potential rate of change of signal	Slow	Fast	Fast
Ability to go past obstacles	Good	Good	Poor
Locatability	Variable	Medium	High
Energetic cost	Low	High	Variable

be transferred. For example, in arboreal environments visual information may be less effective than auditory information. A densely foliated arboreal environment would provide few opportunities for close-proximity intraspecific interaction, and so we would not then expect a large facial expression repertoire. In contrast, arboreal environments may favour the evolution of a complex vocal repertoire (Maestripieri, 1999; Marler, 1965). A **semi-gregarious**, nocturnal species may be expected to exploit chemical signalling that is preserved over a long period of time, rather than auditory and visual signalling that disperses in the environment quickly. Table 1.1 gives an overview of the different modalities of communication and how they compare on these various parameters.

In terms of the receiver, the senses and cognition available to accept the information may limit the characteristics of the signal. Thus, primate signals may have evolved in relation to the receiver. Only those signals that were easy to perceive, discriminate and remember were likely to affect receiver behaviour and were therefore selected for (**receiver psychology**; Guilford and Dawkins, 1991; Rowe, 1999). Auditory signals outside the range of conspecific hearing, for example, are unlikely to prove fruitful. Likewise, the sensory capacity of the receiver can be a powerful force in shaping signals, as how the receiver interprets and uses the information will determine the salience of a signal. The sensory capacity of primate species, therefore, is important to consider when analysing the characteristics of their communication (see also Chapter 2).

1.3.1   Communication across different primate social systems

Communication is clearly influenced and constrained by a host of proximate and ultimate factors both at the individual and group level. Being a social behaviour, though, the evolution of communication must be framed against a backdrop of the group and its social dynamics.

As section 1.2.2 demonstrated, most primate species live in some form of group, however, primate social systems can vary in regard to their social organization and structure as well as their mating system. Communication between group members plays an integral part in the maintenance of such social systems

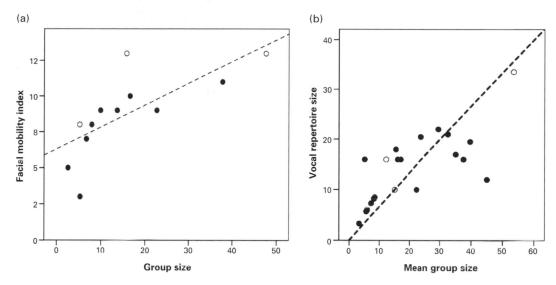

Figure 1.4 Group size and communicative repertoires. (a) Dobson's facial mobility index (Dobson, 2009) plotted against group size. The outliers are: gorilla (top left, open symbol), cotton-top tamarin (lowest solid symbol). (b) Vocal repertoire size plotted against mean group size for different genera. The outliers are: tamarins (upper left), talapoin (lower right). Repertoire size from McComb and Semple, 2005; group sizes from Dunbar (1992, 1995, 2011). In both cases, the correlations are significant (facial mobility: r = 0.726, n = 12, p = 0.007; vocal repertoire: r = 0.746, n = 22, p < 0.001, respectively); O = apes, ● – monkeys (from Dunbar, 2012).

(Altmann, 1962). Clearly, the nature of these social systems influences the communicative interactions within a species. For example, group size is an important factor, since several studies demonstrated that an increase in group size results in an increase in both facial and vocal repertoires (Dobson, 2009; McComb and Semple, 2005) (Figure 1.4).

Furthermore, a relationship between the social structure of a species and its communicative repertoire was proposed by several authors. Preuschoft and van Hooff (1995) suggested that the degree of flexibility in the production of communicative signals is partly determined by the social structure of the species, with more flexibility present in species with more relaxed dominance structures (see also Chapter 7, section 7.2.1 and Chapter 10, section 10.3, for more information on the *Power Asymmetry Hypothesis of Motivational Emancipation*). Maestripieri investigated several macaque species (1996a, b, 1997, 1999) and showed that species with a more despotic social structure, like rhesus macaques, in which the outcome of most social interactions is predetermined, use comparably few signals because of their strict dominance structure resulting in stable and predictable social relationships between group members. In contrast, species with a more egalitarian social structure, like stump-tail macaques, have more symmetrical and less kin-biased interactions that require more negotiation between individuals, resulting in larger,

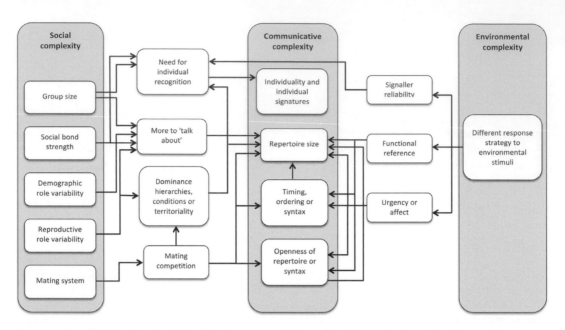

Figure 1.5 Possible functional relationships between attributes of social complexity and attributes of communicative complexity. Both sociality and communication have multiple attributes. These attributes affect one another in complex ways and different attributes of social complexity are likely to drive different attributes of communicative complexity (from Pollard and Blumstein, 2012).

more complex signal repertoires. Finally, one of the very few comparative studies that focused on olfactory communication found that the degree of complexity of female odours differed across several species of lemurs as a function of their social structure or mating system, with chemical signals being more complex in multi-male, multi-female species than pair-bonded species (delBarco-Trillo *et al.*, 2012).

Thus, social complexity seems to be an important factor for the evolution of complex communicative systems (Freeberg, Dunbar and Ord, 2012). There is a variety of possible functional relationships between different attributes of social complexity and communicative complexity, in addition to the influence of different ecological factors (Figure 1.5). A potential problem, however, is how to define and assess the complexity of such communication systems, which might be evident, for example, in the variety of signals used by a species and/or in complex structural properties of a particular signal (Pollard and Blumstein, 2012). Whilst the evolution of primate communication is discussed in more detail in Chapter 10, the following section will describe some examples to demonstrate the variety of forms of communication with particular focus on the social systems of different primate species.

In nocturnal species that are either solitary or form a noyau, the development and maintenance of their relationships may best be served by intensive use of chemical signals and vocalizations rather than visual communication. Both chemical signals and vocalizations can travel relatively long spatial distances and can therefore be perceived over larger distances and reach distant receivers. Olfactory signals vary greatly in the duration they persist in the environment. Some may last for many days, allowing an individual to mark and maintain a territory even when conspecifics will not come into direct contact (see Chapter 1, Table 1.1). During aggressive interactions between males, the use of loud, boisterous vocalizations in addition to physical contact is reported for several species including Mohol bushbaby, mouse lemurs and aye-ayes (Atsalis, 2008; Martin and Bearder, 1979; Sterling and Richard, 1995). Because of their nocturnal and often solitary lifestyle, however, it has been suggested that such species rarely use alarm calls to warn other conspecifics when discovering predators, but rely more on crypsis to escape the predator (Rahlfs and Fichtel, 2010).

Monogamous species often have very loud and ritualized vocalizations that seem to function both in a territorial display and in maintaining contact among group members. For example, pair-bonded males and females in gibbons and indris engage in 'dueting' vocalizations at specific times of the day that travel considerable distances (Bartlett, 2008; Powzyk and Mowry, 2003). In gibbons, such songs are not only used to defend territories, but also to strengthen pair bonds or to attract new mates (Colishaw, 1992; Geissmann, 1999). Within groups, gibbons seem to use a rather limited repertoire of gestures, facial expressions and vocalizations, most likely because of the small and stable nature of these family groups (Chivers, 1976).

In polyandrous species, chemical signalling is vital to maintain the group (Abbott, Barrett and George, 1993; Epple, 1986). While aggressive interactions within groups are relatively rare and mild in form, intergroup interactions may be aggressive in the contexts of defending food resources or mates. Defence typically includes vocalizations but may escalate into chases or physical aggression (Garber, 1988, 1993; Peres, 1989). However, loud long-distance vocalizations do not seem to be common in any inter- or intragroup interactions.

In polygynous mating systems, one-male groups tend to be highly territorial and much of the maintenance of the range boundaries is achieved via the use of loud, long-distance vocalizations uttered by the males, but females may also join in (Cheney, 1981; Hill, 1994). While olfactory signals may be important to communicate the reproductive status of females, they also signal their receptivity through particular postures and through facial expressions (Cords, 1987).

Many of the social life characteristics of multi-male species are similar to those of one-male species such as female interactions and territoriality. In some species of macaques, though, strong female dominance hierarchies exist as **matrilines**, in addition to the male hierarchies (Thierry, 2007). Although in these species vocal communication is important, as it is in one-male species, a variety of visual signals like facial displays are frequently used, particularly in the contexts of appeasement and aggression (Maestripieri, 1999; Parr et al., 2010) (Figure 1.6).

In fission–fusion species, communication involves vocalizations that are used when parties re-group and in both aggressive and affiliative encounters (Laporte and Zuberbühler, 2010; Nishida and Hiraiwa-Hasegawa, 1987; van Roosmalen and Klein, 1988). Spider monkeys possess pectoral glands located in the chest region, which are used in

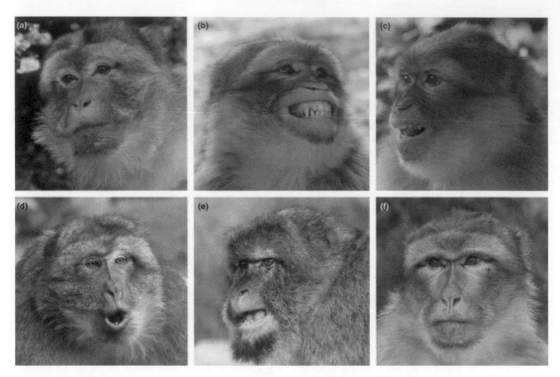

Figure 1.6 Different types of facial expressions in Barbary macaques: (a) 'alert' facial expression of an adult male mainly characterized by the raised eyebrows; (b) 'submissive' expression of a subadult female: the corners of the lips are fully retracted and the upper and lower teeth are shown; (c) 'affiliative' facial expression in a subadult female: the mouth is half open and the lips slightly protruded. This expression involves a chewing movement and clicking or smacking of the tongue and lips; (d) 'threat' face in an adult male: the eyebrows are raised, the animal stares intently and the lips are protruded to form a round mouth. This facial expression may be accompanied by a 'head bob'; (e) 'commenting' facial expression in an adult female: the mouth is half opened and the corners of the lips are slightly retracted; (f) neutral facial expression in a subadult male (from Teufel *et al.*, 2010).

greeting behaviour between males and in sexual contexts. In both of these contexts, individuals perform chest-rubbing and pectoral gland sniffing. In addition, female spider monkeys have an elongated clitoris that they use to deposit drops of urine as scent marks (van Roosmalen and Klein, 1988), while chimpanzees or bonobos do not seem to use olfactory communication to any significant extent. In the wild, chimpanzees use a variety of body postures, visual and tactile gestures as well as facial expressions in a variety of contexts, such as play, sexual behaviour, grooming and aggressive encounters (Goodall, 1986; Hobaiter and Byrne, 2011b; Roberts, Vick and Buchanan-Smith, 2012a) (Figures 1.3 and 1.7). Very little is known about nonvocal communication in spider monkeys and visual signals like facial expressions or gestures might not be suitable in their arboreal and thus densely foliated environment where visual signals may not be as rapidly recognized.

Figure 1.7 Playful interactions of chimpanzees that involve 'playfaces' (Photos: Katie Cronin, Katja Liebal, Cat Hobaiter).

Summary

This chapter provided the basis for this book by introducing the term communication and showing that communication can be understood at different levels. This is important since these levels are often confused, related to the fact that biologists are traditionally more interested in the ultimate level of communication (survival value and evolution), while psychologists mostly focus on the proximate level (causation and ontogeny). It is important to keep this in mind when reading this book, since the chapters differ in regard to their approach to communication. However, all chapters share the fundamental definition of communication, which involves some form of social behaviour that transmits information from one to another individual.

This chapter also introduced primates who (in contrast to many other mammals) are highly social throughout all life stages. Primates vary in regard to their social organization, structure and mating system and a complex combination of ecological and phylogenetic factors influences the evolution of their different social systems. To maintain such complex social systems, communication within and between groups is essential. As a result, primates use a variety of different forms of communication, which might differ between species as a result of their different social systems, but also life styles and habitats.

2 The morphology of primate communication

Like most mammals, primates possess the special senses of olfaction, taste (gustation), vision and hearing (audition) as well as the general sense of touch (somatosensation).[1] These senses represent the interface between an animal and its environment. Indeed, the survival and fitness of an animal depend upon its ability to accurately and adequately gather sensory stimuli. Olfaction and taste are grouped together as chemical senses because the stimuli they detect are molecules or compounds that either travel through the air or are settled on a physical substrate. Hearing, vision and somatosensation are grouped together as physical senses because the stimuli they detect are physical forces (e.g. sound waves, light, pressure, temperature). While taste is a critical part of an animal's ability to judge aspects of its environment (such as food quality), it does not play a direct role in social communication and thus will not be included in the ensuing discussion (Dominy, Ross and Smith, 2004).

While we discuss these senses separately from one another for the benefit of conceptualizing their roles in primate communication, we recognize that there is usually a mix of sensory modes that are locked together in any potential interaction. In the following sections, we describe the morphology of each of these sensory systems, how they function to gather and interpret external stimuli and how they are used in primate communication.

2.1 Olfactory communication

Elaboration of the visual system has long been considered a hallmark of primates and their evolutionary history (Cartmill, 1972; Elliot Smith, 1927; Martin, 1990; Ross, 1995). In most mammals, though, olfaction is the dominant sensory mode. Primates have been historically described as **microsmatic** (having a reduced olfactory sense) relative to many other mammals (Elliot Smith, 1927; Le Gros Clark, 1959; Martin, 1990). However, characterizing primates as microsmats has recently been challenged, since all primates still use scent in daily life (Barton, 2006; Hoover, 2010; Smith and Bhatnagar, 2004).

---

[1] 'Special senses' are distinguished from somatosensation because they use specialized sensory receptors in the eye, ear, nose and taste buds to transmit external stimuli via cranial nerves to brain centres where they are registered as sensations. Touch as a sense involves general sensory receptors in the skin and external stimuli are carried to brain centres via spinal nerves (Schultz, 1969; Young, 1957).

Primates use odours (volatile compounds) and pheromones (non-volatile compounds) that communicate a wide range of information to receivers, including sex of the signaller, reproductive status and receptivity in females, kin identity and even the individual signature of the signaller (see **Box 2.1**). Pheromones are relatively heavy hydrocarbon molecules produced by a wide range of animal species. It has been suggested that secretion of these pheromones is energetically efficient, since only a small volume needs to be secreted in order to be effective (Knapp, Robson and Waterhouse, 2006). In mammalian species, pheromones come in several varieties. One classification system for pheromones groups intraspecific signalling molecules into primer pheromones and releaser pheromones (Swaney and Keverne, 2009; Tirindelli *et al.*, 2009). Primer pheromones induce delayed, long-term behavioural or physiological responses that are mediated through activation of the **neuroendocrine system**. Releaser pheromones are responsible for individual odour signatures and trigger immediate, short-latency behavioural responses in conspecifics such as aggressive attacks, mating behaviour or territory marking.

The ability to discriminate between individuals is an essential foundation in primate social behaviour such as territorial marking, defence of the social status or resources, parental behaviour and mate choice (Halpin, 1980; Porter, 1998; Smith, 2006). These functions are completed by interdependent and parallel function of both the main and accessory olfactory systems (see section 2.1.1). Auditory and visual communication also fulfil these functions but olfactory communication has several advantages: the signal maintains its integrity in the absence of the sender, no light source is needed for successful communication, and the signal itself can span a long time range before it breaks down (see Chapter 1, Table 1.1 comparing such parameters of all modalities). In terms of development, olfaction becomes functional earlier than vision or hearing in most mammals (Gottlieb, 1971) and olfaction as a sense remains stable throughout life, providing a reliable, long-term identity label (Meredith, 1991; Porter, 1998).

Nocturnal primates seem to rely on olfactory communication more than diurnal primates and prosimians are thought to rely more on it than anthropoids (Ankel-Simons, 2000; Martin, 1990; Smith, Rossie and Bhatnagar, 2007; Smith, Siegel and Bhatnagar, 2001). However, many platyrrhines heavily use olfactory communication. For example,

Figure 2.1 Male chimpanzee inspecting the swelling of a female (Photo: Tobias Deschner).

Box 2.1 The major histocompatibility complex (MHC)

Each mammal produces an odour signal that is its own individual signature. The nature of that particular odour is influenced by a family of genes known as the **major histocompatibility complex** (MHC). MHC genes themselves are found in all vertebrates and they play a role in controlling immunological activity, tissue rejection, autoimmunity and immune responses. The MHC contains the most diverse set of genes located to date in vertebrates (Klein, 1986). Some MHC molecules are found on the surfaces of cells and are important for distinguishing 'self' cells from 'non-self' cells, obviously having great importance in medical arenas, such as as organ transplants and their potential rejection.

Odours associated with the MHC are used, for example, by mice to avoid mating with kin (Penn and Potts, 1999) and individual odour recognition has been demonstrated experimentally in a wide range of mammals including humans (see Beauchamp and Yamazaki, 2005, for a review). These individual odour signatures are located in urine and are at least partially determined by the MHC. It is thought that the MHC genes account for about half of individual odour signatures and odour variation (Beauchamp and Yamakazi, 2005; Eggert, Ferstl and Müller-Ruchholtz, 1999). Primates have a wide variety of MHC genes, and they recognize and respond to MHC-based odour cues (Knapp, Robson and Waterhouse, 2006). Thus, there is mounting evidence to indicate that the MHC is important for mother–offspring recognition and mate choice in a variety of mammals including humans (Jacob *et al.*, 2002; Knapp, Robson and Waterhouse, 2006; Wedekind and Furi, 1997; Wedekind *et al.*, 1995).

---

callitrichids (marmosets and tamarins) rely extensively on olfactory communication to maintain their social groups (Epple *et al.*, 1993). Prosimians have numerous scent glands scattered over their bodies and use urine to mark their territory. Male mantled howler monkeys sometimes perform 'urine-wash' displays in the sexual context, which represent combinations of an olfactory signal with tactile and/or visual signals (Jones and Van Cantford, 2007b). Much less is known about the extent to which Old World primates rely on olfactory signals. The few existing studies provide little evidence that great apes use olfactory cues, e.g. to detect the ovulation of females by inspecting their genitals or sniffing urine (Reichert *et al.*, 2002), although male chimpanzees are frequently observed to sniff sexual swellings in females (Goodall, 1986) (see Figure 2.1).

Determining exactly how much any given species uses olfactory communication is difficult. One way to 'measure' the capacity to receive chemical stimuli is to determine the area of the nasal cavity covered by olfactory neuroepithelium and the density of the receptors, called olfactory sensory neurons, it contains (see section 2.1.1). One way to fit more epithelium into a given nasal cavity and increase the density of olfactory sensory neurons is to have a long snout, like many lemurs. One of the characteristics of anthropoids compared to prosimians is a reduction in snout length leading to the hypothesis that they possess lower density of olfactory sensory neurons, and thereby a decreased olfactory ability relative to prosimians. However, this approach suggesting that the size of the olfactory neuroepithelium and/or the size of the brain areas involved in the processing of olfactory signals are suitable predictors for the olfactory abilities of primates has been repeatedly challenged (e.g. Laska, Seibt and Weber, 2000; Smith and Bhatnagar, 2004).

2.1.1   The perception of olfactory signals

The perception of olfactory signals involves two functionally and anatomically distinct systems that may have evolved in different ways: the **main olfactory system** (MOS) and the **accessory olfactory system** (AOS). The MOS is primarily concerned with analysing general molecular compounds that are airborne (volatile). These odours are functionally related to ecological factors such as food quality, for example state of ripeness, and predator detection. The AOS detects heavier compounds that cannot travel through the air including non-volatile, liquid-based pheromones found in urine or in the secretions of scent glands. This olfactory system is more involved with physiological markers of social and reproductive factors such as mate choice and care of infants.

All extant primates possess a functional, intact MOS. However, although all primates also have an AOS, it seems only to be intact and functional in prosimians and many platyrrhines, while **catarrhines** lack anatomical structures associated with the AOS (Barton, 2006; Dixson, 1998; Martin, 1990; Smith, Siegel and Bhatnagar, 2001; Smith et al., 2011).

2.1.1.1   The main olfactory system (MOS)

This system is composed peripherally of the external nose, the nasal cavity lined by olfactory neuroepithelium, and olfactory nerves that terminate in the olfactory neuro-epithelium as specialized sensory receptor cells, the olfactory sensory neurons (see also Chapter 3, section 3.3.1). The cartilaginous midline nasal septum separates the left and right nasal cavities. The olfactory nerves take information carried by airborne, volatile odours from the sensory receptors to the olfactory bulb of the brain. Generally, prosimian species have a greater density of olfactory sensory neurons than anthropoids and nocturnal species have a greater density than diurnal species, even after controlling for differences in body size (Smith and Rossie, 2006).

2.1.1.2   The accessory olfactory system (AOS)

While only prosimians and New World monkeys possess an intact functional AOS (Aujard, 1997; Barrett, Abbott and George, 1993; Smith et al., 2011), humans and at least some apes possess a **vestigial vomeronasal organ**, the primary structural compo-nent of the AOS. The **vomeronasal organ** is a paired tubular structure found in the lateral aspects of the cartilaginous nasal septum and is lined by a sensory neuroepithelium that contains specialized sensory receptors (see Chapter 3, section 3.3.1), which are primarily involved in the perception of olfactory signals in the context of reproductive behaviour.

If Old World monkeys or apes do not possess a functional AOS, the question remains how they do access such pheromonal information. Odour itself is an important part of reproductive behaviour in these species and is used, for example, to attract males to oestrous females (Keverne, 1999; Restrepo et al., 2004). There is compelling evidence that some of these catarrhine species are able to detect pheromones with parts of the

Figure 2.2 Scent-marking in ring-tailed lemurs (Photo by Rhett A. Butler/mongabay.com).

MOS (e.g. Barton, 2006) indicating that the functional pathways of the MOS and AOS are not completely independent.

2.1.2   The production of olfactory signals

Scents that contain pheromones are released by primates in a variety of ways. Many prosimian species urine-mark their territories, especially in the lemurids (Kappeler, 1998; Perret, 1995). Urine-marking has been shown to be associated with intrasexual competition in many prosimians, both males and females (delBarco-Trillo *et al.*, 2012; Gursky, 2003; Kappeler, 1998; Perret, 1995). Some anthropoids, including callitrichids, howler monkeys, and squirrel monkeys, show urine-marking behaviours (Candland, Blumer and Mumford, 1980; Epple *et al.*, 1993; Milton, 1975; Ziegler, Snowdon and Uno, 1990). There is also evidence that some members of the lemurids scent-mark with saliva during non-nutritive gouging of tree branches (Patel and Girard-Buttoz, 2008; Rasoloharijaona, Randrianambinina and Joly-Radko, 2010).

Prosimians and callitrichids also have a variety of specialized scent glands (Hagey, Fry and Fitch-Snyder, 2007; Schilling, 1974). Many lemurids have glands on their wrists, the circumgenital region and on their throats, which they rub on environmental structures (Figure 2.2) to defend food resources, mates and/or territorial boundaries. For example, male ring-tailed lemurs engage in 'stink fights' with extra-troop males. They scent-mark their tails by rubbing them between the brachial glands on their wrists or by urinating on them and then carry their tails over their bodies and flick them at other males (Schilling, 1979).

Figure 2.3 Gorilla 'chest beat'. Gorillas perform this auditory gesture (a) to invite others to play or (b) as part of their display behaviour (Photos: Simone Pika).

Callitrichids scent-mark in similar ways and seem to use olfactory cues to suppress reproduction in females. These species live in small groups consisting primarily of family members but include only a single breeding female (see also Chapter 1). Although the mechanism is not entirely clear, the reproductive female releases some sort of suppressive olfactory cue, which – together with additional behavioural cues – results in other female group members being typically anovulatory or not engaging in sexual behaviour (Abbot, 1984; Digby, Ferrari and Saltzman, 2007; Epple *et al.*, 1993).

2.2   Auditory communication

Sound plays a large role in primate life both in nocturnal and diurnal species. They rely on hearing to locate predators or (in the case of carnivorous species) to locate prey and they use vocalizations to defend territorial boundaries and food resources, negotiate social interactions and locate kin or mates. Auditory signals, unlike visual and olfactory signals, have the advantage of being able to travel long geographic distances (see Chapter 1, Table 1.1). Like olfactory and visual communication, auditory signals can transmit information on sex, age and reproductive status, and can contain an individual signature. While the production of vocalizations is the most obvious kind of auditory communication, there are sounds created without the use of vocal folds, such as 'kiss squeaks', 'lipsmacking' sounds, or 'chest beats' (Figure 2.3). Therefore, although both vocalizations and sounds convey auditory information, they differ in regard to the structures involved in the production of these auditory signals and most likely also in regard to the underlying neural substrates (see Chapter 3, section 3.3.5). In the following sections, the focus is on vocalizations and thus those signals that involve the use of vocal folds.

2.2.1   The production of vocalizations

The generation of a vocalization involves the expulsion of air from the lungs, regulated by the diaphragm, the chest muscles and ribs. The air needs to pass through the larynx, which is located in the neck. The larynx is bordered by several cartilages and contains numerous small muscles. The interplay of these laryngeal structures together with the air pressure that builds up beneath the larynx results in the rhythmic opening and closing of the vocal folds. This vibration causes sound waves to be created.

The frequency of this sound is controlled in part by the vocalis muscle. When this muscle contracts, the diameter of the glottis, which is the opening between the vocal folds, is constricted, raising the frequency of the sound. When this muscle is relatively relaxed, the glottal diameter is increased, lowering the frequency of the sound. The sound generated within the larynx itself then moves on to the pharynx, oral cavity, and nasal cavity (see Figure 2.4) where it is resonated and amplified.

In all primates except for humans, the **epiglottis**, which is a flap attached to the entrance of the larynx, is in contact with the posterior border of the soft palate.

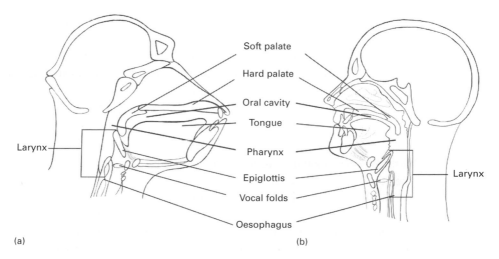

(a)   (b)

Figure 2.4 Vocal tracts of chimpanzees (a) and humans (b).

Figure 2.5 Vocal sacs of siamangs, which are inflated during their characteristic **duetting** songs (Photo: Thomas Geissmann, www.gibbons.de).

This contact reduces the volume of the pharynx available for resonance of sound and hinders much phonation via the oral cavity. Instead, most of the vibrating air passes into the nasal cavity for resonance. Unlike any other living primate, humans have an epiglottis separated from the soft palate. This expands the volume of the pharynx available for resonance and allows air to freely enter the oral cavity for resonance.

The difference in air passage routes is one of the important modifications of the vocal tract that – together with a variety of additional changes in anatomical and neural structures – played an important role in the evolution of human language (Lieberman, 2012). For example, the permanently descended larynx of adults may be an important anatomical modification of the human vocal tract, and has historically been identified as a key feature that may differentiate humans from other animals. It has been argued to represent a key prerequisite for modern speech and thus to have played an essential role in the evolution of language (Lieberman, 1984; Lieberman, Klatt and Wilson, 1969). The descended larynx allows humans to make more extensive modifications to the shape of the vocal tract compared to animals with a higher larynx and as a consequence, humans can produce a variety of speech sounds. The uniqueness of a descended larynx in humans has, however, been challenged by Fitch and Reby (2001), who showed that a descended larynx has evolved independently in different lineages. For example, it is also found in the red deer, where it most likely serves the function of exaggerating the perceived body size, since the elongation of the vocal tract – caused by the descended larynx – results in a decrease of the vocal tract resonant frequencies. Therefore, the descended larynx is most likely not directly related to the ability to produce a more varied sound repertoire, but may represent a preadaptation for speech- or language-specific functions (Fitch and Reby, 2001).

Even though the oral cavity route is reduced in nonhuman primates, some air does enter the oral cavity along with the nasal cavity for resonance. The vibrating air that enters the oral cavity can be modified by the position of the tongue and the position of

(a)

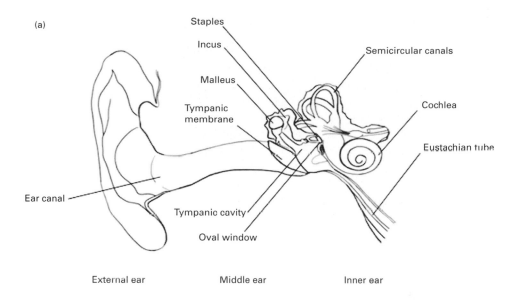

Staples

Incus

Semicircular canals

Malleus

Cochlea

Tympanic
membrane

Eustachian tube

Ear canal

Tympanic cavity

Oval window

External ear   Middle ear   Inner ear

(b)

Inner hair cells

Tectoral membrane

Outer hair cells

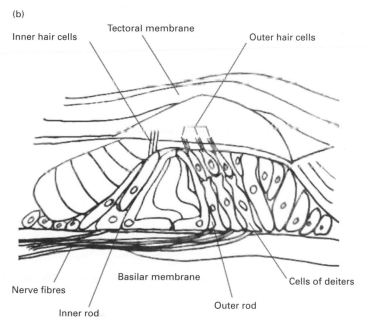

Nerve fibres

Basilar membrane

Cells of deiters

Inner rod

Outer rod

Figure 2.6 The primate ear consists of three major parts, the outer, middle and inner ear (a). The basilar membrane (b) is located within the cochlea and transforms sound waves into mechanical movements.

the lips. Some species, such as gibbons and siamangs, also possess laryngeal air sacs that can further alter the quality of the vocalizations. Males of long-nosed or proboscis monkeys take air modification further by vibrating the air through their pendulous external nose (Ankel-Simons, 2000).

2.2.2   The perception of vocalizations

Across mammals, the ability to hear high-frequency sounds is related to the size of the head: mammals with small heads tend to hear higher frequency sounds than mammals with large heads (Masterton, Heffner and Ravizza, 1969). Therefore, although most primate species are able to hear frequencies from about 35–45 kHz, galagos and ring-tailed lemurs have been reported to detect higher frequencies around 60 kHz (Gillette *et al.*, 1973; Heffner, 2004; Heffner, Ravizza and Masterton, 1969), while humans and chimpanzees are able to detect much lower frequencies, around 17–29 kHz (Heffner, 2004). It is important to note, however, that only a fraction of primate species have been tested for frequency range of hearing.

Sounds are received by the ear. In primates, like in most mammals, it consists of three parts: an external ear, a middle ear and an inner ear. The external ear, or pinna, is composed of elastic cartilage. Most nocturnal prosimians, such as galagos and aye-ayes, have relatively large, semi-membranous external ears, which are highly mobile in a wide range of directions and can be moved independently from each other. In many diurnal species, the outer ear is relatively small and immobile. The external ear acts as a funnel to collect sound waves and to forward them via a bony tube to the tympanic membrane, or eardrum (see Figure 2.6).

The tympanic membrane is a connective tissue sheet attached to a bony ring formed by the tympanic bone, which serves as a separation between the outer ear and the middle ear cavity. The sound waves encounter and vibrate the tympanic membrane. The vibrating tympanic membrane in turn causes vibratory movement in the three middle ear ossicles (bones): the malleus, incus and stapes. The stapes itself is in contact with a membrane separating the middle ear cavity from the inner ear cavity, the oval window (see Figure 2.6a).

The inner ear cavity contains the cochlea and semicircular canals. The cochlea is a coiled bony tube filled with fluid. It contains the **basilar membrane** (Figure 2.6b) that vibrates in response to movement generated from the stapes and oval window and thus transforms the incoming sound waves into mechanical movements. These movements displace hair cells located in the basilar membrane, which triggers electrical impulses in the auditory nerve (see Chapter 3). The semicircular canals, while housed with the cochlea, are not involved in hearing, since these fluid-filled structures are responsible for sensing movement, position and orientation.

2.3   Visual communication

While most mammals depend primarily on olfaction as their dominant mode of communication, living primates are usually referred to as examples of elaboration of the visual system and they rely heavily on visual information (Cartmill, 1972; Elliot Smith, 1927; Martin, 1990). Accordingly, the primate visual system has evolved such that primates have unusually high **visual acuity** relative to most other mammals (Ankel-Simons, 2000; Kirk and Kay, 2004; Martin, 1990). Visual acuity, the ability to continue to see the details

> **Box 2.2** Evolution of binocular vision
>
> The majority of mammals have eyes and bony orbits that face laterally, away from one another. This lateral placement yields mostly monocular visual fields that have a very low degree of binocular overlap. Thus, species with laterally placed eyes and low fields of binocular overlap, such as rodents and ungulates, probably have wide panoramic visual fields. This may be advantageous for species that are typically more preyed upon than are predators themselves. Having the eyes placed forward brings them, necessarily, closer together and reduces the degree of separation between the two eyes. Thus, having **binocular visual fields** is necessary for the development of **stereoscopic** vision, which depends upon central nervous processing of data derived from simultaneous observation of objects with both eyes (Martin, 1990). Primates and anthropoids in particular have evolved a relatively high level of orbital convergence, which is a high degree of the orbits facing the same direction (Cartmill, 1972; Heesy, 2003; Ross, 1995). This convergence is thought to be related to the evolution of a binocular visual field and stereoscopic vision, which is associated with a high degree of depth perception ability and three-dimensional perception. While the factors leading to the evolution of high degrees of orbital convergence, a binocular field of vision and stereoscopic vision are not completely clear, many authors have considered these developments to be tied to an arboreal, predatory lifestyle in early primates, a hypothesis called the *Nocturnal Visual Predation Hypothesis*. This hypothesis was first developed by Cartmill (1972, 1974, 1992) and refined by subsequent investigators (Heesy and Ross, 2001; Kirk *et al.*, 2003). Cartmill argued that the earliest primate adaptations involved nocturnal, visual predation on insects in the lower canopy of tropical forests. This implies that the last common ancestor of extant primates, like some living prosimians, was nocturnal and fed largely on insects that were visually located (not located via olfaction) and manually captured. As a mammal hunting for insects in the terminal branches of trees, it would indeed be advantageous to have highly accurate depth perception abilities and the ability to judge the immediate surroundings in three dimensions.

of an object separately and unblurred as those details are made smaller and closer together, is a feature that living primates share with birds of prey more than with most non-primate mammals (Walls, 1942; Young, 1957). The evolution of this high degree of visual acuity is due in part to the evolution of the eye and optical convergence, the medial or inward migration of the bony orbits and eyes from the primitive lateral position found in most other mammals. This medial migration of the bony orbits and eyes allowed for the development of a binocular visual field and stereoscopic vision (**Box 2.2**) (Ankel-Simons, 2000; Heesy and Ross, 2001). Because of this forward-facing position of the eyes, primates are capable of perceiving depth. However, as opposed to other species with their eyes located on each side of the head, the visual field of primates covers a comparably smaller area. For example, humans have a horizontal field of view of approximately 200 degrees covered with their two eyes, of which a portion of 120 degrees represents the binocular field that is seen by both eyes.

Visual communication depends upon light or electromagnetic radiation, which consists of different types of waves. These light waves travel quickly, therefore transmitting information almost instantaneously from signaller to receiver. However, light waves become disrupted and distorted in dense vegetation. As a consequence, visual signals should be used over short distances only in densely foliated environments or their use should be more pronounced in more open habitats (Maestripieri, 1999). Furthermore, since visual signals like facial expressions, manual gestures or body postures occur so rapidly, their perception depends on the presence and visual attention of the sender (see Chapter 1, Table 1.1).

Most aspects of any primate's lifestyle are driven by what time of day it is awake and active (Ankel-Simons, 2000; Heesy and Ross, 2001). Generally, primates that spend most of their waking time at night are referred to as nocturnal. This includes most prosimians like galagos, lorises, tarsiers, aye-ayes and cheirogaleids (dwarf lemurs and mouse lemurs) and one anthropoid, the owl monkey. Nocturnal species depend heavily upon olfactory and auditory communication, but they employ vision to a great extent as well. They tend to have relatively large eyes, probably to allow more light to enter the eye. Nocturnal primates are referred to as using **scotopic vision**, which works at low light levels and depends primarily on **rods** (Ankel-Simons, 2000; Martin, 1990). Primates that spend most of their waking time in the daylight are referred to as diurnal. This includes some prosimians such as some lemurs (e.g. ruffed lemurs) and most indriids as well as all anthropoids except for the owl monkey. Diurnal primates tend to have relatively smaller eyes than nocturnal species and they are described as using photopic vision, which depends upon high light levels and is dominated by cone light receptors (Ankel-Simons, 2000; Martin, 1990).

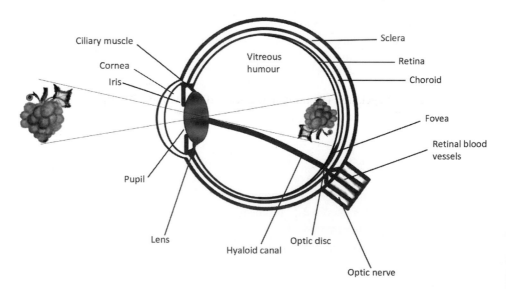

Figure 2.7 The primate eye with the retina (with an example of a reversed image).

2.3.1   The perception of visual signals

Visible light that conveys visual signals to receivers is perceived by the eyes. The light enters the eye through the cornea and then encounters the lens. The lens can refract or change the direction of the incoming light so that it can be focused on the **retina** (Figure 2.7). Through contraction of small, involuntary muscles called ciliary muscles, the lens changes its shape to alter the focal distance of the eye so it can focus on objects at differing distances to produce a sharp image on the retina. This process of accommodation helps produce a sharp image at varying distances (Martin, 1990).

The retina is the innermost lining of the eye and is considered to be an extension of the optic nerve. It contains two main types of photoreceptors, rods and **cones**, that react to light thus initiating several chemical and electrical processes. This in turn triggers nerve impulses, which are then further transmitted via the optic nerve to the corresponding areas in the brain (see Chapter 3, section 3.3.3).

Rods and cones are specialized neurons and they each contain a specific light-sensitive pigment. Rods contain the pigment **rhodopsin**, which facilitates the absorption of low-intensity light. They are good at discriminating shades of grey and function primarily at night. Cones contain the pigment **iodopsin**, which functions primarily in bright light. They are used in colour vision and are characterized by elevated visual acuity. In primates, there are three types of cones: red, green and blue cones, named according to their ability to absorb light of specific wavelengths.

These types of cones work together and overlap in function such that primates are able to perceive many different shades among the red, green and blue spectrum. Nocturnal species like many prosimians are either monochromatic and thus can only recognize shades of grey or they are **dichromatic** and can differentiate blue and green, but not red. Diurnal species generally possess good colour vision, but there is considerable variability between groups. Old World primates like Old World monkeys, apes and humans are **trichromatic** and thus can differentiate between blue, green and red, while most species of New World monkeys are dichromatic (Mollon, 1989). Why trichromatic colour vision has evolved is still debated (Surridge, Osorio and Mundy, 2003), but the capacity to discriminate between green and red is advantageous for finding ripe fruits and young leaves (which are often red in the tropical forest) against a background of dense vegetation (Dominy and Lucas, 2001; Mollon, 1989).

All mammals have a **blind spot** on the retina, which contains no photoreceptor cells. This is the area where the optic nerve fibres exit the eye to transmit the neural impulses to the brain. In primates, the blood vessels that pass from this blind spot to the rest of the retina pass around a lateral area of the retina called the central visual area. Therefore, this area of the retina is free of blood vessels and is thus permitted greater precision in retinal imaging. This unique aspect of the primate retina is referred to as 'retinal centralization'. Some nocturnal prosimians and diurnal lemurs simply have this thickened area of the retina while tarsiers and anthropoids have a small pit, or **fovea**, in this temporal region of the retina, which is densely packed with photoreceptors (Martin, 1990).

The **choroid** is the other lining layer of the eye (see Figure 2.7). This layer contains the majority of blood vessels of the eye, a vascular network called the choroid plexus.

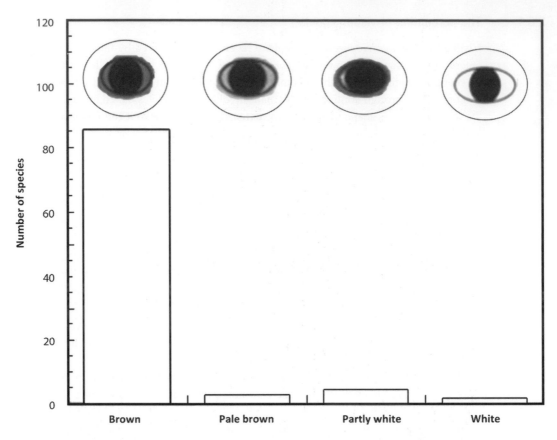

Figure 2.8 Variation of scleral colour across primates. Shaded areas indicate the portion of the eyeball with colour (modified after Kobayashi and Kohshima, 2001). In the majority of species (92%, 85/92), the exposed part of the sclera is uniformly brown or dark brown. Barbary macaques and pigtail macaques possess a pale brown sclera, while some tamarins, marmosets and Goeldi's monkeys have a partly white sclera. A white sclera is unique to humans.

Most mammals have an additional cell layer between the choroid and retina called the tapetum lucidum. This layer reflects light back through the retina to increase the available ambient light and is the structure that causes mammals' eyes to 'glow in the dark' when a light is shined on them. The pattern of distribution among primates is somewhat reflective of their time of activity, since the tapetum lucidum is found in all prosimians except for some diurnal lemurs and tarsiers, while diurnal anthropoid species do not possess it (Ankel-Simons, 2000; Martin, 1990).

The sclera is the outer, protective layer of the eye. Only humans possess a white sclera that does not have any pigmentation and is thus in stark contrast to the darker iris (Kobayashi and Kohshima, 1997). Therefore, the unique morphology of the human eye and the sclera in particular illustrates the important role eyes play in human communication, e.g. in the form of establishing eye contact, gazing at others, or directing others' attention to other events or objects in the environment (see Chapter 8, see section 8.3.2) (Figure 2.8).

Summary

This chapter demonstrated that primates produce and perceive signals of different sensory modalities. However, the occurrence and predominance of each of these modalities differ between species depending on their life styles, habitats or social systems. Traditionally, primates are considered microsmatic, but there is increasing evidence that olfactory signals play an important role in a variety of contexts, such as kin recognition and territorial and reproductive behaviour. Although the use of scent- and urine marking is mostly limited to prosimian species, olfactory signals seem to play an important role in anthropoid species as well, particularly in the form of pheromones and thus in the recognition of kin or of the reproductive status of females.

In regard to auditory signals, the morphological structures of the perception and production of vocalizations in primates generally resemble those of other mammal species. However, as opposed to other primates, only humans possess an epiglottis that is separated from the soft palate. This difference in the anatomy of the vocal tract results in a unique air passage route that enables humans to produce a much more sophisticated modification of the sounds originating from the larynx, which likely played an important role in the evolution of language.

Vision is often referred to as the most dominant sense in primates. This is supported by the fact that only primates possess forward-directed, comparably large eyes, resulting in binocular visual fields, which enable stereoscopic vision. Again, humans differ from other primates in regard to the morphology of their eyes and the sclera in particular. Unlike other primates, humans possess a sclera that lacks any pigmentation, which most likely reflects the important role of eyes and eye contact in particular in human communication.

3 The neural substrates of primate communication

Primates – like any animate being – receive different kinds of sensory input from their environment mediated, amongst others, by olfactory, visual and auditory channels. The processing of these very different kinds of information is mediated by the nervous system. In primates and other vertebrates, it is comprised of two major parts, the central and the peripheral nervous system. The central nervous system consists of the brain and the spinal cord, and is responsible for integrating the sensory information it receives from the body and selecting an appropriate motor response, thus coordinating the corresponding actions. The peripheral nervous system connects the central nervous system with the body. It relays sensory information from the body to the central nervous system and carries nerve impulses from the central nervous system to different body parts, such as limbs or organs.

The focus of this chapter is on those areas of the brain that are specifically involved in the perception and production of communicative signals. First, a brief section will introduce the major parts of the primate brain and will highlight those features that differentiate primates from other mammals with special reference to the structures that are unique to the human brain. Second, the corresponding brain areas involved in the production and perception of signals will be presented separately for each modality, and with special emphasis on differences among species. Third, a final section will focus on those brain areas that are responsible for the integration of multiple sources of sensory information to emphasize the multimodal nature of primate communication at a neural level.

3.1 Anatomy of the primate brain

Primate brains, like those of other vertebrates, consist of five major parts: the myelencephalon, metencephalon, mesencephalon, diencephalon and telencephalon (for an overview, see Table 3.1). This section will only provide a more general introduction to these major brain parts with focus on those areas that are relevant for primate communication (for more detailed information, see Martin, 1990; Nieuwenhuys, Donkelaar and Nicholson, 1998).

The myelencephalon comprises the medulla oblongata, which contains several cranial nerve nuclei, for example, the nucleus ambiguus and the hypoglossal nucleus, which innervate the muscles of the larynx and tongue, respectively.

Table 3.1 Major divisions of the primate nervous system.

Central nervous system (CNS)	Brain (encephalon)	Forebrain	Telencephalon (cerebral hemispheres)	Neocortex Basal ganglia Limbic system
			Diencephalon	Thalamus Hypothalamus
		Midbrain	Mesencephalon	
		Hindbrain	Metencephalon	Cerebellum Pons
			Myelencephalon	Medulla
	Spinal cord			
Peripheral nervous system (PNS)	Somatic (skeletal) nerves			
	Autonomic ganglia and nerves	Sympathic division		
		Parasympathic division		

The metencephalon consists of two major parts. The pons contains several nuclei that relay information from the forebrain to the cerebellum, and nuclei that control eye movements and facial expressions. The cerebellum is the major area for motor coordination, although it does not initiate movements.

The mesencephalon comprises the tectum, which contains several nuclei. The superior colliculus is involved in the processing of visual information and the control of eye movements, while the inferior colliculus is involved in the relay of auditory information to the auditory cortex.

The diencephalon comprises the thalamus and the hypothalamus. While the thalamus relays different kinds of sensory information (auditory, visual and touch/**somatosensory**) from subcortical areas to the cerebral cortex, the hypothalamus contains the pituitary gland, which controls most of hormone secretion.

The telencephalon consists of three major parts. The basal ganglia are important for controlling movements, while the limbic system contains structures involved in a variety of functions such as emotion (amygdala), long-term memory and learning (hippocampus), and olfaction (olfactory bulb). The neocortex is the outermost layer of the telencephalon (see Figure 3.1). Evolutionarily, it represents the newest part of the cerebral cortex, which is unique to mammals and particularly enlarged in primates (see section 3.2) (Finlay and Darlington, 1995). It is involved in the execution of motor movements and the integration of sensory information as well as in higher cognitive functions that seem unique to humans, such as conscious thought or language.

The telencephalon is divided into two hemispheres that are connected via the corpus callosum. Each hemisphere is divided into four different lobes. Figure 3.1 provides an overview of the major cortical divisions of the human brain to facilitate the reader's navigation through the different brain areas involved in the production

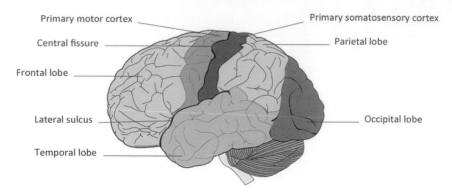

Figure 3.1 Major cortical divisions of the human brain (lateral view of the left hemisphere). The division into these lobes is based on the corresponding bones that cover them and not on structural or functional differences.

and perception of different communicative modalities. The occipital lobe is responsible for the processing of visual information (see section 3.3.3), the temporal lobe contains the auditory cortex (3.3.4), but is also involved in the perception of faces and facial expressions (3.3.3.1), the parietal lobe is associated with integration of different kinds of sensory information (3.4.1), and the frontal lobe mediates executive or cognitive functions such as attention, self-control, planning and working memory.

3.2   Evolution of the primate brain

In general, the primate brain consists of many of the same parts and structures as the brain of other mammals. The distinguishing feature of the primate brain is that it is in general larger than predicted by body size (Dunbar, 1998; Shultz and Dunbar, 2007). This high degree of **encephalization** is commonly ascribed to the enlargement of the cerebral cortex, especially the neocortex. This is mostly achieved by the expansion of the occipital lobe indicating the importance of vision in primates, and the enlargement of the frontal lobe, especially the prefrontal cortex responsible for a variety of higher cognitive (executive) functions. Although the size of primate brains increased over the course of evolution, there is some evidence that some parts might have evolved independently from the rest of the brain. Thus, since the brain consists of functionally distinct areas, natural selection caused selective size changes of some brain areas but not others and as a result, the brain was not enlarged in its entirety (Barton and Harvey, 2000). For example, the size of the olfactory bulbs decreased in anthropoids compared to prosimians, most likely due to their less pronounced olfactory communication (see Chapter 2), while the particular enlargement of the frontal lobes is characteristic for great apes (Rilling, 2006).

The human brain is distinct from other primates' brains in the overall larger proportion of the neocortex. Although there is some debate whether the human prefrontal cortex is disproportionately enlarged given the overall size of the human brain (see Rilling, 2006; Semendeferi *et al.*, 2002), it seems that the prefrontal cortex of

Box 3.1 Neural substrates of language

Language is considered a localized and lateralized function of the human brain: two main areas of the human brain are involved in language processing and both are located in the left hemisphere. Broca's area is located in the inferior frontal **gyrus** and consists of Brodmann areas 44 and 45. Wernicke's area has been traditionally located in the posterior section of the superior temporal gyrus that is anterior to the primary auditory cortex (the anterior part of Brodmann area 22), but there is recent evidence for a greater Wernicke's area that extends to the posterior section of the lateral **sulcus** in the parietal lobe (Brodmann area 39 and 40) (Catani and Jones, 2005). Broca's area is considered the main area for language production, while Wernicke's area is mainly involved in language comprehension. However, this classic dichotomy has been challenged more recently. For example, although, neurons from Broca's area project to the larynx, the tongue, and mouth motor areas in the motor cortex and therefore control the structures that are involved in the articulation of sounds (Liotti, Gay and Fox, 1994), Broca's area is also involved in language comprehension, such as syntax recognition (Embick *et al.*, 2000) and other, less language-related cognitive tasks such as the recognition of hand actions (Decety *et al.*, 1997). Wernicke's area, on the other hand, is usually described as the area of language comprehension and thus responsible for tasks such as auditory word recognition (DeWitt and Rauschecker, 2012). However, there is also evidence that Wernicke's area is at least partly involved in the motor processing of language (Hickok and Poeppel, 2007). Although the left hemisphere is predominantly involved in language processing, there is recent evidence for a 'division of labour' between the two hemispheres, since the right hemisphere contributes to both language production and perception (Jung-Beeman, 2005). For example, the posterior temporal lobe of the right hemisphere is involved in the processing of affective prosody of speech (Friederici and Alter, 2004). Some models even suggest that temporal information of language is processed in the left hemisphere, while spectral information is mostly processed in the right hemisphere (Zatorre and Belin, 2001; Zatorre, Belin and Penhune, 2002).

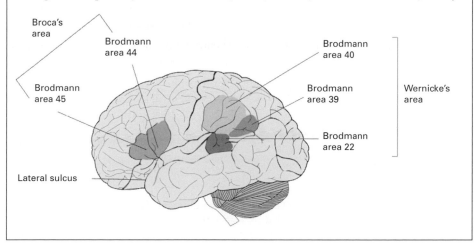

humans shows distinctively richer interconnectivity, and an increase in surface area, caused by a higher degree of **gyrification**. In humans, the prefrontal cortex is responsible for higher cognitive functions such as planning, reasoning, abstract thought and language (see **Box 3.1** for neural substrates of human language).

3.3   The processing of communicative signals

In general, the processing of a signal involves the perception of this information by particular receptor cells in, for example, the retina, ear, skin or nasal epithelium, respectively (see Chapter 2), which relay the sensory information to the corresponding areas in the central nervous system. From there, efferent neurons, which relay signals away from the brain, project to particular effectors such as muscles. This basic process differs of course depending on the kind of signal perceived and is modulated by a variety of factors such as the degree of arousal, attention and also memory. For the purpose of this chapter, we differentiate between production and perception of signals, and focus on the major brain areas involved in these processes. The Figures 3.2–3.8 illustrate the major brain structures involved in the processing of each modality, separately for its perception and production.

3.3.1   Olfactory communication: perception

As described in Chapter 2, all primates have a main olfactory system (MOS), while only prosimians possess a clearly functional accessory olfactory system (AOS). These two structures differ in some aspects related to the neural substrates underlying the perception of olfactory information.

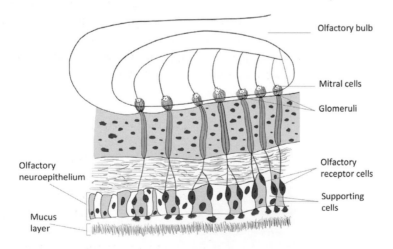

Figure 3.2 Perception of olfactory signals by the main olfactory system. The olfactory epithelium is located in the nasal cavity and consists of olfactory sensory neurons (receptor cells) that project to the olfactory bulb.

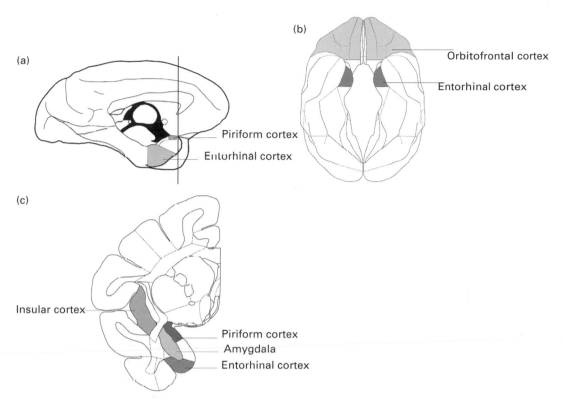

Figure 3.3 Neural substrates involved in the perception of olfactory signals. (a) Medial view, (b) ventral view and (c) coronal slice of the rhesus macaque brain. The vertical line in (a) indicates the location of the coronal slice.

In the MOS of primates, airborne (volatile) odorants are perceived by olfactory sensory neurons in the olfactory epithelium of the nasal cavity (Dulac and Torello, 2003). The olfactory sensory neuron projects its **axon** to the olfactory bulb, where it synapses with mitral cells (Figure 3.2). These synapses between the axons of the olfactory sensory neurons and the dendrites of mitral cells constitute a glomerulus. Each odour activates a different array of glomeruli, resulting in specific 'odour maps' or 'odour images', similar to the spatial patterns in the visual system (see section 3.3.3).

The mitral cells then relay this information to different brain areas including the amygdala that is involved in the affective reaction to odour and two major regions in the olfactory cortex: the piriform cortex in the frontal lobe, which is closely associated with odour identification, and the entorhinal cortex in the medial temporal lobe that is associated with memory (Zald and Pardo, 1997) (Figure 3.3a–c). It is of note that the olfactory cortex is the one area of cortex that receives direct sensory input without an interposed thalamic connection. From the olfactory cortex, information is relayed to the insular and orbitofrontal cortex (Figure 3.3c). The insular cortex also receives taste input and may be the site where olfactory and gustatory input is integrated to produce the sensation of 'flavour'. The orbitofrontal cortex receives these projections as well as projections from the amygdala. This area of neocortex is considered to be associated with forming conscious perception of odour and seems to be involved with decision-making based upon olfactory input.

The AOS, on the other hand, is responsible for detecting non-volatile odorants such as pheromones, which are perceived by the sensory neurons of the vomeronasal organ (e.g. Halpern, 1987; Dulac and Torello, 2003). Similar to the MOS, these receptors project to mitral cells in the accessory olfactory bulb (Swaney and Keverne, 2009). Contrary to the MOS, the mitral cells of the AOS project directly to the amygdala and hypothalamus, bypassing cortical areas (Jia, Chen and Shepherd, 1999). How olfactory information is eventually encoded in the brain and how primates differentiate between different types of odours is still a matter of debate.

3.3.2   Olfactory communication: production

Very little is known about the neural substrates mediating the production of olfactory signals in primates. Most of the existing studies are conducted with rodents to investigate which brain areas are involved in behaviours such as scent-marking or the release of pheromones (e.g. Bamshad and Albers, 1996; Petrulis, Peng and Johnston, 1999; Yahr and Stephens, 1987). These studies indicate the importance of different hormones such as vasopressin stimulating scent-marking in these rodents and thus refer to the importance of the pituitary gland in the hypothalamus regulating the release of these hormones. A study with mouse lemurs found that the removal of the vomeronasal organ in males resulted in a lack of interest in females and a decrease of aggressive interactions with other males, but did not affect their marking behaviour (Aujard, 1997). The author's interpretation was that the changed behaviour of male mouse lemurs is not due to a chemosensory deficit caused by the removal of the vomeronasal organ, but rather represents a functional disturbance of those brain areas connected to the vomeronasal system. In summary, although it is obvious that the hypothalamus and most likely also the amygdala are involved in olfactory communication in primates, it is still largely unknown how exactly the brain relays efferent information to the effectors to produce the corresponding behaviours.

3.3.3   Visual communication

Visual information is perceived by photoreceptors in the retina of the primate eye (see Chapter 2, section 2.3). From the ganglion cells of the retina, the optic nerve consisting of a bundle of axons projects this information to the lateral geniculate nucleus located in the thalamus from where information is relayed to the primary visual cortex in the occipital lobe of the brain. From there, it is hierarchically processed in several other secondary visual areas that are referred to as extrastriate visual cortex. The number of areas involved in the processing of visual information varies across primates, ranging between 14 in galagos to more than 20 in New World monkeys (owl monkeys) and Old World monkeys (macaques), and they cover more than half of the neocortex in those nonhuman primates (for a review, see Sereno and Allman, 1991). Many of these areas are retinotopically organized, which means that visual input from specific regions of the retina is mapped onto individual neurons in the visual cortex. Thus, adjacent points in the visual field are represented in adjacent regions in the visual cortex. In humans, the basic organization of visual areas is similar to those identified in monkeys, but they differ in regard to their position and size. More specifically, foveal representation in humans is larger, as more neurons are available to process images at the centre of a human's gaze (Sereno *et al.*, 1995).

Similar to the dual processing of auditory information (see section 3.3.4), a two-stream hypothesis is suggested for the processing of visual information after it exits the visual cortex (Ungerleider and Mishkin, 1982). While the ventral stream is responsible for object identification (*what*), the dorsal stream processes spatial information (*where*) (but see Schenk and McIntosh, 2010). The following sections, however, will specific-ally focus on those substrates involved in the processing of facial expressions and visual gestures.

3.3.3.1   Faces and facial expressions: perception

The perception of facial expressions relies on the visual channel – as does the perception of visual gestures. Still, these two modalities are considered separately in this chapter, because it is suggested that the perception of faces and facial expressions is one of the most highly developed visual skills in humans and most likely also in other primates (Emery, 2000; Tomonaga, 2010).

In humans, face processing relies on a distributed neural network that includes multiple, bilateral regions in the brain, with the core system consisting of occipitotem-poral regions of the visual cortex (Haxby, Hoffman and Gobbini, 2000, 2002). *Invariant* facial features that characterize a face as such and thus enable the recognition of *face identity* are mediated by the lateral fusiform gyrus, while *changeable* features such as different *facial expressions* that can convey very different information are processed in the superior temporal sulcus. Furthermore, an extended system that includes regions such as the amygdala or the anterior temporal lobe contribute to the processing of faces by mediating emotional perception or recognizing a person's identity, respectively. Thus, in humans, different brain regions that are part of a larger distributed network mediate the perception of face identity and facial expressions separately (Haxby, Hoff-man and Gobbini, 2000, 2002).

There is increasing evidence for a similar distributed network in the occipitotem-poral cortex in both monkeys and apes, which mediates face recognition in these nonhuman primates (Freiwald and Tsao, 2011; Parr and Hecht, 2011; Parr *et al.*, 2009). Similar to humans, face identity and facial expressions are processed in different regions of the brain that include structures in the temporal lobe, such as the inferior temporal cortex and the superior temporal sulcus, as well as the amygdala and the premotor cortex (Figure 3.4) (for a comprehensive review, see Barraclough and Perrett, 2011).

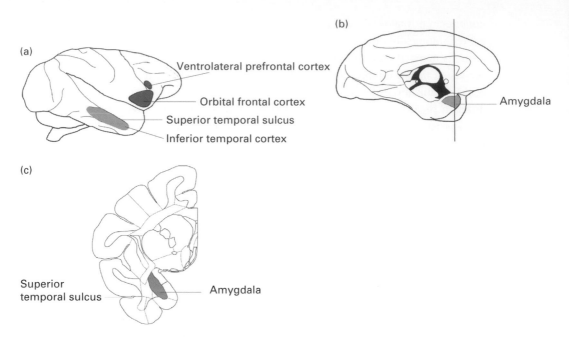

Figure 3.4 Neural substrates involved in the perception of faces and facial expressions. (a) Lateral view, (b) medial view and (c) coronal slice of the rhesus macaque brain. The vertical line in (b) indicates the location of the coronal slice.

In rhesus macaques, the face-selective neurons in these different brain areas differ in regard to how they respond to particular characteristics of the face. For example, neurons in the inferior temporal cortex and the superior temporal sulcus respond to either facial identity or facial expression (Hasselmo, Rolls and Baylis, 1989). In regard to face identity, different populations of neurons respond to and distinguish between different facial identities: while neurons in the inferior temporal cortex respond to physical characteristics of the face, neurons in the superior temporal sulcus respond to the familiarity of faces (Young and Yamane, 1992). In regard to facial expressions, neurons in the superior temporal sulcus of the rhesus macaque not only differentiate between facial expressions as opposed to ingestive mouth movements like chewing (Perrett and Mistlin, 1990), but they also respond to and distinguish between different facial expressions, such as 'open-mouthed threats', 'lipsmacking' and neutral expressions (Hasselmo, Rolls and Baylis, 1989).

In contrast to these neurons in the temporal lobe that respond to either facial identity or facial expression, the amygdala shows a more complex pattern of response. The majority of neurons in the amygdala of monkeys respond to both identity and expression, while other neurons in the amygdala show a specific response – that is an increase or decrease in firing rates – as a function of the type of facial expression. For example, in response to appeasing faces, neuronal firing rates decrease, whereas their firing rates increase in response to threatening faces (Gothard *et al.*, 2007).

In summary, like in humans, face perception in nonhuman primates relies on a distributed network involving multiple brain regions, with some of them being specialized in the processing of specific facial expressions. Because of this similarity, it has been suggested that face perception in nonhuman and human primates is functionally homologous. For example, face-selective neurons are found in several distinct patches (three in each hemisphere) in the superior temporal sulcus in both monkeys and humans, suggesting that they represent homologous structures (Tsao et al., 2006; Tsao, Moeller and Freiwald, 2008). However, there are two major reasons why this needs to be treated with caution. First, the overwhelming majority of studies on face perception are conducted with a very limited number of macaque species (Freiwald and Tsao, 2011), usually rhesus macaques, while there are virtually no studies with other species of monkeys or with great apes (but see Parr et al., 2009). Second, despite the evidence that the involved brain areas in humans and nonhuman primates are homologous in terms of their function, it is not always clear whether they are also homologous in terms of their anatomical features (Parr and Hecht, 2011).

3.3.3.2 Facial expressions: production

Facial movements are largely mediated by two cranial nerves. The trigeminal nerve innervates those muscles involved in ingestion and thus is responsible for movements such as chewing, swallowing or biting. The facial nerve innervates those muscles that attach to the skin of the face and thus mediates the production of facial expressions. The facial nerve itself originates in the facial (motor) nucleus located in the pons of the brainstem (Figure 3.5b). This neural structure mediates spontaneous facial movements – often closely connected to emotional states and then referred to as facial expressions – by projecting efferent information via the facial nerve to the corresponding muscles of the face (Rinn, 1984). This seems to indicate that primates have little cognitive control over the production of their facial expressions (see Chapter 8). However, there is a second route to facial movements mediated by cortical structures, mostly the facial area of the motor cortex (Figure 3.5a), which provides a greater degree of voluntary control over the production of facial expressions (Parr, Waller and Fugate, 2005b). For example, Morecraft et al. (2001) found that in addition to the facial nucleus, various cortical face representation areas and their connections influence the production of facial expressions in rhesus macaques suggesting a much higher degree of voluntary control than previously suggested.

Interestingly, recent research indicates that neural substrates of great apes and humans differ from other primates in some important aspects suggesting that hominoids have more control over their facial movements, specifically orofacial movements (Sherwood et al., 2005). First, in contrast to monkeys and prosimians, the volume of the facial nucleus of great apes and humans is comparably larger than predicted by standard phylogenetic regression, possibly reflecting an increased differentiation of those muscles involved in the production of facial expression (Sherwood et al., 2005). Second, in great apes and humans, the region of orofacial representation in the primary

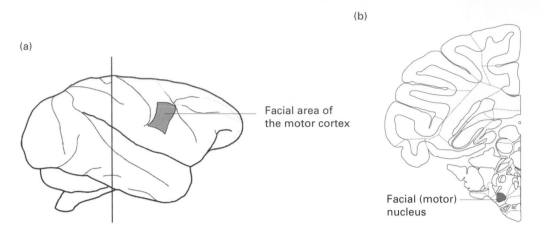

Figure 3.5 Neural substrates involved in the production of facial expressions. (a) Lateral view and (b) coronal slice of the rhesus macaque brain. The vertical line in (a) indicates the location of the coronal slice.

motor cortex is characterized by a greater proportion of neurons that are enriched in **neurofilament protein**. Although its function is not completely understood, this protein seems to play a role in the maintenance and stabilization of the **axonal cytoskeleton**, resulting in heavily myelinated, fast-conducting axons, which hints at a greater degree of control over the production of the corresponding orofacial movements (Sherwood *et al.*, 2004a). Third, particular layers of the primary motor cortex that contain neurons responsible for associations between different areas of the cortex show an increased relative thickness in great apes and humans suggesting that the cortical orofacial representation is involved in more complex processing of a variety of afferent and efferent information in hominoids, but not other primate species. At the same time, there is a greater proportion of neuropil in this area – that is the space between neurons filled with axons, glia cells and dendrites – suggesting that there is more space for a variety of synaptic interconnections between neurons (Sherwood *et al.*, 2004b). Altogether, these differences hint at a greater degree of voluntary control of orofacial muscles, resulting in a greater mobility of the face and a more varied repertoire of facial expressions in great apes and humans compared to other primate species.

There is also evidence for the influence of social factors such as group size on the size of neural substrates involved in the production of facial movements (Dobson and Sherwood, 2011). Thus, species that live in complex social groups have larger facial nuclei and enhanced facial mobility (Dobson, 2009) (see Chapter 1, Figure 1.4). At the same time, they also have larger primary visual cortices indicating that these neural structures are adapted for processing facial signals. Together, these two findings high-light the importance of faces for social communication in primates.

Box 3.2 Neural substrates of action recognition: mirror neurons

Understanding actions by others is a cognitive capacity that is a fundamental component of social behaviours. In searching for the neurophysiological mechanisms underlying this action recognition, it has been suggested that it is most likely realized by the mapping of the perceived visual information onto its motor representation in the brain (see Rizzolatti, Fogassi and Gallese, 2001). To explain these processes on the level of single neurons, much attention has been dedicated to a particular class of visuo-motor neurons that discharge both when an individual performs an action or observes the same action of another individual. These **'mirror' neurons** were originally discovered in the ventral premotor cortex (F5 area) of rhesus macaques while they either performed or observed grasping hand movements (di Pellegrino *et al.*, 1992; Gallese *et al.*, 1996). Since then, mirror neurons were not only discovered in other species, including songbirds and humans (see Keysers and Gazzola, 2009), but also for a variety of other behaviours, including communicative behaviours. In rhesus macaques, for example, some neurons in the F5 area in the ventral premotor cortex respond to the perception of facial expressions, like 'lipsmacking', 'lip protrusion', 'teeth chatter' and 'open-mouth face', but also when the monkeys perform the same communicative mouth actions (Ferrari *et al.*, 2003). The recognition of actions, however, does often not only rely on visual information, but also on auditory cues. Kohler and colleagues (2002) found audiovisual neurons in the F5 area that discharged not only when pigtail macaques either performed or observed a particular action, but also when they heard the corresponding sound of this action, like breaking a peanut or ripping paper. Altogether, these findings indicate that the F5 area in the ventral premotor cortex of macaque monkeys controls both orofacial and hand movements and contains mirror neurons that map particular actions to the motor representation in the brain, with some of them responsible for the integration of multisensory information (see also section 3.4.1). The discovery of mirror neurons in the F5 area in the macaque brain is of particular importance, since this area is homologous to Broca's area in the human brain. In humans, Broca's area is mostly discussed in regard to the production of language (see **Box 3.1**), but there is increasing evidence that it is also involved in a variety of other functions, including the control of arm movements (Grafton *et al.*, 1996; Rizzolatti *et al.*, 1996). These two facts – that the F5 area of monkeys and Broca's area in humans represent homologous structures and that Broca's area in humans is involved in both the production of language and the execution of mouth and arm movements – are often referred to as evidence for an evolutionary scenario suggesting a gestural origin of language (Rizzolatti and Arbib, 1998) (see Chapter 5).

3.3.3.3   Visual gestures: perception

There are a variety of studies that investigate the neural substrates underlying the recognition of biological motion (for a review see Allison, Puce and McCarthy, 2000). In both human and nonhuman primates, the superior temporal sulcus seems to play an important role for the recognition of movements of the body or body parts such as the head, hands or whole body (as well as facial expressions, see previous section). However, to the best of our knowledge, there is no study that investigates how gestures of nonhuman primates (as defined in this book) are processed in the brain.

There are studies that refer to facial expressions as gestures (e.g. Ferrari *et al.*, 2003; Perrett and Mistlin, 1990), but these are not considered gestures in the same sense as used in this book. Other studies classify particular hand movements, such as grasping, as gestures. Such studies showed that particular neurons in the premotor cortex of the F5 area of the macaque brain fire when the monkey executes a grasping movement, but also when it observes a human performing the same movement. These neurons are therefore referred to as mirror neurons (for more details, see **Box 3.2**) (di Pellegrino *et al.*, 1992; Gallese *et al.*, 1996; Rizzolatti *et al.*, 1996). However, according to the definition of a gesture used in this book, grasping an object is not considered a gesture, since this hand movement is an action directed towards an object, while manual gestures are movements used to communicate with another conspecific. Thus, it is still unknown whether the perception of such communicative, manual gestures in primates is mediated by the same brain areas as their non-communicative hand movements.

3.3.3.4   Visual gestures: production

Similar to the perception of visual gestures, rather few studies have investigated the neural substrates underlying gesture production. The little that is known about neural correlates of gesture production is from chimpanzees, a representative of the great apes. Such progress was facilitated by a more recent development enabling the use of different brain-imaging techniques such as **magnetic resonance imaging** (MRI) and positron emission tomography (PET) for nonhuman primates (for more detailed information, see Chapter 4). Such studies often specifically target the question of whether brain areas involved in gesture production in great apes are homologous to areas in the human brain that are involved in language processing, such as Broca's area (e.g. Taglialatela *et al.*, 2008). Since language processing in humans is lateralized to the left hemisphere of the brain, research on neural substrates of gesture production is specifically interested in differences between the two hemispheres concerning the size, cortical microstructure, or connectivity of the corresponding homologous brain areas in nonhuman primates (see **Box 3.3** on the lateralization of brain functions).

There is evidence that chimpanzees preferentially use their right hand to produce pointing gestures when they request food from humans (Hopkins and Leavens, 1998). Based on this observation, researchers have investigated whether the lateralization of this communicative behaviour is reflected in neuroanatomical asymmetries in their

Box 3.3 Lateralized functions in primate communication

In humans, many functions related to language processing are lateralized to the left hemisphere. In searching for the evolutionary origins of language, a substantial body of research is dedicated to the question of whether there is also evidence for a lateralized processing of communicative signals in other primates. Many studies aim at finding precursors of lateralized processing specifically for vocalizations and gestures in order to support an either gestural or vocal origin of human language (Corballis, 2003; Ghazanfar and Hauser, 1999). While research into gestures focuses mostly on the production of these signals (e.g. Hopkins and Leavens, 1998), vocalizations are extensively researched in regard to their lateralized perception (e.g. Petersen *et al.*, 1984). Comparably fewer studies exist for the processing of facial expressions and many of them investigate these signals during the production of vocalizations that result in specific oro-facial movements (Wallez *et al.*, 2012). The lateralization of communicative signals can be assessed by analyzing (1) behaviours (e.g., hand preferences while producing a pointing gesture, asymmetries in facial movements while producing a facial expression or vocalization, or by using the head turn paradigm to investigate which side primates turn their heads to when a vocalization is played 180 degrees behind them), (2) the size or cytoarchitectonical structure of the corresponding neural substrates, and (3) the correlation between the lateralized production or perception of a signal and the corresponding asymmetries in the underlying neural substrates. Since the left hemisphere controls the contralateral (right) side of the body and vice versa, a behaviour that is preferentially produced with the right side of the body indicates the dominance of the left hemisphere in controlling its production.

In regard to vocalizations, there is evidence for a hemispheric bias in the perception of species-specific vocalizations, often indirectly measured based on behavioural responses to auditory information (for a review, see Taglialatela, 2007). For example, Japanese macaques show a right-ear orienting bias in response to 'coo' calls suggesting the dominance of the left hemisphere in processing such species-specific vocalizations (Petersen *et al.*, 1978). While some studies find neuroanatomical support for these behavioural asymmetries, for example, a left-lateralized activity in the temporal pole of the superior temporal gyrus in response to conspecific's vocalizations in monkeys (e.g. Poremba *et al.*, 2004), others found a dominance of the right hemisphere (e.g. Gil-da-Costa et al., 2006) or concluded that the lateralization was not specific to the processing of species-specific vocalizations (Joly *et al.*, 2012). In chimpanzees, lateralization seemed not to vary as a function of whether the call was species-specific or not, but depending on the function of the call (Taglialatela *et al.*, 2009). While there was no evidence for lateralized processing of long-distance vocalizations such as 'pant hoots', proximal conspecific vocalizations such as 'grunts' and 'barks' resulted in a lateralized activation of the right but not the left hemisphere. To explain these unexpected findings, Taglialatela and colleagues (2009) refer to the fact that some functions of human language are also processed in the right hemisphere (Zatorre and Belin, 2001; Zatorre, Belin and Penhune, 2002). Similarly,

Continued

Box 3.3 (*cont.*)

species-specific calls of chimpanzees might be processed differentially in the two hemispheres related to the type of information or function of the different calls.

Research into behavioural asymmetries of vocal production often uses a multi-modal approach since it is based on the analysis of orofacial asymmetries during vocal communication. Different species, including monkeys and great apes, show orofacial asymmetries in vocal production towards the left side of the mouth indicating a right hemisphere dominance, which is interpreted as evidence that their vocalizations are closely linked to emotional states (Hauser, 1993; Hauser and Akre, 2001; Hook-Costigan and Rogers, 1998). For example, in chimpanzees, the right hemisphere is more dominant in the production of (emotional) facial expressions, as evidenced by either a larger left hemimouth area or length for some of their facial expressions such as 'silent-bared teeth face', 'playface' and 'scream face' (Fernández-Carriba *et al.*, 2002) (see figure below). Most interestingly, however, some orofacial movements of chimpanzees that are associated with the production of learned attention-getting sounds are lateralized to the left hemisphere (Losin *et al.*, 2008) possibly indicating that different neural substrates are involved in the processing of species-specific and learned vocalizations.

In regard to gestures, many studies that address **laterality** focus on pointing of primates in interactions with humans (Hopkins *et al.*, 2012, see section 3.3.3.4). Most of these studies found a right-hand bias on a population level, as did some studies that investigated hand preferences for different manual gestures in monkeys and great apes in interactions with their conspecifics. (Meguerditchian and Vauclair, 2006; Meguerditchian, Molesti and Vauclair, 2011). The dominance of the left hemisphere in gesture production is also evident in neuroanatomical asymmetries, particularly in those areas that represent homologues to human areas of language processing like the inferior frontal gyrus and the planum temporale (Hopkins and Nir, 2010; Meguerditchian *et al.*, 2012), but also in differences in the cytoarchitectonical structure, such as an increased number of neurons (Spocter *et al.*, 2010).

Altogether, research into lateralization of primate communication often reveals inconsistent results. For example, the lateralized processing of vocalizations varies depending on the type or function of the call, but factors such as characteristics of the caller or the familiarity with a sound may also play a role (Lemasson *et al.*, 2010b; Leliveld, Scheumann and Zimmermann, 2010). Many studies highlight a right-hand preference for gesture use, but lateralization can differ as a function of context and the type of action (communicative or not), and is sometimes limited to individuals but is not characteristic for a population (Hopkins and Cantalupo, 2005). While some species-specific vocalizations and their corresponding orofacial movements indicate a right hemisphere bias possibly reflecting their emotional component, learned attention-getting sounds and their orofacial movements are lateralized to the left hemisphere. This suggests that such voluntary controlled orofacial and gestural communication might have coevolved into an integrated system since they share the same left-hemispheric specialization (Wallez *et al.*, 2012).

Continued

Box 3.3 (*cont.*)

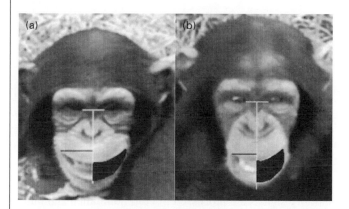

Examples of length and area measures. For the hemimouth length, a straight line was drawn from each outer corner of the mouth to the midline. To measure the hemi-mouth area, a line surrounding the inner side of the mouth perimeter was drawn and the surface calculated. (a) 'silent bared-teeth display'; (b) 'play face' (from Fernández-Carriba *et al.*, 2002).

brains. It is important to consider here that for the majority of functions, the right side of the body receives its neural input from the contralateral and thus left part of the central nervous system. As expected, chimpanzees with a right-hand preference for pointing have a larger inferior frontal gyrus in their left but not right hemisphere, while no asymmetries are evident in those chimpanzees that use their hands ambidextrously (Taglialatela, Cantalupo and Hopkins, 2006). However, it is of note that the relation between hand preference and a larger inferior frontal gyrus in the left hemisphere was specific for manual communicative gestures, while there was no correspondence between hand preference and neuroanatomical asymmetry for simple motor reaching. Further research demonstrates that the inferior frontal gyrus, which is part of the prefrontal cortex, together with other subcortical areas, is active while chimpanzees produce communicative gestures (Figure 3.6) (Taglialatela *et al.*, 2008). These findings are particularly interesting since those areas that are active during gesture production are homologous to the Broca's area of humans. In humans, this brain region is responsible for language production, but there is recent evidence that it is also involved in the recognition of gestures (Gentilucci and Corballis, 2006). The functional significance of the activated additional cortical and subcortical structures for gesture production in chimpanzees, however, is to date still unclear.

3.3.4   Vocal communication: perception

Vocalizations are complex sounds from which the receiver may extract information about the eliciting context, the identity of the caller and the spatial location of the caller. Thus, the perception of a vocal signal can involve extraction of information such as *what, who* and *where* (Wang, 2000).

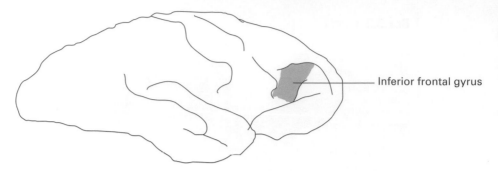

Inferior frontal gyrus

Figure 3.6 Neural substrates implicated in gesture production. Lateral view of the chimpanzee brain.

In general, auditory signals are processed in both nonhuman and human primates in similar ways. Auditory information is converted into neuronal activity at the basilar membrane of the inner ear (see Chapter 2, section 2.2.2). Particular areas of the basilar membrane are each sensitive to very specific frequencies of the incoming auditory signal: high-frequency sounds are localized at the base of the cochlea as opposed to low-frequency sounds that localize near the apex. The corresponding movements of the basilar membrane trigger a specific spatial–temporal discharge pattern of the hair cells that are attached to the basilar membrane. In other words, which of the hair cells are activated differs as a function of the acoustic characteristics of the incoming sound wave. The auditory nerve relays this information to the brainstem, specifically to the ventral nucleus (or medial geniculate complex) in the auditory thalamus, which projects to the auditory cortex in the temporal lobe of the brain (Kaas, Hackett and Tramo, 1999).

The auditory cortex consists of three areas located in the superior temporal plane that correspond to different levels of processing: core, belt and parabelt (Figure 3.7). The core represents primary sensory cortex and consists of tonotopically organized subdivisions or tonotopic maps. Thus, the neurons in the core are arranged according to the frequency of the sound to which they respond best. As a consequence, sounds with similar frequencies are also represented in neighbouring regions in the core area.

While the core is mostly involved in the processing of pure tones, the belt and parabelt are less tonotopic and mostly process more complex sounds, such as vocalizations of other conspecifics that usually are composed of more than one frequency that can also change over time (Kaas, Hackett and Tramo, 1999). How nonhuman primates process such complex sounds is described in the next section.

There is evidence for a 'voice region' in the superior temporal plane, and more specifically in the hierarchically higher anterior regions of auditory cortex of the monkey brain that specifically responds to vocalizations of its own species, but not other vocalizations or natural sounds (Petkov *et al.*, 2008). Although, there is some evidence that a limited number of neurons fire selectively in response to specific call types (Ghazanfar *et al.*, 2005), most neurons in the auditory cortex, especially in the belt and parabelt, respond to more than one frequency and to more than one call type (Ghazanfar and Hauser, 1999; Wollberg and Newman, 1972). The current

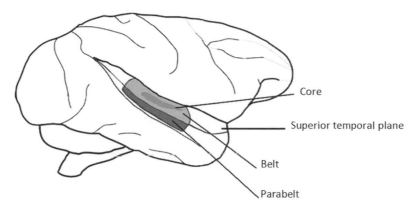

Figure 3.7 Neural substrates of vocal perception. Lateral perspective of the rhesus macaque brain. The lateral sulcus has been partially opened out to illustrate the superior temporal plane.

understanding of the encoding of auditory information refers to so-called neuronal ensembles, comprising some neurons that fire most to a particular frequency (rate coding), while others encode the temporal features of a sound (temporal coding) (Brosch and Scheich, 2003). Both types of neurons are differentially activated depending on the type of acoustic stimulus and as a consequence, particular call types result in a specific neuronal response pattern. Together, the activity of such a neuronal ensemble represents the neuronal encoding of this specific call type.

Some research has been conducted on the neural substrate encoding the location of a caller and thus the *where* of auditory information. Similar to the processing of visual information, a dual cortical pathway has been suggested for the processing of auditory information (Rauschecker and Scott, 2009). While the temporal pathway is responsible for the identification of patterns, such as species-specific vocalizations as described above, the parietal pathway was hypothesized to process spatial information (Rauschecker and Tian, 2000).

3.3.5   Vocal communication: production

The question of whether nonhuman primates are able to control the production of their vocalizations is fiercely debated, particular in regard to searching for the roots of human language (see Chapters 5, 8 and Slocombe, Waller and Liebal, 2011). From an evolutionary perspective, the ability to voluntarily produce vocalizations seems an essential milestone towards the evolution of speech and modern language. This development includes two important components: first, the ability to either produce, or alternatively, to suppress a vocalization depending on the communicative context, and second, the ability to voluntarily control the articulators such as the mouth, tongue or lips (Hopkins, Taglialatela and Leavens, 2007). Together, these abilities enable the flexible use of vocal utterances (Chapter 7). The question of the degree of control over the production of vocalizations is inevitably linked to the underlying neural substrates. The majority of studies suggest that unlike in humans, vocal production by nonhuman primates does not involve cortical motor areas (e.g. Jürgens, 2002). However, recent research indicates a more differentiated picture as summarized in the following section (for an overview of experimental studies with chimpanzees, see Hopkins, Taglialatela and Leavens, 2011).

Overall, the basic anatomy and mechanisms of vocal production are similar in nonhuman and human primates (Ghazanfar and Rendall, 2008). The production of vocalizations involves three components: respiratory movements, laryngeal activity and supralaryngeal or articulatory activity (Jürgens, 2002). The primary sound is produced in the larynx while exhaling air, resulting in a vibration of the vocal folds. The supralaryngeal vocal tract above the larynx then modifies this primary sound resulting in a modulated, more complex sound (Chapter 2, Figure 2.4).

These two basic processes are mediated by different motor nuclei in the pons and the reticular formation in the medulla that integrate the laryngeal, respiratory and articulatory activity (Jürgens, 2002). These motor nuclei include the nucleus ambiguus that projects to the larynx, while three further nuclei innervate articulatory structures such as the tongue (hypoglossal nucleus), the mouth (trigeminal motor nucleus) and the upper face (facial nucleus). The latter two are also involved in the production of orofacial movements related to ingestion as well as facial expressions (see section 3.3.3.2). However, since many vocal utterances are intrinsically linked with movements of the face, the trigeminal and facial nuclei also play a role in the production of vocalizations (Sherwood et al., 2005).

The activity of these subcortical motor nuclei is coordinated by a network in the medulla, while this motor-coordinating network receives facilitatory input from the periaqueductal grey of the midbrain (Figure 3.8b). Altogether, these subcortical neural structures are involved in the production of species-specific, innate vocalizations (Jürgens, 2002).

Voluntary control over the production or suppression of vocalizations requires the input from cortical structures, more specifically, the mediofrontal cortex including the anterior cingulate gyrus as well as the supplementary and pre-supplementary motor area (Figure 3.8a). In general, it is argued that this differentiates humans from other primates since nonhuman primates lack this kind of cortical control. For example, the electrical stimulation of these motor areas in monkeys results in the movement of the articulators such as tongue and lips, but not a vocalization, as opposed to humans that will utter vowel-like sounds after the stimulation of these areas (Ghazanfar and Rendall, 2008). Furthermore, lesions in the primary and premotor cortex in the area of orofacial representation have no impact on the call rate or structure of monkeys (Aitken, 1981; Jürgens, Kirzinger and von Cramon, 1982). However, a similar destruction in the human primary motor cortex results in a complete loss of voluntary control of linguistic phonation, while vocalizations that are linked to particular emotional states such as groaning, crying or laughing are not affected (Groswasser et al., 1988; Mariani et al., 1980).

Overall, findings suggest a lack of cortical control of the production of vocalizations in nonhuman primates. In contrast to humans, there is no direct connection between the motor cortex and the motor nuclei in nonhuman primates, since the information from the cortical areas to the motor nuclei in the pons is always mediated via the reticular formation. Only humans are characterized by a direct connection between the motor cortex and the nucleus ambiguus, which innervates the larynx, suggesting that only

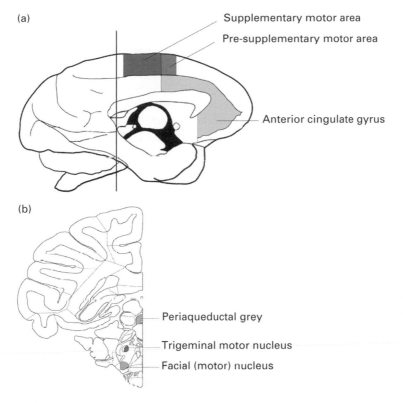

Figure 3.8 Neural substrates of vocal production. (a) Medial view and (b) coronal slice of the rhesus macaque brain. The vertical line in (a) indicates the location of the coronal slice. The nucleus ambiguus and the hypoglossal nucleus are not shown in this figure.

humans can voluntarily produce vocalizations (Jürgens, 1976; Kuypers, 1958). However, there is some evidence that the anterior cingulate gyrus of monkeys is involved in situations that require some form of learning such as the production of a vocalization in order to receive a food reward from a human (Gemba, Miki and Kazuo, 1995). This seems to indicate some voluntary control over vocalizations; however, the activity of the anterior cingulate gyrus is most likely related to motivation rather than to the release of the vocalization itself, as found for both human and monkeys (Aboitiz *et al.*, 2006).

In summary, the majority of studies suggests that unlike humans, nonhuman primates have little cortical control over the production of their vocalizations, although this is also true for largely innate vocalizations in humans, such as crying, groaning or laughing.

It is, however, important to emphasize that the studies reported in this section concern only monkey species, and more specifically mostly squirrel monkeys or rhesus macaques. Therefore, this data might not be sufficient to generalize to monkeys or even to all species of nonhuman primates, particularly since there is some recent evidence suggesting that chimpanzees might differ from monkeys in this regard. Indeed, Taglialatela *et al.* (2008, 2011) used PET and MRI imaging techniques (see Chapter 4) to measure brain activity in chimpanzees after they produced manual gestures alone or in combination with vocalizations. They found an activation of the inferior frontal gyrus only when the chimpanzees used multimodal signals comprising manual gestures and attention-getting sounds, but not if they only used gestures (Taglialatela *et al.*, 2011). Although these results suggest that chimpanzees have some control over the production of their vocalizations, it is important to note that to date the inferior frontal gyrus has only been implicated in the production of attention-getting sounds. Therefore, the question remains of whether the findings by Taglialatela *et al.* (2011) generalize to other vocalizations. Many of these attention-getting sounds also included nonvocal sounds, such as 'raspberries' and 'kisses' that are not produced by the vocal folds and thus might rely on different cortical neural substrates than other vocalizations. Further-more, vocalizations like the 'extended food grunt' are idiosyncratic signals used by only some individuals in some captive populations and thus they do not represent species-typical ones. Since the mechanisms underlying the acquisition of these two types of vocalizations might differ (see Chapter 6), idiosyncratic and species-typical vocaliza-tions might be processed by different neural substrates. There is some evidence that idiosyncratic vocalizations like the 'extended grunt' and sounds that are not produced with the vocal cords like the 'raspberries' or 'kisses' are both socially learned, since chimpanzees are more likely to produce such attention-getting signals if their mothers also used these sounds (Taglialatela *et al.*, 2012). In general, however, very little is known about how these signals are acquired during ontogeny and how this relates to neural substrates.

3.4   Multimodal integration on a neural level

Up to this point of the book, the different modalities have been mostly considered separately. During natural, 'in situ' primate interactions, however, many facial expres-sions, vocalizations and gestures often co-occur, e.g. many vocalizations result in characteristic movements of the face or some gestures have both visible and auditory components. In this section, the aim is to present some examples to show how the integration of different sensory modalities is realized on a neural level.

3.4.1   Perception: integration of multiple sensory modalities

Multisensory integration is the combination of sensory information from different modalities in the nervous system, like smell, sight, touch or sound, and the way these modalities interact with each other and influence each other's processing. The pioneering work by Stein and Meredith resulted in three general principles that characterize multisensory integration (Stein and Meredith, 1993). Thus, the multi-modal response of neurons is stronger than expected when (1) the contributing, constituent unimodal sensory stimuli arise from the same location (spatial rule) or (2) at the same time (temporal rule) and (3) compared to weak unimodal sensory responses (inverse effectiveness) the level of multisensory enhancement is inversely related to the strength of the unimodal responses. Thus, the weaker the unimodal responses, the greater is the multisensory enhancement (e.g. King and Palmer, 1985; Meredith and Stein, 1983, 1986). Often, the neuronal response to the combination of two sensory inputs is greater than the sum of the responses to the two sensory inputs presented alone, which is referred to as supra-additivity representing one marker of multisensory integration (Stein and Meredith, 1993). In humans, this phenomenon can be seen in the integration of visual and auditory information, e.g. while observing others' faces while they are talking (Calvert, 2001). Although several brain areas seem to be involved, the superior temporal sulcus in the cortex plays an important role in this integration of input from several modalities, especially during the perception of audiovisual speech.

For rhesus macaques, Barraclough et al. (2005) found evidence for multisensory integration in neurons in the superior temporal sulcus, including supra-additive integration of visual and auditory stimuli. More specifically, the researchers presented a visual action in combination with the sound of this particular action and focused on how the addition of auditory signals affected the response to visual, biologically meaningful actions, such as particular hand actions, whole body movements, or facial movements. The visually evoked response of the neurons in the superior temporal sulcus was increased by the combined presentation of an action together with sound, but only if the sound matched the particular action. For example, the neurons' response to the visually presented 'lipsmacking' was more pronounced if combined with the corresponding auditory stimulus of 'lipsmacking', while the response was significantly smaller if 'lipsmacking' was combined with other incongruent auditory signals such as 'coos' or 'pant-threats'. These results indicate that the integration of different modalities is even realized at the level of single neurons, since particular neurons seem to be selective for specific multimodal communicative signals.

In addition to multisensory integration in the superior temporal sulcus, there is also evidence that the combination of auditory and visual stimuli together changes the response of the corresponding neurons in other areas of the macaque brain, such as the ventrolateral prefrontal cortex (Romanski, 2012; Sugihara et al., 2006) and the auditory cortex (Ghazanfar et al., 2005). For example, Sugihara et al. (2006) showed that approximately half of the recorded neurons in the ventrolateral prefrontal cortex of rhesus macaques were multisensory. Thus, they were either bimodal and responded to unimodal auditory and visual stimuli, or they showed an enhanced (or decreased) response to the combined audiovisual stimulus (vocalization and face) compared to the response to the unimodal stimuli.

Furthermore, neurons in the auditory cortex showed an enhanced response when monkeys were presented with conspecific facial expressions and matching vocalizations, compared to the response when either the face or vocalization were presented alone (Ghazanfar et al., 2005). Within the auditory cortex, the lateral belt region shows a greater frequency of multisensory integration than the core region. These findings suggest that the multisensory integration of visual and auditory information and more specifically, the face and voice, occurs separately in different areas of the brain, and is likely to include interactions between the auditory cortex and the superior temporal sulcus.

3.4.2   Production of multiple communicative signals

While the integration of multiple incoming sensory information on a neuronal level has been so far exclusively studied in monkeys, much less is known about the production of multiple signals, and the little that is known is from chimpanzees. Furthermore, unlike the research with monkeys, studies with chimpanzees focused on the combination of gesture and vocalizations. Thus, it is difficult to draw any comparisons across species and modalities.

For chimpanzees, Taglialatela *et al.* (2008) showed that the inferior frontal gyrus, a homologue of Broca's area, was active during the production of gestures and vocalizations, which the apes directed towards an experimenter with desirable food items. It remained unclear, however, whether the activation of the inferior frontal gyrus was related to the use of either gestures or vocalizations, or the multimodal combination of both. Taglialatela *et al.* (2011) addressed this issue by repeating this experiment with four chimpanzees all of whom produced gestures to request food, but only two of whom produced attention-getting vocalizations ('kisses', 'raspberries', 'extended grunts') in combination with their gestures (see also section 3.3.5). Overall, activation of the inferior frontal cortex was only seen in the two chimpanzees, who produced vocalizations in conjunction with gestures, but not in the other two chimpanzees that used gestures but no vocalizations. This indicates that multimodal signals consisting of vocalizations and manual gestures selectively activate the inferior frontal gyrus in chimpanzees. Since this brain area is the homologue to the Broca's area in humans, this finding might have important implications for theories of language evolution, thus pointing to a multimodal – rather than gestural – origin of language (Taglialatela *et al.*, 2011). However, strong conclusions should not be drawn on the basis of such few individuals. Furthermore, without a further control condition featuring chimpanzees, who only vocalize without gesturing, it is feasible that these activation patterns represent the neural underpinnings of vocal production, rather than multimodal signal production. It is also questionable whether manual gestures and vocalizations can be considered multimodal signals in this study by Taglialatela *et al.* (2011), because the later PET scans measured any activity related to the use of either gestures or vocalizations over a period of 40 minutes. Interestingly, one subject that produced food-associated calls in the communicative task but no attention-getting calls, showed no activation of Broca's homologue, which seems to indicate that perhaps only these learned vocalizations recruit this area of the brain.

Summary

Although there is a vast body of research on the neural substrates involved in the production and perception of primate communication, two major conclusions should be drawn from this chapter. First, knowledge about the underlying brain structures differs enormously between modalities, and second, much of the current knowledge stems from only a few species, with the overwhelming majority being monkeys and here mostly two species, rhesus macaques and squirrel monkeys. The little that is known about great apes is from research on chimpanzees. Furthermore, although the processing of some modalities such as the perception of facial signals or vocalizations is intensively studied and new findings are rapidly accumulating, comparisons between groups of primates and particularly the comparison with humans remain problematic, since it is often difficult to determine whether the corresponding brain areas represent homologous structures across species. Regarding the production of signals, most evidence is available from facial expressions and vocalizations, while the very few studies on gestures are limited to chimpanzees. Interestingly, great apes seem to be different from monkeys in regard to their control over the production of both facial expressions and vocalizations, as indicated by an increase in size or the complexity of the cytoarchitectonical structure of the corresponding neural substrates or the involvement of cortical areas in the production of these signals. However, given the fact that most research focused on very few species, it is too early to conclude that there are consistent differences between monkeys and great apes. Regarding the perception of signals, there is a substantial body of research on both facial expressions and vocalizations, but virtually nothing is known in regard to the perception of gestures in both monkeys and apes. For multimodal signals, there is evidence of integration of multiple sensory information at a neuronal level. However, these studies focus on the integration of facial and vocal signals (in monkey species), while little is known about combinations that include gestural signals in both monkeys and great apes.

Part II

Approaches to primate communication

4 The methods used in primate communication

Primate communication tends to be divided into four different modalities of behavioural communication – olfactory communication, gesture, facial expression and vocalization – which span four different sensory channels (olfaction, vision, touch and audition). We use the term modality to refer to each type of behavioural communication in this way, but others have used modality to refer specifically to the sensory mode of the stimulus (e.g. Partan and Marler, 1999). We use our broader definition of modality for three reasons. First, one of our aims is to review and examine the classic scientific distinctions made between communicative domains. Second, some types of communication can use more than one sense: for example, gesture can be visual, tactile or auditory. Third, different cognitive mechanisms are thought to underlie both the production and perception of facial expressions, vocalizations, manual gestures and olfactory signals, and so treating them as different modes might be most productive for our purpose.

The aim of this chapter is not to provide a review of what is known about each modality, but instead to give an overview of the methods and general approaches employed. What follows is an outline of the scientific questions, methods for data collection and approaches to analysis within each area of primate communication research. For each modality, there is a section that refers to the brain mechanisms underlying its production and perception. However, it is important to emphasize that the focus here is on the main methods used to study the neural correlates of primate communication, while Chapter 3 discusses the corresponding brain areas and neural circuits in more detail. Importantly, it will soon become clear that the methods (and perhaps more importantly, the scientific questions) can differ between modalities.

4.1 Olfactory communication

Olfaction is the oldest of the senses among vertebrates (with the exception of taste) and persists throughout the primate order. Unlike the visual, auditory and somatosensory systems, the fundamental morphology of the olfactory system has been conserved for more than 500 million years (Stoddart, 1980; Young, 1962). It is clear, though, that each vertebrate species employs unique olfactory strategies in response to its own evolutionary history. Early research into primate olfaction was almost exclusively done with captive laboratory populations heavily used in medical research, for example, rhesus macaques, galagos and callitrichids, as an effort to improve husbandry and reproduction rates in breeding colonies (Abbott, Barrett and George, 1993; Clark, 1982; Epple and Katz, 1984; Goldfoot *et al.*, 1978; Savage, Ziegler and Snowdon, 1988). These studies helped to establish general properties of the olfactory systems, and also improved understanding of species-specific olfactory-related behaviours, such as the importance of olfaction in the mating behaviour of callitrichids (Abbott, 1984).

The olfactory system is activated when olfactory sensory cells within the nasal cavity become excited, initiating a process of neural recognition that influences behaviour and hormonal state (Munger, Leinders-Zufall and Zufall, 2009). As described in Chapters 2 and 3, the olfactory system is often distinguished as two specific subsystems, the MOS and the AOS. The AOS is often thought to be a more narrowly tuned sensory instrument that focuses preferentially on pheromones that are important in social behaviours such as mate-finding, kin recognition and individual signatures (Firestein, 2001; Swaney and Keverne, 2009). As this book is concerned with social communication, the olfactory signals we will focus on are pheromones, as these are the social signals within this modality.

4.1.1   When do primates produce olfactory signals?

Primates produce pheromones automatically in scent glands, and so species that have numerous scent glands are typically the target of investigation as researchers often assume that this will correlate with heavy use of olfactory signals. Pheromones have therefore mostly been studied and analysed in prosimians and callitrichids, which are taxa with numerous scent glands. Pheromones are contained in secretions from these scent glands but have also been detected in some species' saliva, urine and vaginal secretions (Dixson, 1998; Phillips *et al.*, 2011; Schilling, Perret and Predine, 1984). These compounds can contain information related to reproductive status, parental care, kin identity and individual identity.

Observational studies on scent-marking or urine-washing behaviour can be used to determine when primates use pheromone-based olfactory signals. Scent-marking behaviours can take many forms, depending on the scent gland being used. Prosimians, for example, possess anogenital glands, brachial (wrist) glands, sternal (chest) glands and glands clustered around the head. These glands are rubbed against a substrate, such as a tree limb, or different parts of the body (Ankel-Simons, 2000). Observational studies suggest that scent-marking among prosimians commonly occurs at the boundaries of individual and group territories (Gursky, 2003; Kappeler, 1998). Among the anthropoids, only callitrichids have been shown to actively scent-mark (Digby, Ferrari and Saltzman, 2007). Several primate species have been observed to engage in urine-washing behaviour where an individual urinates into the palm of the hand and then rubs the lower limbs and tail. Urine from many mammalian species contains chemical and pheromonal information, such as the sexual steroids testosterone and oestrogen that may be important in social interactions (Dixson, 1998; Ma and Klemm, 1997; Rajanarayanan and Archunan, 2011; Rekwot *et al.*, 2001). The exact function of this behaviour is not fully understood, but observational studies indicate that urine-washing may serve as a signal for female sexual receptivity, convey individual signatures or it may serve an appeasement function (Colquhoun, 2011; Phillips *et al.*, 2011).

While most studies into primate pheromones have been done on prosimians and callitrichids, behavioural observations of some catarrhine primates suggest that pheromones are also important to these species, particularly in the context of sexual behaviours. For example, observations of males indicate that they alter their behaviour towards females in oestrus between the time of nascent sexual swellings and maximal swelling (Deschner *et al.*, 2004; Dixson, 1983). Interest in copulation in male chimpanzees intensifies around the period of female ovulation and some males increase mate-guarding activities at this time. It has been postulated that these behavioural changes are due to the male's ability to detect ovulation from olfactory signals produced by the female at the time of ovulation.

Observational studies, like those outlined above, are complemented by chemical analysis of secretions to investigate the pheromones they contain. For instance, studies of female chimpanzees have revealed that vaginal secretions during oestrus contain pheromones that signal reproductive capacity of females as well as physiological status (Fox, 1982; Matsumoto-Oda *et al.*, 2003). There is evidence that vaginal secretions of female rhesus macaques and chacma baboons also contain pheromones that alter sexual behaviour in conspecific males (Clarke, Barrett and Henzi, 2009; Michael and Zumpe, 1982). This suggests that they are strong olfactory cues to reproductive behaviour in catarrhine primates, which do not possess a functional AOS, thus adding further evidence that the MOS/AOS subsystem division is not entirely warranted.

4.1.2   Structure and form of olfactory signals

The chemical composition of a pheromone or scent-gland secretion can be measured using **gas chromatography** and **mass spectrography (GC-MS)**. The tools for chemical analysis are difficult to use for quantitative analysis in the field, unlike the study of visual or auditory signals, where more portable field instrumentation is available (Epple, 1986; Heymann, 2006). In addition, the nature of volatile chemicals can make it difficult to collect samples for later analysis. As a result, the structure and form of olfactory signals are usually studied in a controlled laboratory setting.

GC-MS can be used to study primate pheromonal composition by profiling the chemicals contained in secretions. For example, one study of ring-tailed lemurs identified up to 300 different chemical compounds in each brachial gland secretion that was collected (Knapp, Robson and Waterhouse, 2006). Knapp, Robson and Waterhouse (2006) demonstrated that each individual had a specific odour profile (i.e. an individual signature) as determined from GC-MS analysis (for a typical read-out from GC-MS, see Figure 4.1). Similar results on both the structural complexity of secretions and individual odour signatures have been obtained from analyses of secretions from other prosimians and callitrichids (Belcher *et al.*, 1988; Hayes, Morelli and Wright, 2004; Smith, Siegel and Bhatnagar, 2001; Yarger *et al.*, 1977).

One difficulty for all modalities, but perhaps particularly olfactory signals, is that in the absence of data on the sensitivity of the receiver to the components or parameters of

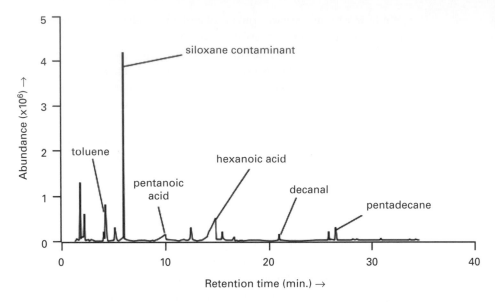

Figure 4.1 Chromatogram generated by GC-MS analysis showing a typical odour profile of an odour sample from a ring-tailed lemur (modified after Knapp *et al.*, 2006). Such an analysis can reveal up to 300 different chemical compounds, but only some of them are shown in this figure.

the signal that are measured, we do not know that we are identifying 'meaningful' characteristics in particular signals. For instance, although the GC-MS may reveal high levels of a certain chemical compound in a urine sample, we do not necessarily know that this compound is detectable by the receiver or whether it has a specific influence on the receiver's behaviour.

4.1.3   What mechanisms underlie the production of olfactory signals?

Olfactory communication is mostly studied in regard to the underlying physiological mechanisms, however, as opposed to studies with humans and non-primate species such as rodents, comparably less research is conducted with nonhuman primates (Heymann, 2006). The signals involved in olfactory communication are produced by glands, which are epithelial structures housed in various places in the body. The activity of these glands is mediated largely by hormones, with no evidence for cognitive control over the production of olfactory signals. Many studies focus on the perception of olfactory signals and investigate the genes that encode olfactory receptors (Whinnett and Mundy, 2003). In regard to production, the role of olfactory signals as individual signatures and the link to the MHC receives increasing attention (Knapp, Robson and Waterhouse, 2006) (see Chapter 2, **Box 2.1**). However, developmental aspects of primate olfactory communication have rarely been studied, partly because of methodological problems in regard to recording and measuring olfactory signals (Heymann, 2006).

4.1.4   What does a sniffer understand from olfactory signals?

While numerous experimental studies of visual and auditory perception in primates have been conducted using a variety of methods, the systematic measurement of receiver responses to olfactory signals remains rare, despite the possibility of applying modified versions of behavioural paradigms commonly used in other modalities. For instance, a recent study with captive ring-tailed lemurs used preferential sniffing paradigms to assess discrimination of urine from group members and strangers (Palagi, Dapporto and Borgognini Tarli, 2005). Researchers do, however, face difficulties in quantifying the 'strength' of the olfactory signal it provides the receivers in such behavioural experiments, as this requires costly laboratory analysis. Furthermore, it is also generally unknown what effect both storage durations and techniques (e.g. freezing) have on the signal. Thus, the best current proxy for measuring the MOS olfactory acuity in any mammal is the density of olfactory sensory neurons embedded within the olfactory epithelium lining the nasal cavity. Similarly, measuring AOS olfactory acuity (i.e. ability to detect pheromones) is best determined by measuring the density of the neuroepithelium lining the vomeronasal organ, the major functional feature of the AOS (Munger, Leinders-Zufall and Zufall, 2009; Smith and Bhatnagar, 2004). In species that do not possess an AOS, such as most catarrhines, there is no established way to measure their ability to detect pheromones.

4.1.5   Brain mechanisms underlying olfactory signal production and perception

It is well known that odour information is coded in the brain of the awake, behaving animal. However, how this information is relevant to behaviour and which aspects are salient to the animal is less well known. These questions have led to many investigations, often on non-primate species such as rodents that centre on gene-targeting experiments and gene expression experiments as well as electrophysiological recordings of awake animals (Rinberg and Gelperin, 2006; Rinberg, Koulakov and Gelperin, 2006; Smear et al., 2011) (see Chapter 3, for a more detailed discussion of the neural substrates of olfactory communication).

4.2   Gestural communication

Although the term *gesture* is frequently used both in the scientific world and in everyday life, it nevertheless evades consistent definition. Human gestures are usually very broadly referred to as 'a manner of carrying the body and movements of the body or limbs as an expression of feeling' (Simpson and Weiner, 1998, p. 476). According to Kendon (2004), a gesture is a form of nonverbal communication in which visible bodily actions communicate particular messages, either in place of speech or closely intertwined with spoken words. In humans, gesturing occurs very early in life: before children start to speak, they show or request objects, and the onset of pointing is a major milestone in their cognitive development (Bates et al., 1979; Butterworth, 1998).

As a result, the criteria used to define a gesture in nonhuman primates are largely adopted from research into gestures of preverbal children (Leavens, 2004; Leavens, Russell and Hopkins, 2005). Call and Tomasello (2007) define a gesture as a behaviour that, unlike an action, is motorically ineffective, is produced in the presence of an audience and is tailored to the attentional state of the audience. Furthermore, gestures are characterized by the sender waiting for the recipient's response and his persistence and elaboration when his communicative attempts fail (Leavens, Hopkins and Bard, 2005). This definition is intrinsically tied to intentional communication, characterized by the voluntary and directed use of a particular signal to influence the behaviour of a specific recipient (see Chapter 8 for a more detailed discussion of intentionality).

Unlike research into human gestures, which is limited to the visual channel, research into gestures of apes and monkeys includes both auditory gestures that produce sound from body parts other than the vocal cords, and tactile gestures that include physical contact between the two interacting partners (see Chapter 1, Table 1.1). Visual gestures may not be suitable for use in dense vegetation or if the recipient is not attending, while tactile and auditory gestures do not necessarily require the visual attention of the recipient. Tactile and auditory gestures are most often not 'pure' modalities in terms of the perceptual channel, since they can also contain a visual component. For example, 'chest beat' is an auditory gesture that also has a salient visual component.

Gesture research into nonhuman primates is not restricted to the use of hands but also considers movements of limbs, the head and even body postures. This already hints at two problems of gesture research. First, a gesture is rather broadly defined in terms of the involved body parts. As a consequence, some studies consider orofacial movements or facial expressions as facial gestures (Ferrari *et al.*, 2003; Maestripieri, 1999), while other scholars refer to facial expressions as a separate mode of communication (Liebal *et al.*, 2004c; Pollick and de Waal, 2007). Second, and even more importantly, the investigation of tactile gestures automatically raises the question of how instrumental actions can be differentiated reliably from gestures. More specifically, researchers often struggle to differentiate communicative gestures from actions that can also serve to initiate interactions with other conspecifics, but lack the characteristics qualifying them as intentionally used gestures (Liebal and Call, 2012).

The resulting ambiguity leads to the notion that research into gestural communication of nonhuman primates suffers from a great deal of subjectivity given the lack of an objective, standardized method to identify and describe gestures as there is for facial movements of monkeys and apes (see section 4.3.2, for the **Facial Action Coding Systems (FACS)** available for nonhuman primates). Scholars design their individual coding schemes and the resulting differing terminology hinders systematic comparisons across studies. However, Roberts and colleagues (2012b) proposed a coding scheme that is adopted from research into speech-accompanying gestures and which is based on structural properties of gestures (e.g. form or position of the arm or hand while performing a manual movement) (see also Forrester, 2008). Thus, similar to FACS used to identify facial movements, this coding system is based on form features and not on functional aspects of manual movements (see also section 4.2.2).

Since the current definitions are adopted from developmental psychology, research into nonhuman primates' gestures entails a rather anthropocentric approach. For example, when talking about human gestures, people think of gestures like 'thumbs up', 'waving goodbye', 'blowing a kiss', etc. In contrast, the overwhelming majority of nonhuman primate gestures look more like concrete actions than these abstract, highly conventionalized human gestures. One solution might be to dispose of the term 'gesture' and replace it with 'nonvocal or bodily behaviour'. However, in doing so we may miss the interesting distinction between behaving (action) and communicating (gesture), which may reflect different levels of complexity.

Seminal studies of primates did not use psychological definitions for the body postures and hand movements they observed (e.g. Carpenter, 1940; Kummer, 1968; Rijksen, 1978; van Hooff, 1973), but in the 1980s, Michael Tomasello, Josep Call and their colleagues initiated a new, psychological approach to gestural communication in nonhuman primates. They specifically focused on the intentional use of gestures in order to differentiate gestures from more stereotypical, phylogenetically ritualized displays during their studies of two captive groups of chimpanzees (Tomasello et al., 1985, 1994, 1997). The same approach was later extended to captive populations of the other great apes, siamangs and Barbary macaques (for a summary, see Call and Tomasello, 2007) and, more recently, to wild populations of chimpanzees (Hobaiter and Byrne, 2011a, b; Pika and Mitani, 2006) and gorillas (Genty et al., 2009). A more qualitative approach to primate gestural communication with detailed descriptions of the individual gestures is applied by Tanner and King (King, 2004; Tanner, 1998, 2004).

4.2.1   When do primates produce gestures?

In general, there are two main approaches used to capture situations when primates produce gestures: observations of gestural interactions between conspecifics and experimental settings of a primate interacting with a human experimenter. Both will be explained in the following section, along with the main questions of interest within each approach.

4.2.1.1   Observations of interactions between conspecifics

In its simplest form, researchers observe and record an individual or a group of individuals and aim to record the occurrence of gestures alongside the context of production and the responses of recipients. Gestures are usually captured by using video cameras, which allows the identification of gestures to be verified by an independent coder, but they are also sometimes coded in real time by researchers. The very first studies that referred to behaviours that could qualify as gestures (but often were not labelled as such) focused primarily on wild populations of monkeys and apes, for example baboons (Kummer, 1968), gibbons (Carpenter, 1940), orangutans (MacKinnon, 1974; Rijksen, 1978), gorillas (Schaller, 1963), chimpanzees (Goodall, 1986) and bonobos (Kuroda, 1980). Van Hooff was one of the first to study gestural communication in a captive group of chimpanzees (van Hooff, 1973), followed by de Waal, who later compared their repertoire with that of bonobos (de Waal, 1988). Gestures were mostly reported as part of the normal behaviour of a species (**ethogram**) and usually consisted of detailed descriptions of those behaviours and the contexts of their use.

The majority of studies on gesture are conducted on great apes in captivity (Slocombe, Waller and Liebal, 2011, see also Chapter 5). This is partly due to the practical advantage of access and visibility in captive settings as compared to the wild. Although there are virtually no systematic comparisons of wild and captive populations, there are two studies that indicate that captive baboons and siamangs use a slightly greater repertoire but also use gestures more frequently than wild individuals (Fox, 1977; Kummer and Kurt, 1965).

Observation times are either distributed over the daily activity period of the primates to capture the range of gestures they use across different contexts or alternatively, if gestures used in a particular context are of special interest, observations take place at certain times of the day, such as feeding times. Some studies have focused on when primates use their first gestures in ontogeny (see Chapter 6), in an effort to ascertain the mechanisms and necessary conditions for acquisition. They either use a longitudinal design to observe (mostly few) individuals over longer periods of time to capture the onset of gestural communication (e.g. Gómez 1990; Schneider, Call and Liebal, 2012a) or a cross-sectional design to investigate the gesture use across different age classes (e.g. Tomasello *et al.*, 1997).

There are very few studies on monkey gestures and most of them refer to facial expressions as gestures (Hesler and Fischer, 2007; Laidre, 2008, 2011; Maestripieri, 1996a, b, 1997). Currently, it is unclear whether the lack of published studies concerning monkey gestures is due to the fact that research has preferentially focused on great apes, following fruitful early studies on these species, or whether monkeys are less likely to use gestures. The focus on great ape gestures is reflected in the greater number of gestural repertoires reported for these species as well as some lesser apes, compared to monkeys (white-handed gibbons: Baldwin and Teleki, 1976; siamangs: Fox, 1977; Liebal *et al.*, 2004c; orangutans: Liebal, Pika and Tomasello, 2006; gorillas: Genty *et al.*, 2009; Pika, Liebal and Tomasello, 2003; bonobos: Pika, Liebal and Tomasello, 2005; chimpanzees: Tomasello *et al.*, 1994, 1997, 1985; Tomasello, Gust and Frost, 1989). However, the complexity and breadth of an observed repertoire depend on the methods of data collection and level of analysis, and result in varying degrees of overlap of individual gestural repertoires. For example, Genty *et al.* (2009) report on more than 100 gestures for gorillas, while Pika, Liebal and Tomasello (2003) mention only about 30 different gestures. Since they observed different groups, one explanation might be that the gorillas studied by Genty and colleagues use a greater variety of gestures. A more likely explanation is that Genty and colleagues obtained a more representative picture of the full gestural repertoire through observation of different populations in both captive and wild settings for relatively long periods of time, compared to the relatively short study period performed by Pika and colleagues on only two captive groups. It is also possible that Genty and colleagues used a more detailed coding system to differentiate their gestures. Given the lack of consensus on the definition of a gesture, different authors often use different labels for one and the same gesture. As a consequence, it is not only difficult to compare results across different species, but also between different studies on the same species.

One way to elicit certain types of gestures in interactions with conspecifics is to manipulate the observational conditions, for example by excluding particular individuals from the group and then observing how this influences the interactions of the remaining individuals, by transferring individuals into another group, or by introducing objects such as a monopolizable food source. These are methods that could capture gestures that are characteristic for a particular context, such as submissive gestures directed towards high-ranking individuals or those used to beg for food.

4.2.1.2   Interactions with humans in experimental studies

A different approach investigates gesture use by monkeys and apes in interactions with a human experimenter. The experimental paradigm usually requires a primate subject to point to the food item the human is holding or to a location where desirable food is out of reach.

Based on this general setting, there is now a great variety of studies that investigate how great apes and monkeys adjust their communicative means to their interactions with a human experimenter and in particular to the human's attentional state (Cartmill and Byrne, 2007; Leavens, Hopkins and Bard, 1996; Liebal et al., 2004c; Poss et al., 2006; Povinelli and Eddy, 1996).

This particular experimental setting with the desirable food items being out of the primates' reach often results in the use of pointing gestures to request the food from the human experimenter (see Chapter 8, **Box 8.3**; Chapter 9). Thus, in contrast to their interactions with conspecifics, chimpanzees use pointing gestures to beg for food (Leavens and Hopkins, 1999; Leavens, Hopkins and Bard, 2005; Theall and Povinelli, 1999). Furthermore, these interactions with humans also elicit auditory gestures or vocalizations to attract the attention of humans if they are turned away from the chimpanzee. Similar results were also reported for orangutans and gorillas (Cartmill and Byrne, 2007; Poss et al., 2006).

There are also some studies on pointing behaviour of capuchins, squirrel monkeys and baboons (Anderson et al., 2001; Anderson, Kuwahata and Fujita, 2007; Kumashiro et al., 2002; Meguerditchian and Vauclair, 2006; Mitchell and Anderson, 1997), but subjects often received special training, which gives rise to the conclusion that pointing by monkeys is less flexible than pointing in great apes. However, to date there are few published studies that actually compare the use of pointing gestures in monkeys and great apes (see Gómez, 2007).

4.2.2   Structure and form of gestures

Research into primate gestures (as opposed to vocalizations and facial expressions) tends to focus on how specific gestures are used, but rarely considers their form or structural properties. In contrast to olfactory, facial and vocal signals, that can all be classified using objective techniques based on the structure of the signal, there is no widely accepted analogous system for gestures. As outlined above the lack of tools to objectively map and discriminate different gestures based on structural properties creates difficulties, as researchers tend to adopt idiosyncratic labels and descriptions for gestures, making comparisons even within the same species difficult. One tool used to describe the structural properties of gestures is adapted from human gesture research, specifically from sign language studies: simultaneous structures describe parameters of one particular gesture at a given time, such as hand shape, position of the hand, palm orientation, location and movement. This approach is adapted from sign language studies, modified for speech-accompanying gestures (Bressem, 2008) and is now used to describe structural properties of apes' manual gestures. For example, Roberts and colleagues used a coding system, which is based on morphological features of manual movements to describe the gestural repertoire of a community of wild chimpanzees (Roberts *et al.*, 2012b). Based on this approach, they were able to discriminate 30 morphologically distinct manual gesture types and suggest that this focus on structural (rather than functional) properties of gestures enables a systematic comparison across different studies and species.

4.2.3   What mechanisms underlie the production of gestures?

In this section, both the proximate mechanisms that produce gestures and the developmental mechanisms that influence the manifestation of a gesture are described. Both are covered in more detail in Chapters 6–9, but here these issues are discussed specifically in relation to the methods used to study the mechanisms of gesture production.

4.2.3.1   Proximate mechanisms

Gestures are generally thought to be underpinned by complex cognitive mechanisms and the result of voluntary intentional cognitive processes, an approach that is highly tied to the definition of gestures (Call and Tomasello, 2007). Thus, the methods used to tackle the question of proximate mechanisms tend to focus on unravelling the skills needed to produce a gesture in a certain context. For example, both observations of interactions between conspecifics as well as experimental methods that include monkeys' or apes' interactions with humans are used to investigate whether they adjust their gestures to the behaviour of the recipient, and, for instance, whether their gesture production is linked to an understanding of the recipient's attentional state.

4.2.3.2   Developmental mechanisms

In contrast to vocalization research, very few cross-fostering experiments have been used to identify gestures that are present from birth. There is one report that a gorilla raised by humans that never observed or interacted with other conspecifics still used the 'chest beat' gesture (Redshaw and Locke, 1976). Other studies transferred adult individuals into groups of another species. For example, a wild female savanna baboon was moved from her group to a group of hamadryas baboons. Female savanna baboons usually respond to male approaches by running away, whereas hamadryas baboons present their anogenital region. The newly introduced savanna baboon female first responded with her species-typical behaviour, but soon started to produce the hamadryas-specific behaviour (Kummer, 1997). Longitudinal studies can be used to follow the same individuals over extended periods of time, to identify the onset of gesture use and to investigate if individual repertoires change over time (Schneider, Call and Liebal, 2012a, b). In gesture research, there are only a few studies that use this method and most have used qualitative descriptions of the individual's gesture use, given the small sample sizes available (e.g. Gómez, 1990). Some studies use a combination of a cross-sectional design and longitudinal observations. Thus, several individuals or groups are repeatedly observed over intervals ranging from weeks to even years, to compare the gestural repertoires used across the different points in time (e.g. Tomasello *et al.*, 1994, 1997) and to investigate the mechanisms that underlie gesture acquisition (Chapter 6, section 6.2).

4.2.4   What does a receiver understand from gestures?

The majority of gesture studies focus on the different types of gesture and the contexts in which they are used, but much less is known about the effect they actually have on the recipient's behaviour. In part, this may be due to the dominant theoretical approach to gestures. What a recipient understands from a gesture is closely linked to the question of whether gestures have a particular meaning, and whether they are functionally referential (see Chapter 9). Most gestures seem to have no specific meaning because of their flexible nature – since one gesture can be used in different contexts and several different gestures can be used to achieve the same goal, which is also termed means-ends dissociation (Bruner, 1981). Therefore, the information that a gesture conveys might vary depending on the context. For example, an 'extend arm' gesture can function as a submissive signal when a subordinate individual uses this gesture towards a higher-ranking individual while approaching them, but it can also serve as an invitation to follow when used by a mother towards her offspring. Again, subtle changes in form that are not yet considered in gesture research might reveal subtle differences between gestures, but so far it is proposed that the majority of gestures – as opposed to many vocalizations and facial expressions – do not carry a context-specific meaning.

Some studies have focused on the change of recipient's behaviour after a gesture was directed towards them, to try and explore what the receiver understands from the gesture (Liebal *et al.*, 2004b). However, it is usually only reported *that* the recipient reacted, but not exactly *how*, but an individual can react to a gesture in very different ways. For example, the recipient can show no obvious change in behaviour, or can respond with a particular action such as starting to play or by producing another gesture in response to the sender's communicative attempts. Finally, the recipient can respond by leaving, a behaviour usually recorded as different from showing no response at all. As opposed to research on vocalizations and facial expressions, however, very little is known about what primates understand about gestures of other conspecifics.

Experimental settings enable a more controlled way to investigate what apes understand about gestures. However, the disadvantage is that this investigation is often limited to interactions with humans and to a particular type of gesture, namely pointing. There is a vast variety of studies that investigates the understanding of human pointing gestures in monkeys and apes, but also in non-primate species (for a review, see Miklósi and Soproni, 2006). Many of these studies are based on the object-choice paradigm, which consists of a human placing food under one of at least two cups with the baiting process invisible for the subject. Then the human indicates where the food is hidden, usually by establishing eye contact and then extending an arm towards the baited cup. However, primates repeatedly failed to use this information to find the food (e.g. Call, Agnetta and Tomasello, 2000, Povinelli, Bierschwale and Cech, 1999). While some authors argued that these results are due to fact that primates do not understand the human's pointing gestures as means to inform them about the location of the food and thus as a cooperative communicative act (Hare and Tomasello, 2004), others emphasize methodological problems and the inconsistent use of this paradigm across different primate and non-primate species (Mulcahy and Call, 2009; Mulcahy and Hedge, 2012). To address these methodological issues, some researchers have changed the basic setting of the object-choice task, either by manipulating the distance between the human experimenter and the food (Mulcahy and Call, 2009) or by using a competitive rather than cooperative setting (Hare and Tomasello, 2004; Tempelmann, Kaminski and Liebal, 2013).

4.2.5   Brain mechanisms underlying gesture production and perception

Research methods to investigate the neural basis of behaviour are developing at a fast rate. Here, we outline the current methods available to scientists, and how they have been used to investigate gesture specifically.

4.2.5.1   Laterality

In most humans, language comprehension and production is lateralized in the left hemisphere. Human manual (and some oral) movements are also lateralized, suggesting a similarity between these manual gestures and spoken communication (Corballis, 1992; Kimura, 1993; MacNeilage, Studdert-Kennedy and Lindblom, 1987). Thus, one common way scientists investigate the neural substrates underlying the production of nonhuman primate gestures is by investigating whether gesture use is similarly lateralized (see Chapter 3, **Box 3.3**). Commonly, researchers simply record whether there is a preference for hand use while performing manual gestures. Although it is not possible to identify any specific brain regions by using this non-invasive, observational method, the method does enable tentative conclusions about the role of the two hemispheres for the production of manual gestures.

4.2.5.2   **Magnetic Resonance Imaging (MRI)**

MRI can be used to gain a greater understanding of the neural areas involved in the production of behaviour. MRI provides a detailed anatomical image of the brain, and allows measurement of subtle differences in size and structure between regions. For MRI, animals need to be anaesthetized to ensure their head remains still. Bill Hopkins, Jared Taglialatela and colleagues have used this technique to investigate whether the preferential use of one hand while gesturing correlates with asymmetries of different brain regions between the two hemispheres (Hopkins, Cantalupo and Taglialatela, 2007;

Pilcher, Hammock and Hopkins, 2001; Taglialatela, Cantalupo and Hopkins, 2006). However, those studies do not locate the brain regions responsible for gesture production directly, but instead demonstrate an association between laterality and anatomical asymmetries in the brain.

4.2.5.3   Positron Emission Tomography (PET)

PET is a nuclear imaging technique that allows investigation of functional processes in the brain. Thus, PET can be used to investigate which specific brain regions are involved in the execution of a specific behaviour. Before animals participate in an experiment, they first must ingest a **radioactive tracer isotope** that is incorporated into a biologically active molecule, called a tracer (e.g. fluorodeoxyglucose). They then engage in the behaviour of interest for a period of time (e.g. 30–40 minutes) and the tracer accumulates in those neurons that are active during the behaviour as the glucose attached to the tracer is metabolized. The tracer is trapped in those tissues and as the tracer decays it emits gamma rays, which can later be detected in the PET scan. The animals need to be anesthetized after they finish the behaviour session, during transportation to the imaging facility and for the scanning itself. PET has been used to demonstrate the neural underpinnings of gesture production in chimpanzees (Taglialatela *et al.*, 2008).

4.2.5.4   Single-cell recording

Single-cell recording is a technique used to study the activity in specific neurons during experimental tasks. This method has been used to observe mirror neurons in the monkey brain, which are active when a monkey either performs or observes actions such as grasping an object (Rizzolatti and Arbib, 1998). In this context, grasping is often referred to as gesture. However, given our definition at the beginning of this chapter, we would not consider this to be a communicative gesture, since it is an action directed towards an object and not used towards another individual to influence their behaviour.

4.3   Facial expression

Most mammalian species are capable of changing the structure and appearance of their faces using facial musculature (Diogo *et al.*, 2009). These facial movements can form meaningful and adaptive components of the animal's behavioural repertoire, and are often termed facial expressions. Primates have particularly mobile faces, and the majority of social primate species exhibit some form of facial expressions, involving movements of the jaws, lips, ears, eyelids and other facial landmarks. Some expressions (such as the 'bared-teeth' display, where the lips are retracted and the teeth exposed) can be seen in many different species, implying that they have a common ancestral root. Some expressions are also common to many other mammal species, as well as the repertoires of different primate species (e.g. the 'playface'; canids: Bekoff, 1974; bears: Egbert and Stokes, 1976; rats: Panksepp and Burgdorf, 2003). Other facial expressions are unique to certain species, or a group of related species, and do not have obvious counterparts in humans or other animals. The 'lipsmack' expression, for example, is a dynamic display found in many Old World monkeys, particularly macaques and baboons (van Hooff, 1967). Facial expressions (such as the 'lipsmack') often include auditory components, but these sounds are only considered vocalizations if they involve the vocal folds or larynx (see section 4.4).

An ethological approach is usually employed to study the facial expressions of nonhuman animals. Within such an approach, adaptive function in terms of information transfer is considered (Smith, 1977), and facial expressions are conceptualized as stereotyped displays (Huxley, 1914). A consideration of the adaptive function often involves a strong focus on the receiver (receiver psychology: Guilford and Dawkins, 1991). In contrast to the study of other animals, however, the study of primate facial expression often attracts a less functional approach, and is instead more similar to that used for human facial expression, with a strong focus on the feeling state of the sender. Darwin's early accounts of facial expressions as behavioural counterparts to felt emotion (Darwin, 1872), as opposed to communication adaptations, have been highly influential in this emotional approach to facial expression. Thus, many scientists focus on facial expressions as part of an emotional system when studying primates.

4.3.1   When do primates produce facial expressions?

The vast majority of studies examining the production of primate facial expressions are observational. Early scientists produced several comprehensive descriptions of the facial repertoires of many species (e.g. Andrew, 1963; Chevalier-Skolnikoff, 1974; van Hooff, 1967), in both captive and wild settings. Similar to gestures, many descriptions of facial expressions are also embedded within broad ethograms of behaviour. The meaning of the expression is often inferred from the other behaviours that accompany the expression (in the sender). For example, a 'playface' often occurs with play behaviours, so it is largely considered a signal of play. Inevitably, this has also led to inferences about the feeling state of the sender, that the sender feels playful. Thus, the context of the expression is often bound up with feeling state and meaning. Context is also often described in terms of the actions of other individuals, so if the facial expression tends to occur as a response to others, then the action of the other may be seen as a stimulus event (i.e. an aggressive attack).

The use and frequency of facial expression is also commonly analysed in relation to individual differences such as sex, age and dominance, the specific relationship between individuals, and compared between populations. Although such analyses are not experimental, they are often approached quantitatively by comparing characteristics of the facial expressions between different conditions. Researchers now often take video recordings of primate interactions in order to code facial expressions in detail. Fully experimental approaches to record the variables that affect the production of specific facial expressions are relatively rare. In some experimental studies, however, facial expression is often recorded as a behavioural measure of a particular emotional state, e.g. yawning as an indicator of stress (Paukner and Anderson, 2006) or as a measure for empathy (Campbell and de Waal, 2011).

4.3.2   Structure and form of facial expressions

Early research employed no standardized measure for describing the structure and form of primate facial expressions. Ethograms included detailed descriptions of facial expressions, sometimes with drawings and photographs (e.g. chimpanzees: Andrew, 1963; Chevalier-Skolnikoff, 1973; van Hooff, 1973; van Lawick-Goodall, 1968; Parr, Cohen and de Waal, 2005), but descriptions can be subjective and so it is difficult to build direct comparisons between species or populations. Researchers were often using different terms, resulting in a large body of descriptions for each expression type. It is sometimes unclear, therefore, whether these reflect genuine differences. Faces can pose particular problems for observational description, as humans process faces in a highly specialized manner. Configural processing of faces (Young, Hellawell and Hay, 1987) streamlines our attention to important information in the face of others (identity, attention, emotion, etc.) and can render our conscious perception of detail unreliable (Seyama and Nagayama, 2002). This specialized processing is a very useful adaptation for complex social living – to respond to relevant information only – but as a consequence, we must be vigilant with scientific observations when we are concerned with the detail of the signal. Waller *et al.* (2007) demonstrated this problem by asking human participants to make comparisons between chimpanzee and human facial expressions after viewing the two sets of images. Comparison of features and judgements of similarity were associated with how the images were interpreted emotionally. Specifically, if the chimpanzee faces were interpreted as fearful, comparisons with human faces were underestimated, even though the two image sets depicted homologous expressions (the human 'smile' and the chimpanzee 'bared teeth'). Standardized observational methods are needed to help minimize these problems.

Scientists studying human facial expressions have overcome the problems inherent in trying to measure faces, through development of The Facial Action Coding System (FACS: Ekman and Friesen, 1978; Ekman, Friesen and Hager, 2002). FACS, although ultimately concerned with facial expressions, is a facial *movement* coding system. The components of facial expressions, as opposed to composite expressions, are identified and recorded individually. This approach is particularly helpful when observing faces, as it forces the observer away from the whole face (which can cause biases) and towards the details. FACS is based on careful consideration of the surface appearance of specific facial movements, informed by an understanding of their muscular basis. Each facial movement (termed *action unit* or AU) is identified by a set of minimal observable criteria relating to the underlying muscle contraction. FACS is heavily influenced by the seminal work of Duchenne (1862) who first identified the muscular basis of facial movements through surface electrical stimulation. Later experiments have built on this work with intramuscular stimulation techniques (Waller *et al.*, 2006). An important strength of FACS lies in standardization: to use the system, researchers must learn to identify the appearance of facial movements and pass a test to become certified FACS coders. This process is designed to minimize individual coder biases.

Modified FACS systems have been developed for use with primates, and have all been designed to allow direct comparison with the human FACS (Figure 4.2). Modification to

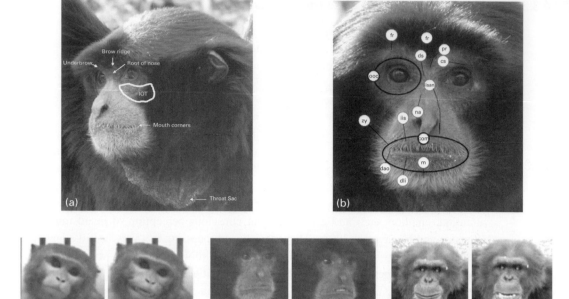

Figure 4.2 Facial Action Coding System (FACS). (a) Examples for key facial landmarks (IOT = infra-orbital triangle) and (b) facial musculature in the siamang face (lines represent approximate muscle location and labels refer to specific facial muscles: fr = *frontalis*, ooc = *orbicularis oculi*, pr = *procerus*, ds = *depressor supercillii*, cs = *corrugator supercillii*, llsan = *levator labii superioris alaeque nasi*, na = *nasalis*, zy = *zygomaticus major*, lls = *levator labii superioris*, dao = *depressor anguli oris*, dli = *depressor labii inferioris*, m = *mentalis*, oom = *orbicularis oris*). (c–e) show an example of an Action Unit (AU 16) across different primate species: (c) rhesus macaque: neutral face (left) and AU 16 (right, with AU 12); (d) siamangs: neutral face (left) and AU 16 (right); and (e) chimpanzees: neutral (left) and AU 16 (right). (Photos: Katalin Gothard (c), Paul Kuchenbuch (d), Manuela Lembeck (a, b) and Lisa Parr (e).)

each species has been fairly straightforward as the facial musculature is broadly similar (Burrows, 2008) (see Chapter 10, Figure 10.1). The same codes can be used to identify each muscle movement. FACS was first modified for use with chimpanzees (ChimpFACS: Vick *et al.*, 2007). This was followed by MaqFACS for rhesus macaques (Parr *et al.*, 2010), GibbonFACS for hylobatids (Waller *et al.*, 2012) and OrangFACS for orangutans (Caeiro *et al.*, 2013). The development of each system was informed by dissections (Burrows *et al.*, 2011; Burrows, Waller and Parr, 2009; Burrows *et al.*, 2006; Diogo *et al.*, 2009) and intramuscular electrical stimulation experiments where possible (Waller *et al.*, 2006; Waller, *et al.*, 2008). Video footage of the study species is required in order to perform detailed FACS coding of their facial expressions. Similar to the human FACS, descriptions and coding instructions for each AU are published in scientific coding manuals, which are available online and users are required to pass a final test to become certified coders.

There are clear practical benefits to using standardized FACS systems: descriptions can be translated between studies to make the growing database of research on primate facial expressions accessible to a wider audience of researchers (including human facial expression researchers). Comparing facial expressions using FACS can also be helpful in identifying whether facial expressions that are similar in appearance have arrived at that similarity through **homology** (stem from the same ancestral display) or **analogy** (look similar but do not stem from the same ancestral display). Assessing the muscle substrate underlying facial movements may not demonstrate homology conclusively, but it does allow scientists to assess similarity on more than one level (Preuschoft and van Hooff, 1995).

4.3.3   What mechanisms underlie the production of facial expressions?

The proximate mechanisms underlying the production of facial expression (i.e. the psychological and physiological mechanisms) and the developmental mechanisms (i.e. the processes of acquisition) are of interest to understand how facial expressions are produced, and how they develop in an individual's lifetime.

4.3.3.1   Proximate mechanisms

Emotional experience, or feeling state, is often assumed to underlie facial expression, and as such is sometimes seen as a causal mechanism for production (e.g. Darwin, 1872). Underlying emotion is rarely studied directly (if this is ever possible), although in other animals, neural activation is sometimes taken as a proxy for feeling state (see Panksepp, 1998). Instead, an emotional underpinning is inferred from the behaviour that accompanies the facial expression (e.g. Bard, 2003). Volitional or cognitive processes are rarely considered to underlie the production of facial expressions, and so few studies employ methods to investigate this. The physiological mechanisms underlying production (facial muscles, neural control) are investigated using facial dissection (e.g. Burrows *et al.*, 2006) and quantitative analysis of size and structure of neuroanatomical areas (e.g. Sherwood *et al.*, 2004b; see section 4.3.5).

4.3.3.2   Developmental mechanisms

Scientists are also interested in the level and type of experience necessary and sufficient for facial expressions to manifest in the individual (see Chapter 6 for a detailed discussion of acquisition). For example, studies have examined differences in the onset, frequency and form of facial expression between different captive rearing conditions (Bard, 2003) and between groups (van Hooff, 1973).

4.3.4   What does a receiver understand from facial expressions?

Scientists have explored how receivers understand the facial expressions of conspecifics both observationally and experimentally.

4.3.4.1   Observational data

Preuschoft (1992) was one of the first to investigate facial expressions in terms of how they affect social interaction between the sender and receiver. Preuschoft used quantitative observational analysis in the form of pre-post-event histograms to assess the social function of a facial expression in terms of the receiver's response. Pre-post-event histograms plot the frequency of events immediately prior to a facial expression, compared to events immediately following the expression. The results of several of these sequences are aggregated, and the resulting frequency distribution of events relative to the facial expression is represented as a histogram. These studies set the scene for much of the observational work assessing the social function of facial expressions. This paradigm has been an influential framework for a great many studies of facial expression in both monkeys and apes and a focus on the receiver has been included in many subsequent studies (e.g. Palagi and Mancini, 2011; Visalberghi, Valenzano and Preuschoft, 2006; Waller and Cherry, 2012; Waller and Dunbar, 2005). Although these studies do not focus on cognitive understanding per se, they do attempt to demonstrate how the behaviour of the receiver may have changed as a result of the signal. Seyfarth *et al.* (2010) conceptualize information transfer as 'a reduction of uncertainty in the recipient', so in this sense, the information salient to the receiver is being documented by these studies.

4.3.4.2   Experimental data

Parr, Hopkins and de Waal (1998) used a computerized match-to-sample paradigm to demonstrate that chimpanzees can discriminate between different conspecific facial expressions (see Figure 4.3 for the general procedure). The basic task requires animals (who are trained to operate a computer using a joystick or touch-screen interface) to match pairs of expression photographs and avoid non-matching expressions. Such studies demonstrate that these facial expressions are perceived as meaningful, discrete stimuli to chimpanzees. As such, unlike the above observational approaches, these

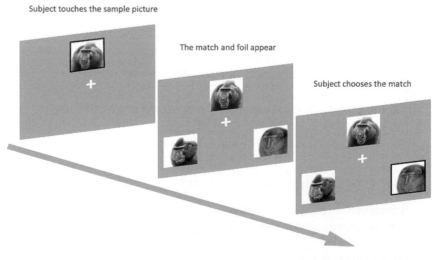

Figure 4.3 General procedure of the match-to-sample task. When the subject touches the sample picture on the touch screen, two additional pictures appear: the match and the foil. Only if the subject touches the correct match, do they receive a food reward (figure provided by Jérôme Micheletta).

studies examine the cognition underlying receiver interpretation of facial expression more directly. Parr, Waller and Heintz (2008) conducted similar experiments, only they used computerized model chimpanzee faces morphed into different anatomically correct facial expressions (based on ChimpFACS) instead of photographic images. This allowed the scientists to analyse the chimpanzees' patterns of discrimination between expressions, in relation to specific component movements. The findings showed that chimpanzees were using both configurational and featural cues in facial expression processing, which is similar to human processing. To understand how the receiver understands facial expressions in terms of positive and negative associations, Parr (2001) asked chimpanzees to match facial expressions to images of positive and negative items (e.g. food vs. veterinary procedures). To date, this is one of very few studies explicitly addressing how nonhuman primates understand expressions in terms of emotional valence.

Miller, Caul and Mirsky (1967) studied receiver understanding of facial expression in a rather different experimental setting. Captive rhesus macaques were first conditioned to avoid an aversive stimulus by pressing a lever when they saw a visual stimulus (flash). The monkeys were then placed in conspecific pairs, but the flash could only be seen by one monkey (the stimulus monkey) and the lever only available to the other (the responder monkey). The only information available to the responder monkey was the face of the other animal. Monkeys were able to avoid the aversive stimulus by pressing the lever when they saw the stimulus monkey responding to the flash. The authors concluded, therefore, that the monkeys were using the information present in the face (presumably, negative facial expressions) to predict the forthcoming aversive stimulus. More recently, Morimoto and Fujita (2011) examined receiver responses to facial expressions, but using a less invasive paradigm. Captive capuchin monkeys witnessed another individual (demonstrator) reacting either positively or negatively to the contents of one of two containers, and were then allowed to choose one of the containers. The observer's choice of container was related to whether they witnessed specific facial expressions in the demonstrator, implying that they have some understanding of what these expressions mean.

4.3.5 Brain mechanisms underlying facial expression production and perception

4.3.5.1 Neuroanatomical approaches

The neural structures underlying facial expression production can be investigated using detailed quantitative neuroanatomical techniques. For example, Sherwood et al. (2005) compared the number of motorneurons in the facial motor nucleus between species. More motorneurons (relative to brain size) result in greater dexterity of facial muscles, so these analyses can quantify the extent of neural control over facial muscles, and assess species differences in relation to socioecological and phylogenetic factors. Similarly, Sherwood et al. (2004b) measured the volume of the region corresponding to orofacial representation of the primary motor cortex and compared them across different species. Histological studies have also been conducted on the specific arrangement and density of neurons in this same region (Sherwood et al., 2004a). The relationship between facial nucleus volume (relative to brain size) and the visual cortex has also been examined (Dobson and Sherwood, 2011), to assess whether production and perception of facial expressions are related (see Chapter 3, section 3.3.3.2).

Earlier studies often used lesioning techniques to examine the neural underpinnings of facial expression. In such studies a specific brain area is surgically damaged and then the behaviours of the animal are recorded and compared to its previous behaviours in order to isolate the functions of this particular brain area (e.g. investigation of social behaviours after amygdale lesioning: Rosvold, Mirsky and Pribram, 1954).

4.3.5.2   Single-cell recording approaches

Electrodes can be inserted directly into specific neurons to record their activity during the perception of facial expressions. For example, this has been carried out in rhesus macaques to measure the neural activity in the amygdala while they were looking at faces and facial expressions of monkeys and humans (Gothard *et al.*, 2007). Another study examined the response of mirror neurons in the premotor cortex during performance and observation of communicative mouth actions such as 'lipsmacking', 'lip protrusion' and 'tongue protrusion' (Ferrari *et al.*, 2003). (Note: the authors referred to these movements as communicative gestures rather than facial expressions.)

4.3.5.3   Functional magnetic resonance imaging (fMRI)

Several studies have used **fMRI** on awake rhesus macaques to explore the neural correlates of the perception of facial expression (e.g. Hadj-Bouziane *et al.*, 2008) (for more details, see Chapter 3). This technique maps dynamic changes in blood flow to different brain regions and increased blood flow to an area is taken to indicate high levels of activity in that brain region. Thus researchers can correlate performance of certain tasks with the use of certain brain areas. As monkeys have to be restrained during fMRI scanning, the technique is usually used to investigate the brain areas involved in the perception of communicative signals rather than their production.

4.4   Vocal communication

Vocalizations are ubiquitous in primates, and are defined as sounds produced from the vibration of the vocal folds within the larynx. Other sounds, such as 'whistling', 'lipsmacking' and 'raspberry blowing', are made by the mouth but do not engage the larynx. Such sounds, therefore, are not termed vocalizations. The terms vocalization and call are often used interchangeably. Calls can vary greatly in acoustic structure (frequency, amplitude and temporal pattern of the sound) and variation in these acoustic parameters can relate both to the target of the call, and the habitat it needs to travel through (Maciej, Fischer and Hammerschmidt, 2011). Sounds with certain acoustic properties (e.g. high pitch) propagate through different habitats with varying success. Thus, over the course of evolution, the sound propagation properties of a species' habitat will have acted as a selection pressure, shaping the acoustic structure of the calls of the species to ensure they reach their target audience. Some primate calls can propagate over very large distances (e.g. over a kilometre) and many allow communication with individuals who are out of sight. In visually dense habitats such as forest, vocal communication allows communication with spatially disparate group members, where visual forms of communication (e.g. gestures and facial expressions) may not be possible (Maestripieri, 1999; Marler, 1965). In contrast to olfactory signals, which can persist over long periods in the environment, vocalizations are momentary signals (see Chapter 1, Table 1.1).

Primate calls are commonly separated into signals that communicate with individuals within a social group (intragroup calls) and signals for communicating with other groups or species (intergroup calls). Intragroup calls are generally lower in amplitude and acoustically more varied than intergroup calls. Intragroup vocalizations are commonly produced by primates in a variety of ecologically and socially important contexts, such as foraging, travelling, maintaining group cohesion, copulation, affiliation, play and agonistic interactions. The structure of these softer vocalizations is likely to be less dependent on habitat-specific propagation than the long-distance calls. The structure of all calls is, however, shaped by receiver psychology: over evolutionary time signals that are easy for the receiver to detect, discriminate and remember are most likely to be selected (Rowe, 1999).

Intergroup calls or 'loud calls' tend to travel relatively long distances through the environment to reach their target audience. Intergroup calls include territorial vocalizations, such as gibbon or indri song and howler monkey roars. Such calls function to announce the location of the caller within their territory and to deter competing groups or individuals from entering. Predator alarm calls are also usually classified as 'loud calls' that travel some distance. The target of the calls varies with species, but can include the predator themselves and potentially spatially scattered group members. Due to the commonly loud nature of these calls, other sympatric species often eavesdrop on these signals and learn to react adaptively to them (Zuberbühler, 2000a).

Humans also produce non-speech sounds in similar contexts to non-human primates. Some have argued that only these sounds, such as 'laughter' and 'crying', may have evolved directly from primate vocalizations and that language is a separate system, without its evolutionary roots in primate vocal communication (Burling, 2005). Burling draws parallels between the largely involuntary, affective nature of these human non-speech sounds and primate vocalizations. Human non-speech sounds are, however, rarely investigated in a similar way to primate vocalizations.

4.4.1 When do primates produce calls?

4.4.1.1 Observational data

In order to establish when a primate produces a specific call (e.g. in which context, and in response to which events) observational data is usually collected. The researcher will sound record the vocalizations and then take detailed notes on the context of call emission. Information such as caller identity, caller behaviour, the behaviour of other individuals as well as the presence and nature of salient stimuli that may have elicited the call (e.g. predator, food, social interaction) may be recorded. This contextual information in concert with the acoustic measurements taken from the sound recordings can help answer questions about whether calls given by different individuals or in different contexts are acoustically distinct signals.

4.4.1.2   Experimental data

In some instances, the event that elicits a certain call can be very rare and thus it can be difficult to collect enough systematic observational data to conclude exactly what elicits these calls. In this case experimental procedures can be used to simulate the event in question so that vocal responses to it can be clearly recorded. Researchers often present primate groups with predator models in order to elicit and record alarm calls (e.g. Arnold, Pohlner and Zuberbühler, 2008; Schel, Tranquilli and Zuberbühler, 2009). Both visual (e.g. a model leopard) and auditory (e.g. a broadcast of leopard growls) stimuli can be used successfully. Such experiments also allow researchers to systematically test predictions (usually formulated from observational work), concerning why certain calls are produced. It is not just the presence of predators that can be experimentally simulated: playbacks of group members' calls can also be used to simulate the presence or indicate the behaviour of a group member in order to test the social variables that affect call production in the subject. For instance, in baboons, playbacks of 'contact barks' from family members were used to test whether individuals reply to these calls to inform the other of their location, or whether they only reply when they themselves are separated from the group (Cheney, Seyfarth and Palombit, 1996). More recently, observational work indicates that both ecological and social factors mediate production of food-associated calls in chimpanzees (Slocombe *et al.*, 2010b). In order to examine the social factors affecting call production in more detail, playbacks of individually distinctive 'travel pant-hoot' calls were used to simulate the arrival of specific individuals into the vicinity of a lone feeding individual (Schel, Machanda *et al.*, in press). Chimpanzee males were most likely to produce calls when the playback simulated the arrival of a high-ranking 'friend' into the area (Schel, Townsend *et al.*, in press).

4.4.2   Structure and form of vocalizations

We have quite a good understanding about how vocal signals are produced in primates (see Chapter 2, section 2.2.1) and this allows us to analyse their structure in some detail in an objective and quantitative manner. In order to examine the acoustic structure of a vocal signal, researchers perform **acoustic analysis** on the signal.

Acoustic analysis is invaluable as it provides us with an objective measure of the structure of a call. The first step in this procedure is to capture the vocal signal using sound equipment. Specialist sound recording equipment usually includes a directional microphone to allow high-quality recordings from a distance and a recording device to record the signal in an uncompressed format. Unfortunately some acoustic measures are distorted if applied to compressed signals, meaning that vocal signals captured and recorded by some camcorders are not suitable for acoustic analysis.

Once the vocal signals have been recorded, researchers make use of specialist acoustic software (e.g. Avisoft, Raven, Adobe Audition, Praat) to visualize their sounds

(a)

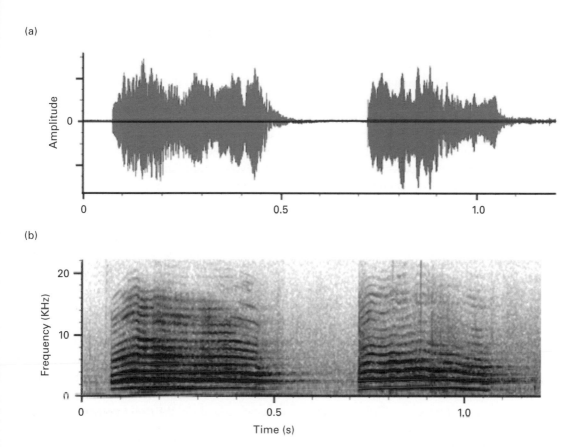

(b)

Figure 4.4 Part of a chimpanzee 'scream' bout illustrated in (a) a waveform and (b) a spectrogram.

and to measure the structure of the calls. The sound can be visualized in several ways, but the two most helpful ways are the waveform and the time-frequency spectrogram (see Figure 4.4). Once the sound is visualized, measurements can be taken to provide an objective description of the structure of the sounds. There is no agreed set of measurements that should be taken on primate vocal signals, which does bring an element of subjectivity into the process. In general, they should (as a minimum) include measurements that pertain to the temporal structure of the call (e.g. duration, call rate) and the spectral structure of the calls (e.g. **fundamental frequency**, peak frequency, change in frequency). Measurements are either conducted by hand, or processed automatically via scripts written for the acoustic software.

Once the acoustic measurements of the calls are complete, the researcher has an objective measure of the structure of the signal. This quantitative description can be used to identify and distinguish between different call types and to map the repertoire of a species. In addition, the measurements are crucial for answering questions about whether calls produced in different contexts (or by different groups or different individuals) are acoustically distinct signals. In order to tackle these questions, inferential statistics can be applied to the measurements to establish if specific measurements (e.g. call duration) differ significantly between the different samples of calls. As there is no agreed set of measurements for primate calls, it is possible, however, for researchers to take a very large number of measurements and to claim that calls are 'different' when they may not be. The danger here is that if enough acoustic variables are measured, at least one or two are likely to differ between samples. However, this may not mean that the overall acoustic structure of the calls is sufficiently distinct to allow discrimination of calls from different samples.

To tackle the issue of discriminating between call types, it is common for researchers to employ a specific statistical technique. **Discriminant function analysis** illustrates how accurately calls from different samples can be discriminated and classified given information on their overall acoustic structure. The analysis considers the overall acoustic structure of a call (based on all the acoustic measurements) and examines whether there are sufficient systematic differences in the structure of the call for them to be discriminated. For example, 25 calls are recorded in context A (e.g. seeing a snake) and 25 calls are recorded in context B (e.g. seeing a leopard). A set of 10 acoustic measures is then taken from the 50 calls. A discriminant function analysis would take the 10 acoustic measures to investigate if these combined measures can account for a significant amount of the acoustic variation in the calls recorded in these two contexts. The statistical function that best explains the variation between calls is then used to identify how many of the calls could be correctly classified according to context. If the function can correctly classify 40 out of the 50 calls (80%), this is above that expected by chance (50% correct classification into one of two groups). One could conclude from this that there is sufficient systematic acoustic variation in the structure of the calls for them to be classified according to context and thus calls given to snakes and leopards are acoustically distinct call types.

Although acoustic analysis is a powerful tool for vocal research that enables objective description and classification of calls, it does have several drawbacks. First, acoustic analysis requires high-quality recordings where a single individual is calling without interference from other primates, commentary from the observer or excessive background noise. These data are difficult to collect, particularly in situations when many individuals in a group tend to call together. Some calls are also very difficult to capture, particularly if they are short, soft calls that occur infrequently: often the call has finished by the time the researcher has started the recording equipment or oriented the microphone towards the caller. Second, acoustic analysis is very labour intensive and it is a lengthy and repetitive process. These two drawbacks may be one of the primary reasons that very few vocal repertoires have been mapped based on acoustic classifications (e.g. Bouchet, Blois-Heulin and Lemasson, 2012). For most species, calls in the repertoire are differentiated based on the context of production and subjective descriptions of the sounds (e.g. chimpanzee vocal repertoire in Goodall, 1986). This is problematic as the human ear can be unreliable and shows great individual variation. It also means there is often an unsatisfactory degree of circularity in terms of the function of the call. For instance, food-associated calls are sometimes defined as calls produced when feeding (e.g. chimpanzees in Goodall, 1986), then it is proposed that food-associated calls seem to refer to food and function to announce the presence of food to others in this species (e.g. Hauser *et al.*, 1993).

4.4.3   What mechanisms underlie call production?

There is great interest in the proximate and developmental mechanisms underlying call production in primates, not least because the production of calls seems to be very different from the flexible, generative (i.e. capacity for generating new meaning) nature of human language. The topic of developmental acquisition is covered in detail in Chapter 6, but the main methods to investigate questions related to this topic will be briefly outlined here. The proximate and developmental mechanisms are connected, of course, and proximate mechanisms can sometimes be inferred from studies of development.

4.4.3.1   Proximate mechanisms

The physiological mechanisms underlying call production are often investigated via anatomical research (inferred from the structure of the vocal tract, Lieberman, Klatt and Wilson, 1969; Riede *et al.*, 2005) and neuroscience research (see Chapter 3 and this chapter, section 4.4.5). Laboratory experiments have also been conducted to investigate the causal psychological mechanisms underlying the production of a call. Voluntary control over vocal production has been tested in the laboratory using conditioning techniques (e.g. Sutton *et al.*, 1973). Although there have been failures, both monkey and ape species have been successfully trained through operant conditioning techniques to reproduce species – typical vocalizations (reviewed by Pierce, 1985). The extensive training required in most studies indicates, however, that vocal control is challenging for most primates (Jürgens, 2002). More recently, the first attempts to apply behavioural markers of intentional signal production, that have been used heavily in gestural research, have been successfully applied to call production (Schel, Townsend *et al.*, in press).

4.4.3.2   Developmental mechanisms

Cross-fostering experiments have been conducted as one way to test the fundamental developmental questions ('nature vs. nurture') concerning primate vocal production (e.g. Owren *et al.*, 1992, 1993). For instance, infant macaques have been raised by a foster mother of a different macaque species, which exhibits a different call repertoire. The calls of the infants are then examined to see if their calls resemble the structure of the foster species (nurture) or their genetic species (nature). Another method to examine the extent to which primate calls are modified by experience (**vocal learning**) is by conducting acoustic analysis on existing calls in different populations. Acoustic analysis of calls can reveal group-specific 'dialects' within a call type (e.g. Crockford *et al.*, 2004), which is indicative of vocal learning if other genetic and environmental differences between groups can be ruled out. Changes in call structure as a result of a changing social environment (e.g. Elowson and Snowdon, 1994) also test the extent to which vocalizations are genetically specified or can be modified through experience.

4.4.4   What does a listener understand from the calls?

4.4.4.1   Observational data

Observational data on the responses of other individuals to a vocalization are vital for generating ideas about the kind of information that may be extracted from the call by the listener. There are, however, numerous difficulties with this kind of data collection. It is often difficult to collect data on all receivers who heard a call, particularly in low visibility environments, and to know the time period after the call in which responses should be recorded. It is also very challenging to control for the receiver's response being influenced by the reactions of other group members, or direct perception of the event that evoked the call in the first place. It is also difficult to establish to what extent the response of the receiver was due to hearing a vocalization, or due to processing cues gained from visual or olfactory signals given by the call producer. One of the biggest challenges that applies to both observational and experimental work on receivers is that we are reliant on overt behavioural changes that we can categorize as a 'response': animals may hear and understand a call, but may decide not to act on it immediately. In this case the researchers may incorrectly assume the receiver had not processed the call.

4.4.4.2   Experimental data

Vocal research has an advantage over other modalities in terms of examining the receiver, as some of the issues highlighted above can be dealt with through the use of playback experiments (in both captive and naturalistic settings). The basic idea underlying playback experiments is to examine what the receiver can process from the call alone, when the original eliciting event and all other communicative cues are removed. If, in the absence of the eliciting event, receiver responses to the played back calls mirror those given to the event that originally elicited the call, this indicates that the call alone allows the receiver to infer the likely cause of the call and act appropriately. Taking the seminal work on vervet monkey alarm calls, this means in the absence of a leopard, if a group member's leopard alarm call is broadcast from a speaker, the receiver will respond to the call in the same way as he would respond to an actual leopard (Seyfarth, Cheney and Marler, 1980a). Thus, conclusions about the type and level of information the receivers can obtain from the call can be made.

Although playback experiments can be powerful tools to examine receiver responses to calls, there are limitations to their use. Playback experiments work best if the animals give qualitatively different responses to the original events that elicit the calls. For instance, vervet monkeys give qualitatively different anti-predator reactions to leopards, eagles and snakes. This allows for very clear hypotheses to be formulated regarding the expected reaction to the playback of calls related to these events and the interpretation of these responses is relatively straightforward. Unfortunately, there are many events to which primate responses are only quantitatively different. For instance, the reaction of many primates to hearing the social calls of others is to turn and look in the direction of the calls, therefore, researchers may need to look for quantitative differences in this basic orienting behaviour to examine if primates can meaningfully distinguish between call variants. So whilst differential duration of attention or interest in the speaker to played back calls can tell you the listener can distinguish between calls (Gouzoules, Gouzoules and Marler, 1985; Slocombe, Townsend and Zuberbühler, 2009), it is more difficult to say what information the receiver has extracted from the calls. It is also difficult to control for low-level acoustic salience of the vocalizations driving the differential orienting responses in receivers, making interpretation of such results without appropriate controls very difficult.

Two different paradigms have been successfully applied to primate playback experiments to try and increase our understanding of receiver processing of the vocal signals in the absence of qualitatively different responses to the calls in question. The first is the violation of expectancy paradigm, which has been an important technique in assessing the cognition of preverbal human infants (e.g. Onishi, Baillargeon and Leslie, 2007). In this paradigm, calls are presented to the receiver that represent either a congruent or incongruent sequence of events. The expectation is that primates, like human infants, will look longer at the incongruent stimuli. For instance, chimpanzees look longer at a sequence of calls that simulate an agonistic interaction where the dominance hierarchy is violated compared to an interaction in keeping with the existing hierarchy (Slocombe *et al.*, 2010a). This showed that the listening chimpanzee must be able to identify individuals from their calls and understand the direction of aggression from the 'scream' variants broadcast to them. This paradigm enables researchers to make strong predictions about the kind of information that the listener needs to be able to extract from the calls in order to show the violation of expectancy response. Second is the **habituation–dishabituation** paradigm, where the researcher repeatedly presents the subject with one type of call until they are no longer responding to it (habituated). The test stimulus is then presented to see if the subject resumes responding (dishabituation) or if they continue to ignore it. For instance, Hauser (1998) habituated rhesus macaque monkeys to one call type given to high-quality food and found they dishabituated when a call given to low-quality food was presented, however, they maintained habituation to an acoustically different type of 'high-quality' food call. This indicated that they readily transfer habituation over acoustically different calls with the same message, and dishabituate to calls that provide a different message. This paradigm is powerful in showing how primates categorize calls and helps to understand the messages they extract from the calls they hear.

There are numerous practical issues involved with conducting playback experiments, which makes them difficult to undertake. Overall, in order to collect meaningful responses from listeners, the experimental set-up has to be psychologically and ecologically realistic (e.g. the researcher must fool the subject into thinking the call actually came from another individual). Negative results from a playback experiment are very difficult to interpret, as it is impossible to rule out the possibility that the animal just did not perceive the set-up as realistic and thus did not give a meaningful response. Often great lengths are necessary in order to make the playback realistic (e.g. Slocombe, Townsend and Zuberbühler, 2009), but at the very least a high-quality speaker capable of broadcasting the appropriate frequencies and high-quality stimuli are necessary. It is often difficult to obtain recordings of calls suitable for playback stimuli that have no interference from human commentary or other group member calls and low background noise. Many studies therefore do not provide a unique stimulus for each subject, and thus pseudoreplicate (Kroodsma *et al.*, 2001; see section 4.6 for a discussion of **pseudoreplication**).

4.4.5   Brain mechanisms involved in vocal production and perception

4.4.5.1   **Neuroanatomical approaches**

Anatomical dissections of brains allow researchers to map lateral asymmetries in the volume of certain brain areas thought to be relevant for vocal production, allowing for interesting comparisons to humans to be made (e.g. Cantalupo and Hopkins, 2001). Historically, lesioning was one of the principal methods for working out the necessary brain circuitry for vocal production. By comparing the vocal behaviour of the subject before and after the removal of certain areas of the brain, either with physical surgery or chemical lesion, researchers could assess whether the brain area was necessary for vocal production (e.g. Kirzinger and Jürgens, 1985).

4.4.5.2   Neurophysiological approaches

Electrical stimulation of different brain areas has also revealed areas of the monkey brain that are required for vocal production and areas that control certain aspects of the vocal behaviour, such as phonation (e.g. Dressnandt and Jürgens, 1992). Whilst electrical stimulation techniques and lesioning are capable of informing us about the brain areas necessary for vocal production, most other techniques only provide correlational evidence. They identify areas of the brain that are active during vocal production, but that may not be necessary for vocal production to occur. In order to examine the neurochemical mechanisms involved in vocal production, experimental studies examining the effect of different drugs and neurotransmitters on vocal production are also conducted (e.g. Glowa and Newman, 1986). Single-cell recording enables researchers to identify areas of the monkey brain that are active during call production (e.g. Lüthe, Häusler and Jürgens, 2000). Due to the invasive nature of this method and the other techniques outlined so far, most of this work has been conducted on monkey species commonly housed in laboratories, such as squirrel monkeys and macaques. Therefore, comparably little is known about the brain mechanisms underlying call production in ape species. New technological advances are, however, hopefully changing this situation.

4.4.5.3   Brain-imaging approaches

PET imaging is now being used with great apes to correlate brain activity with the production of vocal as well as gestural communicative behaviours (e.g. Taglialatela *et al.*, 2008). To examine brain areas involved in the perception of vocalizations, fMRI has been used successfully with monkeys (e.g. Petkov *et al.*, 2008), due to the relative ease of providing calls for the primate to listen to, compared to the difficulty of eliciting call production during testing sessions.

4.5   General data collection and coding methods

One way to record communicative events is to directly enter the data into spreadsheets or palmtop computers (PDAs) while observing the social interactions. To do so, the observer needs to be clear about which behaviours are of interest before starting with data collection. Alternatively, a dictaphone can be used to verbally document and describe the observations. Today, with a range of recording equipment and storage media available, many researchers use video recording for their data collection. Different recording and sampling rules can be used, depending on the scope of the study. In communication research, the measurement of events (as opposed to states and durations) is most commonly used and often all occurrences of a given behaviour are recorded (Martin and Bateson, 2007). To record communicative signals, **continuous behavioural sampling** is most often applied, since the gestures, facial expressions and vocalizations usually are of short duration and would not be captured by periodical **time sampling methods**. Two main sampling methods are used: **Focal-animal sampling** is applied to capture the kind of signal and frequency of signal use by following an individual for a defined period of time. The length of these sampling periods should be consistent within a study, but can of course vary between studies, mostly in the range of several minutes up to an hour. After a sampling period has elapsed, the next focal animal is selected randomly and observed for the corresponding period. This method enables the comparison of signal repertoires and their frequency of use between different individuals, since each individual is observed for the same amount of time. If the scope of research is on rather rare behaviours, such as food-sharing or sexual behaviours, then behaviour sampling is used to capture such events. Here, a group of individuals is constantly observed and as soon as the corresponding behaviour (or the context of interaction) occurs, attention is directed towards the interacting individuals (Altmann, 1974).

Further procedures depend on the medium used for data collection. A crucial step is to divide the continuous stream of behaviours into discrete units and categories, a process referred to as coding. However, if spreadsheets or palmtop computers were used, observations were already assigned to categories. If dictaphones were applied, then the information would have to be transcribed, which is a very time-consuming procedure. Video footage may need to be digitalized and/or converted into a different video format, depending on the requirements of the software used for coding and analysis of the data. There are different ways to code video footage, depending on the scope of the study and the level of analysis desired. In the simplest form, the corresponding behaviour is extracted while watching the video and entered into an Excel spreadsheet or directly into software used to analyse that data, such as **SPSS**. **ELAN** is a free software used to watch and code video footage and is particularly suitable to parallel descriptions of several behaviours. **Interact** and Observer are commercial software packages that support both the coding and analysis of video (and audio) data. In general, the use of such software is very helpful for rather fine-grained analyses based on frames per second, which often require slow playback functions.

4.6   Pseudoreplication[1]

Regardless of the sampling technique used or modality of communicative signal investigated, one problem that is unfortunately common in primate communication research is pseudoreplication (Waller *et al.*, 2013). According to Machlis, Dodd and Fentress (1985, p. 201) pseudoreplication or pooling multiple observations from each individual 'reflects a fundamental error in the logic underlying random sampling, since it implicitly assumes that the purpose of data gathering in ethology is to obtain large samples of behaviour rather than samples of behaviour from a large number of individuals'.

---

[1] This section has been reproduced from Waller *et al.*, 2013, with permission from Elsevier.

The goal of the majority of behavioural scientists is to draw conclusions about a specific group of individuals (usually a species) by examining a sample of individuals from this group. Scientists seek the largest samples they can achieve, in order to increase the reliability of extrapolating from their sample to the population mean, and thus increase the accuracy of their conclusions. Reliability, however, can be increased *only* by increasing sample size in terms of the number of individuals that make up the sample, *not* by taking multiple samples from each individual and pooling them to create a larger data set. In the 1980s, two key papers highlighted this problem (terming it pseudoreplication or the pooling fallacy) (Hurlbert, 1984; Machlis, Dodd and Fentress, 1985), and it is now widely acknowledged as a serious problem to be carefully avoided.

The simplest sampling error that can lead to pseudoreplication is extracting more than one data point per individual, and then adding these data to the main dataset without using appropriate repeated measures statistical techniques. This pooling procedure results in (1) an artificially inflated sample size, which increases statistical power and increases the chances of making a type 1 error (a false positive: claiming a statistically significant effect where there is not one), and (2) data points that are not independent, which violates the assumptions of many statistical tests and renders the outputs from such tests as unreliable. Pseudoreplication can also occur when data points are not statistically independent from each other as they result from the same stimulus treatment and/or temporal, spatial or social group. This latter type of pseudoreplication can get more difficult to identify (and avoid), but the same principle applies – the replicates are not independent from each other, and thus do not increase the reliability of extrapolating from the sample to the whole population.

The pervasiveness of this issue within primate communication research has only recently been explored (Waller *et al.*, 2013b). It seems likely, however, that studies of communication are particularly prone to artificially inflating the data set as the unit of communication (the facial expression, the call, the gesture) is a tempting unit of analysis. Using the unit of communication as the unit of analysis could often lead to pseudoreplication unless each individual contributes one data point derived from their sampled communication, or individual level membership is taken into account by the statistical test. Waller *et al.* (2013b) reviewed 553 papers published in scientific journals from 1960–2008 (the same papers used in Slocombe, Waller and Liebal, 2011; see Chapter 5) and examined whether the data had been pseudoreplicated. The study searched only for the simplest form of pseudoreplication (taking more than one data point from each individual), but 38% of the studies that used inferential statistics presented at least one case of pseudoreplicated data. An additional 14% did not provide enough information to confirm presence or absence of pseudoreplication. Although the modality (gesture, facial expression or vocalization) did not affect the probability of pseudoreplication, observational studies in particular were more likely to report findings from pseudoreplicated data (Waller *et al.*, 2013b).

The high prevalence of pseudoreplication in the research articles examined shows that more work needs to be done to avoid this problem. In addition to avoiding pseudoreplication during data collection, there are also some ways to consider this problem during the analysis of the data. First, even if there are multiple samples of behaviour from a single individual to represent its behaviour more appropriately, it is important to use the average of these multiple data points in the analysis of the data. Furthermore, it is possible to conduct certain statistical tests, such as Generalized Linear Mixed Models (GLMM), that can be fitted with 'individual' as a random factor to control statistically for the fact that individuals may contribute numerous data points to the data set. Scientists should aim to avoid pseudoreplication and should not underestimate the problems it leaves us with: 'all conclusions in the ethological literature, which are based upon pooled data should be viewed with caution' (Machlis, Dodd and Fentress, 1985, p. 213).

Summary

This chapter reviewed and explained the methods commonly employed in research on olfactory, gestural, facial and vocal communication of primates. Most importantly, it showed that the main research questions and subsequent methodology differ widely between the different modalities. Research on olfactory communication tends to focus on observational methods to capture the sender's behaviour when producing olfactory signals, and the precise measurement of the chemical components of olfactory signals in laboratory settings. However, very little research has been conducted to determine what sniffers might *understand* from olfactory signals. Research on facial expressions has a long history of an ethological approach, but is heavily influenced by research on humans and thus assumptions about the emotional meaning of facial expressions are often made. However, this research has recently employed FACS to describe facial movements in a standardized, objective way across primate species. The investigation of structural properties of primate calls through the use of detailed acoustic analysis is a key aspect of research on vocal communication. This research also benefits from the playback paradigm where receiver responses to vocalizations can be measured in both wild and captive environments. In contrast to the other modalities, research on gestural communication generally lacks a focus on structural properties and detailed systems of measurement are only just starting to appear, while the behaviour and contexts surrounding the production of gestures are intensively studied. Methods to examine brain mechanisms underlying the production and perception of signals are mostly used to study monkeys, while there is some recent progress to employ less invasive methods to also investigate great apes. In sum, each modality lends itself better to some methods than others, resulting in an inextricable link between how a modality is studied and the corresponding research questions and assumptions made (see also Chapter 5, for a more detailed discussion). Finally, the high prevalence of pseudoreplication in primate communication research – regardless of the modality investigated – shows that studies using observational methods are more likely to pseudoreplicate. As a consequence, both the methods employed and the treatment of data may influence the conclusions of a particular study.

5 A multimodal approach to primate communication[1]

5.1 What kind of research has been done on primate communication?

When the authors decided to write this book our impression was that primate communication studies overwhelmingly focused on a single communicative modality, be that vocalizations, manual gestures, olfaction or facial displays. Each of us had expertise in each of these four areas, yet we agreed we were irritatingly ignorant as to the methods and findings of scientists operating outside of our own modalities. In particular we discussed how difficult it was to directly compare and integrate findings from our respective modalities to create a coherent understanding of primate communication as a whole. These conversations were not only the catalyst for this book, but also for a systematic review of the current primate communication research, in order to examine the patterns and biases that are present in the literature as it stands. This systematic review is published as an essay in *Animal Behaviour* (Slocombe, Waller and Liebal, 2011), and the key points will be reviewed here.

In order to gain an accurate picture of the current state of the primate communication literature, we conducted a systematic search for research conducted on primate communication from 1960–2008. This search was not designed to be exhaustive, but to provide a representative snapshot of the kind of empirical research being published on primate communication. We used a standard set of search terms in three major relevant databases (Web of Science, Science Direct and PrimateLit; see footnote[2] for search terms). As this is a large and greatly varied field, we applied a set of criteria to all studies returned by the searches: we excluded all non-empirical studies (e.g. reviews, essays, opinions), all non-peer-reviewed material (e.g. books, book chapters, conference proceedings) and as we were interested in spontaneous communication, we also excluded all studies where primates were asked to interpret human signals or use artificial language systems. As one original aim of this review was to examine how primate communication literature is used in debates about language evolution, we focused on the three main modalities that are most relevant to this topic (vocalization, gesture, facial) and we thus excluded studies that focused solely on olfactory communication. Finally, we included only those studies that had been published in English, in order for

---

[1] This chapter has been adapted from Slocombe, Waller and Liebal, 2011, with permission from Elsevier.
[2] Search terms entered into search engines were:

> 'facial communication' OR 'facial expression*' OR 'facial display*' OR 'display AND behaviour' OR 'gestur*' OR 'gestur* communication' OR 'gestur* display*' OR 'vocalization*' OR 'vocalization*' OR 'call' OR 'vocal communication' OR 'vocal*'AND 'primate*' OR 'ape*' OR 'monkey*' OR 'macaque*' OR 'gorilla*' OR 'orangutan*' OR 'baboon*' OR 'vervet*' OR 'chimpanzee*' OR 'gibbon*'

us to be able to assess it properly. A total of 553 studies returned by these search terms met these basic criteria. Each of these studies was then assessed on different measures:

1. **The primary modality of communication investigated** (vocal, gestural, facial or multimodal). We classified a study as multimodal if it assessed relationships or inter-action between communicative signals from different modalities in an integrated fashion.
2. **The nature of the research** (observational or experimental)
3. **The species class of primate investigated** (great ape, lesser ape, monkey or prosimian; a single study may investigate multiple species classes)
4. **The research environment** (wild or captive)
5. **The focus of the research** (producer or receiver)

5.2   What are the findings of the systematic review?

Most studies only investigate a single modality. Only 28 out of 553 studies (5%) examined two or more modalities in an integrated fashion. This illustrates that multi-modal research is rare in primate communication. The combinations of modalities investigated are shown in Table 5.1.

Table 5.1 The combined modalities examined by the 28 multimodal studies found by the literature search (from Slocombe, Waller and Liebal, 2011).

Modalities examined in an integrated fashion	Number of studies
Facial, Vocal, and Gestural	9
Facial and Gestural	4
Facial and Vocal	9
Vocal and Gestural	5
Vocal and Olfaction	1

Systematic differences between vocal, gestural and facial research. The remaining 95% of studies only examined a single modality and differences between these unimodal studies were then investigated in more detail using statistical tests. A summary of the descriptive findings can be found in Table 5.2.

Table 5.2 Percentage of the studies within each modality that were conducted in different research environments, using different approaches, foci and species classes. Single studies may have used multiple research environments, approaches, foci and species classes, therefore the sum of these percentages sometimes exceed 100% (from Slocombe, Waller and Liebal, 2011).

		Vocal (N = 352)	Gestural (N = 51)	Facial (N = 122)	Integrated multimodal (N = 28)
Research environment	Wild	38.4	7.8	8.2	10.7
	Captive	62.8	92.2	95.1	89.3
Approach	Observational	46.3	52.9	65.6	50.0
	Experimental	62.2	49.0	36.0	50.0
Species class	Great apes	8.5	78.4	23.8	39.3
	Lesser apes	4.5	2.0	3.3	3.6
	Monkeys	83.2	19.6	77.9	53.6

Table 5.2 *(cont.)*

		Vocal (N = 352)	Gestural (N = 51)	Facial (N = 122)	Integrated multimodal (N = 28)
Research focus	Prosimians	4.3	0	4.9	3.6
	Producer	73.9	100	88.5	67.9
	Recipient	37.5	19.6	39.3	46.4

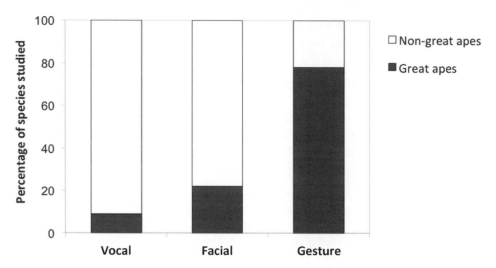

Figure 5.1 Percentage of great ape and non-great ape species (monkeys, prosimians and lesser apes) studied in vocal (n = 354), facial (n = 134) and gestural (n = 51) modalities. Each study could focus on multiple species classes thus the percentage is in terms of the total number of different species classes studied in each modality (from Slocombe, Waller and Liebal, 2011).

1. **Modality investigated**

   There was a great imbalance in the numbers of studies conducted across modalities with vocal research accounting for 64% studies, facial 22% and gestural only 9%. This indicates that gestures in particular have received little overall research effort.

2. **The nature of the research**

   A Chi square test revealed that the relative number of observational and experimental studies varied significantly across modalities ($\chi^2_2 = 25.80$, p < 0.0001) with vocal studies being the most experimental (62%) and facial expressions the least (36%; see Table 5.2).

3. **The species class of primate investigated**

   There were considerable differences in the species class most commonly used across the three modalities, with vocal research heavily focused on monkey species and at the other extreme gestural research heavily focused on great ape species. A Chi square test revealed a significant difference in the use of great ape species across the three modalities ($\chi^2_2 = 144.79$, p < 0.0001) and this pattern is illustrated in Figure 5.1.

4. **The research environment**

   The majority of all the reviewed primate communication work was conducted in captivity. The research conducted in the wild was distributed unequally across modalities with significantly more than expected focusing on vocal communication (89%; $\chi^2_2 = 55.26$, p < 0.0001). Studies using experimental methods in the wild focused exclusively on vocalizations.

5. **The focus of the research**

   A significantly higher proportion of the facial (39.3%) and vocal (37.5%) studies considered the role of the recipient, compared to gestural studies (19.6%; $\chi^2_2 = 6.86$, p $= 0.032$). All gesture studies examined behaviour in terms of the producer, indicating that signal production was a central focus for gesture studies.

   To summarize, gesture studies tended to use both observational and experimental methods in captive settings to investigate the communication of great apes. The behaviour of the signal producer was examined in all studies, with the role of the receiver being relatively neglected. Facial expression data were mostly observational and generated from captive monkeys but with a focus on both recipients and producers. Vocal studies tended to focus on monkey species in both wild and captive settings. Experimental methods were favoured and both the roles of producer and receiver were considered. Thus, there was a relative lack of vocal research on great apes, gestural work conducted in the wild and experimental facial research. Across all modalities there were fewer studies that focused on the behaviour of the recipient compared to the producer

5.2.1 Why do vocal, gestural and facial research differ systematically in their approaches?

Considering the great diversity of methods used in different modalities as described in Chapter 4, section 4.4, the differences outlined above may not come as a great surprise. For instance, whilst receiver responses to vocal signals are possible to test with carefully controlled playback experiments, similar techniques are not widely available in the visual domain. Although some video touch-screen technology has been successfully used to conduct experiments in perception of facial expressions (e.g. Parr, 2004), such equipment can be expensive, is only usable in captive contexts and requires considerable training of the primates before it can be used. In contrast, auditory playbacks can be successfully conducted with unhabituated individuals in the wild. As Chapter 4 outlines, however, even playback experiments are difficult and time consuming to complete and this likely contributes to the relatively low level of consideration for the receiver in research.

Available methodology within a modality will also affect the places and contexts in which data can be effectively gathered. Gestural and facial research relies heavily on video recording of behaviour for subsequent coding: obtaining high-quality footage suitable for such analyses is reliant on good light and clear unobscured views of the animals relatively close to the observer. Whilst these conditions are readily found in captivity, the arboreal lifestyle of many primates, the low visibility of many natural habitats and low light levels associated with dense canopy cover in many forests all provide substantial barriers to collecting such data in the wild. In addition, data on visual signals are very difficult to collect accurately during contexts associated with fast movement of the animals. For example visual signals would be difficult to identify or analyse from footage of a primate fleeing from a predator or involved in an agonistic encounter, whereas the vocal signals from these contexts have been readily analysed. Gestural research therefore tends to focus on relaxed social contexts, where high-quality footage of relatively still or slow moving individuals can be obtained. By contrast, vocal research has generally concentrated on loud calls that are usually emitted in evolutionarily urgent contexts, such as predator encounters, agonistic interactions or food discovery. In order to perform acoustic analysis on a call or use a call as a stimulus in a playback experiment, the call has to be louder than the background noise. Thus, obtaining high-quality recordings of vocalizations is easier with louder calls. Obtaining high-quality recordings of soft calls, given in relaxed social contexts and recorded from a distance is, in contrast, very challenging. It is clear that differences in methodologies across modalities lead to different optimum places and contexts in which to collect data.

5.2.2   What are the consequences of differences in approaches to vocal, gestural and facial work?

The systematic differences in the approaches taken in vocal, gestural and facial research make comparisons of the qualities of the signals in each modality very difficult to directly compare. For instance, later in this book in Chapters 6–9, we examine evidence across modalities for signals being referential, flexible, learnt and intentional, but absence of evidence for one of these qualities within a modality can be driven by methodological constraints and systematic biases in the existing literature rather than necessarily reflecting a true absence of that quality. For instance, gestures are widely cited as being produced intentionally in comparison to vocalizations, which are not (e.g. Tomasello, 2008), yet the evidence on which such an assertion is based is not valid for such comparisons. It is true that captive great ape gestures often have the hallmarks of intentionality, whilst monkey vocalizations do not, but this could simply be one of many cognitive abilities on which great apes outperform monkeys. Until we have comparable evidence in the vocal domain from great apes in similar contexts to directly compare with the gesture data, such a conclusion seems premature. Equally, gestures are often described as being produced flexibly across contexts, in contrast to vocalizations, which are characterized as bound to specific contexts (Tomasello, 2008). Yet again, this comparison is not based on a comparable dataset: captive great ape gestures given in relaxed social contexts such as play are flexibly produced, whereas wild monkey vocalizations produced in response to evolutionarily urgent events such as predator encounters are highly context-specific. There may be a greater evolutionary pressure for unambiguous signals in urgent contexts, such as predator defence, rather than non-urgent social contexts (Tomasello and Zuberbühler, 2002).

Flexibility in communicative signal production (e.g. producing the same signal in a variety of contexts) may only occur in these non-urgent contexts, where the receiver can use contextual information to respond in the most adaptive way to the signal (Smith, 1977). Therefore, gesture work is more likely to reveal flexible signal production than vocal work, due to the biases inherent in the approaches researchers adopt in the two modalities.

There can be an element of circularity in arguments about the qualities demonstrated by signals in specific modalities. The availability of certain methodological approaches influences the findings about communication in a certain modality and these findings then drive assumptions about the modality. These assumptions can add an element of circularity to the research by influencing the way in which findings are interpreted. For instance, Slocombe, Waller and Liebal (2011) highlight that the availability of playback methodologies has enabled the growth of a large body of evidence indicating that vocalizations function referentially (see Chapter 9). This assumption of **functional reference** can then constrain the aim of vocal studies and the interpretation of results. For example, many vocal researchers may look for context-specific call production (an essential aspect of functional reference) and if they fail to find it, the investigation may simply not be published. It is rare that such a finding would be presented as evidence for flexible use of a communicative signal across contexts, as is often reported for production of gestures (e.g. Pollick and de Waal, 2007). It is plausible that if a gestural and vocal researcher were given identical sets of data on the production of communicative signals across contexts, they would analyse and present them in very different ways aligned with their expectations of what they would find in their respective modalities. Equally, the historical characterization of facial expressions being purely emotional displays may prevent researchers from asking cognitive questions within this modality (see Chapter 8, **Box 8.1**).

One major implication of these consequences is that many of the assumptions that are currently held about how vocalizations, gestures and facial expressions compare may not be true. Given the biases and large gaps in the current primate communication data set, absence of evidence for a certain trait may not necessarily reflect absence of ability. This has very serious ramifications for theories of language evolution (see **Box 5.2**), but before these are discussed in more detail, the following section explains how research into the communication of nonhuman primates contributes to our understanding of language evolution.

5.2.3   How does evidence from primate communication inform our understanding of language evolution?

It is often proposed that language is one of *the* features separating humankind from other animals and so researchers have long tried to establish how, why and in what ways language is different from other animal communication systems (e.g. Hockett, 1960). Although the term language is commonly used, it is rather difficult to find a universally accepted definition that covers the multiple facets of this term. In its broadest sense, language is a cognitive faculty that enables humans to learn and use systems of complex communication. Amongst a variety of approaches, language can be defined with focus on its structural properties, the cognitive mechanisms involved, or the functions it serves (see **Box 5.1**).

Box 5.1 What is language?

Language as structural system

Language represents a system of signs (sounds, gestures, letters or symbols, depending on whether language is spoken, signed or written) and is governed by particular rules that relate a specific sign to a specific meaning (de Saussure, 1983). The most basic linguistic units of (spoken) language are meaningless phonemes (vowels and consonants) that form longer meaningful sequences (morphemes, words). These morphemes and words can be combined into longer, meaningful phrases or sentences based on a set of combinatorial rules called syntax.

Hockett (1960) defined at least 13 design features of language with the aim of identifying similarities and differences between animal communication and human language. Those that are highly relevant for a comparative approach to human communication include:

- *Arbitrariness*: There is no direct connection between the signal (word) and what is being referenced (meaning).
- *Discreteness*: Language is composed of discrete units that are distinct from each other.
- *Displacement*: Ability to communicate about things that are not present spatially, temporally, or realistically.
- *Duality of patterning*: Combination of a finite number of meaningless parts creates a potentially infinite number of meaningful utterances such as words, which in turn are combined into an unlimited number of sentences.
- *Productivity*: Ability to create new meanings by combining already existing utterances.
- *Semanticity*: A specific signal can be linked to specific meanings.

In searching for the evolutionary roots of these structural properties of human language, comparative research is particularly interested in features like semanticity, syntax and productivity. Research into nonhuman primates thus centres on whether nonhuman primates (1) use signals that have a particular meaning (Chapter 9: referentiality), (2) combine single signals into longer sequences based on particular combinatorial rules (Chapter 7: flexibility), and thus (3) create new meanings (Chapter 7: flexibility).

Language as social-cognitive faculty

Language consists of acts of social, collaborative communication, which are intentionally directed towards others, and are governed by social conventions with the aim of achieving social ends. An important psychological mechanism underpinning language is shared intentionality, referring to the shared understandings, aims and purposes of those individuals using language. Shared intentionality includes several socio-cognitive skills (e.g. joint attention) to create a common conceptual ground and the prosocial motivation to help and share information with others. Another

Continued

Box 5.1 (*cont.*)

essential cognitive skill is the ability to attribute mental states such as knowledge or ignorance, desires or false beliefs to others (**Theory of Mind**). This is an essential prerequisite for using language, since knowing what others know and do not know is important for deciding which information needs to be communicated to others (Cheney and Seyfarth, 2005). Language is acquired by cultural learning in inter- actions with others, with **imitation** playing a major role in passing on these conven- tions to the next generation (Tomasello, 2008; Tomasello and Carpenter, 2007).

In searching for the evolutionary roots of these social-cognitive foundations of human language, comparative research on nonhuman primates centres on (1) the intentional use of signals implying voluntary control over the production of these signals (Chapter 8: intentionality), which is closely related to (2) the flexibility of use in terms of the use of signals across different functional contexts (Chapter 7: flexibility), and (3) the mechanisms underlying signal acquisition to investigate whether repertoires are largely genetically determined or whether signals are acquired in interactions with other conspecifics (Chapter 6: acquisition).

How language evolved is the matter of an ongoing debate that still leaves us with many more questions than answers (see Christiansen and Kirby, 2003; Hauser, Chomsky and Fitch, 2002; Wilcox, 1999). Within the scope of this book, it is impossible to provide a comprehensive overview of existing theories to language evolution and the corresponding controversies, since there is an almost endless list of books and papers suggesting a variety of scenarios from the perspectives of many different disciplines (e.g. Arbib, 2005; Bickerton, 1992; Christiansen and Kirby, 2003; Corballis, 2003; Seyfarth, Cheney and Bergman, 2005). Although it is widely accepted that language as a whole is unique to humans, there is much less agreement in regard to whether language evolved from scratch in the human lineage or whether it built on existing cognitive and communicative skills humans share with other animals (King, 1999).

Individuals who propose that human language evolved from scratch uniquely in the human lineage clearly dissociate animal communication from human language and argue that language builds on very specific cognitive skills that are unique to humans (Chomsky, 1966). Animal communication is considered qualitatively differ- ent from language, since it is generally holistic, and lacks grammatical items as well as the infinite productivity that is characteristic for human language (Bickerton, 1992). In contrast, individuals who propose that language built on skills already present in the primate lineage argue for continuity in the evolution of human language and suggest that precursors to human communication are present in other primate species as well. Because of our shared phylogenetic history, it is assumed that skills present in extant related species would also have been present in the common ancestors of those species. A comparative approach therefore attempts to identify which skills could have served as building blocks for language by

comparing humans with other nonhuman primate species. As language seems to
have evolved uniquely in humans, one cannot explicitly identify components of
language in other species. Nevertheless, the search for homologous traits in our
primate cousins can identify potential precursors to human linguistic abilities. Since
many primate species use gestures, vocalizations and facial expressions to communicate, most comparative work focuses on their communicative behaviour and the
cognitive skills underlying primate communication as potential precursors to our
own linguistic abilities.

Amongst the many controversies in the field of language evolution (see Christiansen
and Kirby, 2003; Hauser, Chomsky and Fitch, 2002), one major debate concerns the
communicative modality language built on and thus the question whether language
originated from gestures or vocalizations. Both gestural and vocal theories of language
evolution rely heavily on comparative evidence of the communicative abilities of living
primates. As outlined in **Box 5.2**, both vocal and gestural theories cite 'positive'
evidence of commonalities between their modality and human language (e.g. vocalizations are referential; gestures are intentionally produced). Both theories also, however,
use the apparent absence of certain traits in the opposing modality to support their case.
Gestural researchers will thus highlight vocalizations as inflexible, unintentional and
non-generative, whilst vocal researchers will draw attention to gestures being meaningless signals that are never produced in rule-governed sequences. Often such arguments
against the 'other' modality seem to be central lines of argument in these theories. As
highlighted previously, we argue – given the significant differences in many aspects of
research between modalities – these comparisons are not based on a comparable data set
and as such may lead to invalid conclusions.

5.3   The way forward

We suggest several ways in which the field can change in order to help in building a
more complete understanding of primate communication. This will not only aid our
understanding of primate communication systems, but also provide a more solid basis
for theories of language evolution.

First, due to methodological constraints, unimodal research will continue to be the
only option in some circumstances, but if this research focused on the areas that are
currently understudied, this would help to fill the gaps in our knowledge. Researchers
should consider adopting similar approaches, study species and research environments
to other modalities, so that more meaningful comparisons can be made. Equally,
researchers should try and be open-minded about their goals and consider trying to
analyse their data in a way similar to that of a different modality.

Second, more integrated multimodal research is urgently needed. Most non-human
primate and human communication is produced and perceived in a multimodal format,
so by focusing on only single modalities we may be overlooking the complexity
inherent in the system or key functions of communicative acts (Partan and Marler,
1999). In the discussion of multimodal communication that follows it is necessary to

---

Box 5.2 Theories of language evolution

Gestural theories of language evolution

Gestural theories suggest that human language originated from a gestural stage that involved the use of voluntarily produced manual gestures (Hewes, 1973). Although this seems to be counterintuitive since gestures and spoken language rely on different sensory modalities, such theories emphasize evidence from both human and nonhuman primate gestural communication that supports the claim of a gestural origin of human language. First, gestural theories refer to the discovery of mirror neurons and the apparent link between manual gestures and homologous areas of language production in the monkey brain (Rizzolatti and Arbib, 1998). Second, attempts to teach apes spoken language failed, while several projects succeeded in teaching American Sign Language to great apes; a language that relies on manual gestures and body movements (Gardner and Gardner 1969; Hayes and Hayes, 1951; Patterson, 1978, see also Chapter 10, Table 10.1). Third, research into gestural interactions between conspecifics emphasizes that gestures are voluntarily produced signals, that are flexibly used across different contexts and adjusted according to the recipient's behaviour and are characterized by the potential to learn and generate novel gestures (Arbib, Liebal and Pika, 2008; Corballis, 2002; Tomasello, 2008). Finally, even before children acquire spoken language, they use gestures such as showing or pointing to communicate with others. While they learn to speak, gestures are not simply replaced by words, but remain as speech-accompanying gestures, which are closely intertwined with spoken language (Goldin-Meadow, 2003; Kendon, 2004; McNeill, 2002). Gestural theories often emphasize the absence of these language-like qualities in primate vocal or facial communication and specifically highlight the unintentional, inflexible nature of vocalizations and facial expressions (Tomasello 2008). Furthermore, such theories emphasize that all individuals of one species usually share the same vocal or facial repertoire, while there is a much higher degree of individual variability in gestural repertoires. Related to this, there is limited evidence for the production or learning of novel calls or facial expressions resulting in rather constrained and closed repertoires as opposed to gestural repertoires that can be expanded by learning or creating new signals.

Vocal theories of language evolution

Vocal theories suggest that language built directly on the vocal abilities of our ancestors, because both spoken language and vocalizations share the same sensory modality (Zuberbühler, 2005). First, in support of the commonalities in primate vocalizations and human language, vocal theories highlight that primate calls can refer to external objects and events (Seyfarth *et al.*, 1980a,b) possibly representing a precursor to human referential abilities. Second, vocalizations can be combined into meaningful sequences that are produced in response to specific contexts, indicating that simple rule-based combinations may exist in primate vocal behaviour (Ouattara, Lemasson and Zuberbühler, 2009a, b). Third, there is evidence that

Continued

Box 5.2 (*cont.*)

primate vocalizations can be perceived as discrete signals (Fischer, 1998), and thus have the potential to be combinatorial, a key feature of language. Thus, although there seems to be less flexibility in regard to the production of vocal utterances such as learning new calls, flexibility in vocal communication is much more evident in the perception of vocalizations, since primates flexibly assign meaning to calls and call combinations (Arnold and Zuberbühler, 2008; Cheney and Seyfarth, 2005). It is thus in terms of call perception rather than call production that the greatest similarities with human language are apparent (Cheney and Seyfarth, 2005). In contrast to vocalizations, gestures rarely refer to external objects and events and their meaning or communicative function often depends on the context they are used in. Furthermore, although gestures are combined into longer sequences, they usually do not represent meaningful combinations as found for vocal sequences.

Facial origins of human language

A growing body of literature points to the important role that 'lipsmacking' facial expressions may have had in the evolution of language (Bergman, 2013; Ghazanfar *et al.*, 2012). It is suggested that the rhythmic facial movements required in the production of speech evolved from existing facial movement patterns in a primate ancestor. Studies of rhesus macaques show that the temporal patterning of their 'lipsmacking' signals is similar to the temporal patterns of speech (Ghazanfar, Chandrasekaran and Morrill, 2010). Indeed, the developmental trajectory of 'lipsmacking' in macaques mirrors the ontogeny of speech development in humans, both starting with a highly variable 'babbling' phase, before a faster, less variable pattern emerges (Morrill, Paukner, Ferrari, Ghazanfar, 2012). Speech is, however, just one modality through which language can be expressed, whilst language itself can be considered a multimodal cognitive capacity.

Very few theories examine the role that facial expressions might have played in the evolution of language and those that do treat facial expressions more as a component of gesture or vocalization rather than a different modality. For example, mirror neurons are responsive to mouth-communicative movements in monkeys as well as manual actions and these neurons have been implicated in theories of language evolution (Fogassi and Ferrari, 2007). Recent evidence has also shown facial expressions to be perceived by monkeys in an integrated fashion with vocalizations (Ghazanfar and Logothetis, 2003). When examined alone, facial expressions are usually assumed to be tightly linked to the expression of basic emotions (Ekman, 1992), and indeed the expression of emotional state is often an implicit component of common definitions of facial expression (e.g. Chevalier-Skolnikoff, 1973; Darwin, 1872; Ekman, 1994). However, Ekman and Friesen (1978) saw the dangers of relying on (assumed) emotional correlates when describing the physical features of facial expression and developed the standardized coding system based on muscle movements (The Facial Action Coding System: FACS; see section 4.3.2, Chapter 4).

Continued

Box 5.2 (*cont.*)

The connection between emotion and facial expression has probably influenced the suggestion that facial expressions are largely innate and involuntarily produced, and therefore unimportant for understanding language evolution (e.g. Tomasello, 2008). In some ways this might reflect a legacy left by Darwin (1872), who adopted a conversely 'anti-Darwinian' approach by suggesting that facial expressions are not communicative adaptations, but inevitable byproducts of internal, emotional states (Fridlund, 1992). Only through the introduction of a behavioral ecological approach did the communicative nature of facial expression start to sit within an adaptationist paradigm, and has sometimes been considered a 'paralanguage' (e.g. Fridlund, 1991).

---

first examine this theoretically and in non-primate species where it has been better studied, before discussing some examples of multimodal research in primates.

5.3.1   What is multimodal communication?

There is no overarching consensus as to how multimodal signals should be defined and operationalized. Seminal studies have focused on the senses employed by the receiver to detect multimodal signals and thus define multimodal communication as the use of signals that are simultaneously received through two or more sensory channels (e.g. Partan and Marler, 1999, 2005). Other researchers have focused on the method of production and thus have included the combination of signals within a sensory channel (e.g. visual gesture and facial expression) as a multimodal signal (e.g. Pollick and de Waal, 2007). As one of the aims of this book is to address methodological differences employed by researchers of different modes of communication, we adopt an inclusive definition and use both types of multimodal definition. Thus, we use the term *multimodal* to refer to simultaneous combinations of signals from two or more modalities (gestural, facial, vocal and olfactory signals), and/or any signals requiring sensory integration by the receiver.

5.3.2   Why do animals produce multimodal signals?

There are many potential benefits to producing multimodal signals as opposed to unimodal signals. However, it is important to emphasize that most of the available evidence summarized in this section is on non-primate species. First, there is the potential for more complex information to be extracted by the receiver from a multimodal signal: following the *Multiple Message Hypothesis* different information may be conveyed in each channel, increasing the total amount of information available (Partan and Marler, 1999) and receivers may rely on the multiple components to make more accurate judgements about the meaning of a potentially ambiguous signal (Partan and

Marler, 2005). For instance, Beletsky (1983) showed that female red-winged blackbirds produce a similar visual display in courtship and aggression contexts, but combine the visual display with different songs in the different contexts.

Second, signals in different channels may not only serve different functions, but may also be directed at different recipients, so that the overall efficacy of the communicative act is enhanced. Variability in receiver sensitivity to different signals means that multi-modal signals may ensure a signal is received and processed by a wider number of recipients or different classes of recipients than a unimodal signal alone (Hebets and Papaj, 2005). Whilst this has not been shown explicitly to be the case for multimodal signalling, it has been shown with regard to multicomponent signals, such as the call of the Coqui frog. In this species, playbacks have shown that rival male and receptive female recipients respond to different notes of the multicomponent advertisement song (Narins and Capranica, 1976). In addition, variable female preferences have been suggested to have driven the evolution of complex multifaceted displays by male satin bowerbirds. Thus, by developing a complex display, the males are more likely to be attractive to a wider range of females (Coleman, Patricelli and Borgia, 2004).

Although multimodal signals may allow for more complex or multifunctional messages to be transmitted, much empirical work has revealed the components of multi-modal signals to be redundant; that is the different components serve the same function and thus act as a 'back-up' (Partan and Marler, 1999). Production of redundant signals in multiple modalities can still have some important benefits, however. Signallers may be imperfect in their encoding of information (Bradbury and Vehrencamp, 1998) and signals can be degraded as they are transmitted through the environment. Both of these factors contribute to the uncertain landscape that receivers have to operate in. Perception of multimodal signals may allow receivers to make more accurate decisions in such an uncertain environment, as independent estimates of the same event can be made from each sensory modality and the total amount of information available to the receiver is increased (Munoz and Blumstein, 2012). Psychological experiments have found that in comparison to unimodal signals, receivers are more likely to detect multimodal signals and to be able to react faster to them (Rowe, 1999). In situations where the costs of missing a signal are high (e.g. alarm signal that alerts receivers to the presence of a deadly predator), the production of a multimodal signal that increases the probability of detection and speed of response may be highly beneficial (Munoz and Blumstein, 2012). Multimodal signals are not only more likely to be successfully detected by receivers, but also there is evidence that multimodal signals are easier to discriminate, localize and remember than unimodal signals (Rowe, 1999; Partan and Marler, 1999). For example, a recent study investigated whether rhesus macaques combine faces and voices to enhance their perception of vocalizations. Both humans and monkeys were trained to detect unisensory (auditory, visual) or multisensory (audiovisual) vocalizations that were embedded in noise (Chandrasekaran et al., 2011). Similar to humans, rhesus macaques reacted faster in the multisensory condition that combined auditory and visual information compared to the conditions where either auditory or visual components of a vocalization were presented. Chandrasekaran et al. (2011) conclude that the faster recognition of multimodal signals compared to unimodal signals likely confers a behavioural advantage. Thus, multimodal signals are also likely to be processed more successfully by receivers than unimodal signals.

Although there are benefits associated with multimodal signalling, given the widespread existence of unimodal signalling, there must also be costs involved, with the cost–benefit ratio determining the contexts and species in which multimodal signalling is an adaptive behaviour. Partan and Marler (2005) outline several potential costs of multimodal signals. First, whilst multimodal signals can improve a receiver's ability to localize the caller, this also means that predators may also be able to eavesdrop on this signal and also localize the signaller with higher accuracy. This combined with the signaller becoming more conspicuous due to production of signals in multiple channels is likely to increase predation risks for the signaller. Indeed Roberts, Taylor and Uetz (2007) showed that predatory jumping spiders localize playbacks of wolf spiders producing multimodal (seismic and visual) courtship displays faster than unimodal displays. It is not just signallers who face enhanced predation costs from the use of multimodal signals: receivers may also face greater predation risks as attention in more channels is directed towards conspecific signals, rather than detection of predators. It is also not just predators that may be better able to 'eavesdrop' on multimodal compared to unimodal signals: competing conspecifics may also add to the costs of producing multimodal signals. For instance, rival males are quicker to approach and attack a subordinate male fowl producing a multimodal tid-bitting display to females than a unimodal display (Smith, Taylor and Evans, 2011). In this species, males switch between unimodal and multimodal displays as a function of the attentive state of the dominant male in the group and thus dynamically change their signalling strategy to reduce costs and maximize benefits according to the specific social context. Finally, the production of multimodal signals may be energetically more costly than unimodal signals. It may also be difficult to produce some combinations of signals as the production of one signal may interfere with the production of another (Hebets and Papaj, 2005). For instance, the production of high-quality song may be compromised by vigorous movements required for a visual display and thus male brown-headed cowbirds are rarely observed to combine their most complex visual displays with song production (Cooper and Goller, 2004).

In sum, multimodal signalling has clear costs and benefits and these are likely to change with the ecological and social context (Munoz and Blumstein 2012). Thus, the adaptive value of multimodal signalling is likely to be highly varied across species and contexts, but to date comparably little research has been conducted with primate species.

5.3.3   How can multimodal signals be classified?

Partan and Marler (1999) offer a comprehensive way for classifying multimodal signals, as illustrated in Table 5.3. One key distinction is whether or not components of a multimodal signal are redundant. If the separate components of a composite signal elicit the same reaction in receivers, then they are redundant (e.g. they convey the same information). The combination of composite parts may either result in a response that is

Table 5.3 Classification of multimodal signals. Redundant signals are depicted above, non-redundant signals below. Responses to two separate components (a and b, left) and combined multimodal signals (a + b, right) are represented by characters (x, y and z). The same character indicates the same qualitative response; different characters indicate different responses. Different sizes of the same character indicate different intensities of the same qualitative response: an enlarged character indicates increased, a reduced one decreased intensity (modified after Partan and Marler, 1999).

| | Separate components | | | Multimodal composite signal | | | |
	Signal		Response	Signal		Response	Category
Redundancy	a	$\rightarrow$	x	a + b	$\rightarrow$	x	Equivalence
	b	$\rightarrow$	x	a + b	$\rightarrow$	X	Enhancement
Nonredundancy				a + b	$\rightarrow$	x and y	Independence
	a	$\rightarrow$	x	a + b	$\rightarrow$	x	Dominance
	b	$\rightarrow$	y	a + b	$\rightarrow$	X (or $_x$)	Modulation
				a + b	$\rightarrow$	z	Emergence

similar to (equivalence) or much greater in magnitude (enhancement) than the response elicited by the separate elements in isolation. Multimodal signals with redundant components can be seen as 'back-up' signals (Johnstone, 1996), as they are more robust and likely to be received by the recipient than unimodal signals. They are therefore likely to be associated with evolutionarily urgent contexts, where successful perception and processing of the signal by the receiver is vital and produced in noisy environments with many competing cues that may mask detection of unimodal signals.

If separate components of a multimodal signal elicit different responses (non-redundant), then a more complex array of possible outcomes can arise from the composite signal (see Table 5.3). Non-redundant components provide the possibility of greater informational transfer and for novel responses to be given to the combination of signals (emergence).

Partan and Marler (2005) encourage researchers to investigate multimodal signals by comparing the receiver responses to the composite signal as well as the unimodal component signals, in order to fully understand the type of composite signal that is being observed. They do, however, highlight the need for researchers to check that the realism of any experimental methods used to test receiver responses is equal across the different modalities. For instance, if a robotic model is used to test responses to the visual elements of the signal, and a speaker is used to broadcast the auditory element of the signal it is important to test that the responses receivers give to these experimental playbacks are similar to that given to naturally occurring unimodal signals. If one experimental technique is more realistic than the other, the relative strength of response to the two components of the signal will not be accurately captured.

5.3.4   Multimodal signals in non-primate species

Multimodal communication research has been successfully pursued with a number of non-primate species. This section reviews some of this work that shows the importance of examining signals in concert rather than in isolation for fully understanding the function and complexity of animal signals.

Multimodal signals have been studied extensively in a number of arthropods. Insects commonly combine a visual colouration-warning signal with an olfactory or auditory signal to produce a multimodal warning signal to predators. Research has shown these multimodal signals to be effective in increasing avoidance by insect eating birds (Rowe and Guilford, 1996). A number of arthropod species produce multimodal courtship displays, with wolf spiders producing complex visual and seismic signals in combination. Female wolf spiders show faster detection of males producing multimodal rather than unimodal signals (Uetz, Roberts and Taylor, 2009). Males also show behavioural flexibility in the production of multimodal signals, increasing the frequency of visual signals if located on a substrate that conducts seismic vibrations poorly (e.g. rock; Gordon and Uetz, 2011).

Several species of frog produce multimodal courtship and territorial displays that combine visual and auditory signals. Vocalizations are combined with the distinctive visual signal of the inflating air sac. Exemplary studies investigating the effect of the visual and auditory components of the multimodal signal on receiver responses have used robotic frog models. Narins et al. (2005) placed a male robotic frog inside the territory of male dart-poison frogs in their natural rainforest habitat and recorded the reaction of the resident male to the simulated invasion of another male. Resident males responded to either the auditory or visual playback alone with approach and investigation of the model, however, the two signals needed to be combined in a multimodal signal in order to elicit a fully aggressive attack on the model. Narins et al. (2005) also tested the thresholds in the receiver for integrating the auditory and visual cues from the model frog by varying the spatial disparity between the model and the auditory playback and the temporal synchrony of the signals. Resident males only attacked the model frog if the auditory and visual signals were synchronized within a time window of half a second: with greater asynchrony of signals the resident frog did not attack and left the area quickly. Equally, resident frogs spent more time in the vicinity of the model and attacked it more frequently if the spatial disparity between the visual and auditory signals was small (2–12 cm) rather than large (25–50 cm). Narins et al. (2005) thus elegantly demonstrated the thresholds of the perceptual binding of cues in this multimodal signal in the receiver.

Like the male dart-poison frogs responding to the territorial display of an invading male (Narins et al., 2005), female túngara frogs show a preference for multimodal courtship displays compared to unimodal displays (Taylor et al., 2008). More females approached the call with a model inflating his air sac than the call alone and it seemed that the dynamic nature of the air sac inflation was the important part of this visual cue to enhancing the female's response to the male (Taylor et al., 2008). Similar results were found using video playback technology rather than robotic models (Rosenthal, Rand and Ryan, 2004). A more recent study, however, has shown that the visual signal of the inflating air sac is not sufficient in isolation to attract female attention (Taylor et al., 2011) and that the enhancing effect of the multimodal cue over the auditory signal alone disappears as the amplitude of the auditory signal is increased. Taylor et al. (2011) suggest that the visual component of the multimodal signal is therefore useful for detection and localization of males from a distance, but not in assessing male quality once the female is close to the male and the auditory signal dominates. These experiments demonstrate the potential complexity of the interaction between components of a multimodal signal and illustrate that different methods can be successfully used to investigate multimodal signalling.

Honeybees combine visual and auditory/vibratory signals in their waggle dance that they perform on return to the hive after a successful foraging trip. This multimodal display provides receivers with information about the direction and distance of the food source (von Frisch, 1974; Gould and Gould, 1988). Experiments using robotic models have revealed that the auditory and visual components of the wagging run – which is the essential information-containing element of the dance – need to be present in order to recruit receivers and the main source of information in the dance is the direction and duration of the multimodal wagging run (Michelsen *et al.*, 1992). This study found no evidence for separate roles for the auditory and visual components of the wagging run and the authors suggest that these elements offer redundancy to reduce transmission errors. The bee waggle dance is a rare example of a multimodal display that also functions to refer to an external stimulus (i.e. a functionally referential signal; see Chapter 9) and whose components contain information about different aspects of the stimulus.

As outlined briefly in section 5.3.2, another example of a functionally referential multimodal display is the tid-bitting display produced by male fowl when they discover food. The tid-bitting display consists of individually distinctive food-associated calls combined with a visual display of rhythmic head and neck movements. Careful experiments using video playbacks explored female hen perception of male tid-bitting displays and found that the visual and auditory components of the multimodal signal were both functionally referential and redundant (Smith and Evans, 2008, 2009): similar food-specific responses were given by females to both the multimodal and unimodal signals. Although the multimodal signal seems to confer no advantage in terms of greater responses from females, it is assumed that the combination of redundant signals increases the detectability and memorability of the display. Dominant males always produce the multimodal display, whereas subordinate males produce a silent visual display in approximately one-third of events. Smith, Taylor and Evans (2011) found that subordinate birds were more likely to produce a unimodal display when the dominant males were attending to them: aggression from these eavesdropping rivals was greater to the multimodal display than the silent visual display. These subordinate birds thus seem to be flexibly using multimodal displays when rivals are distracted and thus less likely to respond with aggression.

Video playback techniques have found that, in contrast to the fowl, female pigeons show multimodal enhancement to male courtship displays (Partan *et al.*, 2005). Although the visual and auditory elements of the display were sufficient to elicit some courtship behaviour in female recipients, indicating the component signals are redundant (Partan and Marler, 1999), the multimodal display elicited a much greater reaction than expected by the sum of the responses to the individual components.

Multimodal signalling has been generally investigated in less depth in mammalian species, but good progress is being made with some sciurid species. Californian ground squirrels integrate visual tail-flagging signals with bark vocalizations to communicate with family members about the presence of predators (Hennessy *et al.*, 1981; Owings and Virginia, 1978). Similar multimodal displays are also given by tree squirrels and the contribution of the unimodal components to the efficacy of such anti-predator displays has been recently investigated in the wild. For example, Partan, Larco and Owens (2009) presented robotic models to wild grey squirrels and established that although unimodal visual or auditory signals do elicit anti-predator behaviour in recipients, the strongest responses come from the combined multimodal signal. Thus, in this species, the visual and auditory signals are redundant and the multimodal signal demonstrates multisensory enhancement (Partan and Marler, 1999). Partan *et al.* (2010) compared responses to multimodal and unimodal alarm signals in urban and rural populations of grey squirrels and found an increased response to the visual tail-flagging signal in urban compared to wild populations. This may indicate that urban squirrels are relying more on visual signals in their noisier urban environment where auditory signals may be harder to detect.

This literature highlights the importance of examining interactions across signalling modalities to understand the complexity of a communication system. Research with non-primate species has showcased the importance of testing receiver responses to signal components and composites and indicates that both robotic models and video playbacks can be used successfully in this regard. Animation video playbacks are advocated as a new technique with potential to give the researcher much more control over almost every aspect of the visual signal and this could be an invaluable tool for future multimodal research (Woo and Rieucau, 2011).

5.3.5   Multimodal signals in primates

Although multimodal signalling has not been investigated in as much detail on the behavioural level in primates, we will review here some of the studies that are leading the way in this regard.

Olfaction is often considered the 'neglected sense', particularly in terms of primate communication research, yet studies reveal that olfactory signals can form an important component of multimodal displays. Ring-tailed lemurs actively scent-mark using various scent glands, but they also use urine deposits as olfactory signals to both members of their own and other groups. One difficulty is that potential recipients have to locate the deposit before being able to investigate the olfactory signals it provides. Ring-tailed lemurs combine urine-marking with a salient visual signal, erecting their ringed tail in the air, thus attracting the visual attention of receivers to the location of the urine deposit and increasing their chances of detecting the olfactory signal (Palagi, Dapporto and Borgognini Tarli, 2005). A further study conducted in the wild indicates that these 'tail up' urine-based olfactory signals were produced mainly along borders of territory, and thus were suggested to act mainly as territorial signals for members of rival groups (Palagi and Norscia, 2009). As the rival group members were likely absent at the time the visual signal was produced, it seems one of the target audiences for this multimodal signal will only perceive the olfactory component. This highlights the difficulty in assessing integrated multimodal signals that include olfaction, as the time course of olfactory signals is much longer than either visual or auditory signals that may accompany it. Thus, an olfactory signal that is produced as part of an integrated multimodal signal may only be perceived unimodally by receivers who detect the olfactory component after the visual or auditory signal has already finished.

An observational study on rhesus macaques' visual and vocal signals produced during social interactions provides a very detailed description of the multimodal combinations produced in this species (Partan, 2002). In this study, 30% of the signals catalogued were multimodal vocal-visual signals and careful statistical analyses revealed the most common combinations of vocal and visual signals. Some visual and vocal signals consistently co-occurred (e.g. 'screams' with a facial 'grimace'), some were more flexibly combined (e.g. 'barks' and ears retracted) and some signals were never combined (e.g. ears forward and submissive vocalization). Although Partan (2002) suggests that visual signals, such as staring and ears forward, may function to denote the intended recipient of the accompanying 'threat' vocalization, receiver responses to these multimodal and unimodal signals have not been systematically examined.

Receiver responses have been examined in relation to unimodal and multimodal 'lipsmack' signals in two species of macaques. 'Lipsmacks' represent a dynamic facial expression that functions mainly as a signal of affiliation between individuals. This silent visual signal is sometimes combined with a 'girney' vocalization in rhesus macaques. Partan (1998) found that the signaller and receiver were more likely to approach or start grooming following a multimodal audiovisual signal compared to a unimodal visual signal. A similar pattern has recently been found in crested macaques, with 'lipsmacking' displays combined with a soft 'grunt' resulting in a higher probability of affiliative contact than the visual 'lipsmack' signal alone (Micheletta *et al.*, 2013). Thus, in both these species, the addition of a vocal signal to a visual signal seems to enhance the response of the receiver. In crested macaques, further complexity in this signal was revealed as the basic 'lipsmacking' facial movement was also flexibly combined with additional visual components, such as teeth exposure and a 'head turning' gesture (Figure 5.2). Whilst the addition of a 'head turning' gesture also

Figure 5.2 Multimodal communication in crested macaques. Neutral face and different components associated with the 'lipsmacks'. (a) Crested macaque neutral face; (b) 'lipsmack' with retraction of the scalp; (c) 'lipsmack' with retraction of the scalp and teeth exposure; (d) 'lipsmack' with retraction of the scalp and head turn; (e) spectrogram of a 'soft grunt' produced by an adult female (from Micheletta *et al.*, 2013).

increased the likelihood of affiliative contact, the addition of teeth exposure reduced the likelihood of subsequent affiliation. By examining the composite dynamic display, Micheletta and colleagues have revealed levels of subtle complexity that would have been overlooked if each component had been examined in isolation.

A range of different multimodal signals are produced by male mantled howler monkeys during copulation contexts and although these are primarily audiovisual signals, visual-olfactory and visual-olfactory-tactile signals were also observed (Jones and Van Cantford, 2007b). Subordinate individuals produced higher rates of audiovisual signals before their own attempts at copulation, compared to dominant individuals. Qualitative data indicates that subordinates may have been more successful in gaining copulations after increasing the production of this multimodal signal. Jones and Van Cantford (2007a) illustrate that a range of these signals can be usefully categorized using Partan and Marler's (1999) scheme, although more systematic data on recipient responses to multimodal signals and their unimodal components is required.

Captive chimpanzees produce visual, tactile and auditory signals to communicate with human experimenters and several studies have tested the flexibility with which chimpanzees choose their modality of communication. A desirable food object is typically placed out of reach of the chimpanzees and the experimenter then manipulates their visual attention to be either directed towards or away from the chimpanzee. Chimpanzees preferentially use vocalizations or auditory attention-getting behaviours when the experimenter faces away from them and conversely they produce more visual gestures when the experimenter faces them (Hostetter, Cantero and Hopkins, 2001; Leavens et al., 2004; Leavens, Russell and Hopkins, 2010). Chimpanzees thus flexibly choose or combine these different types of signals as a function of the attentional state of the human. These findings mirror the differential use of auditory or visual gestures in captive chimpanzees communicating with conspecifics as a function of their visual attention (e.g. facing the signaller or not; Tomasello et al., 1994).

Pollick and de Waal (2007) investigated the frequency and efficacy of multimodal signals in captive chimpanzees and bonobos communicating with conspecifics. The authors defined multimodal signals as gestures and vocalization or facial expressions occurring within 10 s of each other, which is a more lax criterion than that adopted by most studies that require temporal overlap of the signals. Nevertheless, chimpanzees were observed to produce significantly more multimodal signals (18% of total signals) than bonobos (7% of total signals). In contrast to the majority of unimodal gesture research, this study examined receiver responses to these signals. Equal response rates were given to gestures alone and multimodal displays in chimpanzees, however, bonobos responded more to the multimodal signals than gestures alone. Unfortunately, the responsiveness of receivers to vocal or facial expressions alone was not examined, so it is not possible to fully characterize this multimodal signalling according to Partan and Marler's (1999) scheme.

In contrast to the research conducted on non-primate species, few primate studies use experimental techniques to examine the receiver responses to multimodal signals. Lisa Parr is one of the few researchers to attempt this and she has successfully used video playback combined with a match-to-sample learning paradigm to investigate the perception of communicative signals in visual and auditory channels. Captive chimpanzees

(a) congruent

(b) incongruent

Figure 5.3 An illustration of a congruent (top) and incongruent multimodal trial (bottom) in an audiovisual match-to-sample task (see also Figure 4.3 for the general procedure). Both examples show a still frame of a 5 s sample video ('pant-hoot' expression) on the left panels and the right panels show the two comparison expressions and the cursor. In the congruent multimodal example (a), the correct response is the 'pant-hoot' expression (left side of right panel). The incongruent multimodal example (b) shows the same sample video with a 'scream' vocalization. Subjects were free to select the comparison expression that matched either the auditory ('scream' face, left side of right panel) or visual component ('pant-hoot', right side of right panel) of the sample (from Parr, 2004).

were trained on a match-to-sample procedure where they were shown a sample video of a communicative signal and then asked to match it to one of two still photos showing different facial expressions (one matching the video; the other not) (Figure 5.3). Chimpanzees showed they could master this paradigm and also succeeded in matching at above chance levels using only visual cues (e.g. silent sample video) or auditory cues (e.g. auditory sample only; Parr, 2004). This shows cross-modal integration in chimpanzee receivers in a communicative context. Parr then presented chimpanzees with incongruent multimodal trials to investigate if one modality was more salient to them in different social contexts. For instance, chimpanzees would watch a video of a chimpanzee face 'hooting' whilst listening to 'screams' and then have the choice to match to either a 'hooting' or 'screaming' picture. Parr found that the dominant or more salient modality depended on the type of communicative signal. For example, long-distance 'pant-hoot' calls were more readily matched in the auditory modality, and 'screams' in the visual modality. Parr's work shows that multimodal perception of communicative cues can be examined successfully in primates and hopefully more researchers will adopt these methods in the future.

Cross-modal integration of communicative cues by primate receivers has also been shown in rhesus macaques. Ghazanfar and Logothetis (2003) presented rhesus macaques with silent videos of two monkey faces: one making a 'coo' vocalization, the other a 'threat' vocalization. Monkeys heard either a 'coo' or a 'threat' call whilst looking at the two silent moving faces and their looking time at each face was recorded. Monkeys looked significantly longer at the face that matched the vocalization they heard, showing that monkeys match auditory signals with the appropriate visual facial display. Cross-modal integration of vocal and visual signals has also been shown in terms of individual recognition in chimpanzees. Using match-to-sample paradigms, chimpanzees can match sample vocalizations to photos of the correct individual (Bauer and Philip, 1983; Kojima, Izumi and Ceugniet, 2003).

Behavioural evidence of cross-modal integration is also supported by research examining cross-modal integration at a neuronal level (see Chapter 3, section 3.4). Single neurons in various brain areas respond to both visual and auditory signals and produce supra-additive responses to multimodal visual-auditory signals (Barraclough et al., 2005). In addition, neuroimaging techniques have recently revealed that areas of the chimpanzee brain that are homologous to Broca's area in humans are active during the production of gestural and vocal signals (Taglialatela et al., 2011).

The findings of multimodal signal production and cross-modal integration in receivers are mirrored by research in humans. Humans constantly augment language with non-linguistic and often nonverbal cues (through face, hands and body), which can fundamentally adjust the message conveyed. For instance, eyebrow movements punctuate speech for emphasis (Ekman, 1979) and gestures automatically accompany speech in the absence of learning (Iverson and Goldin-Meadow, 1998). Perception of speech is inherently multimodal with visual information having the potential to change our perception of auditory stimuli, as demonstrated with the famous 'McGurk effect' (McGurk and MacDonald, 1976). Here participants presented with the sound 'ba' and the facial movement for 'ga', will often report perceiving the sound 'da'. This is an excellent example of multimodal emergence, as defined by Partan and Marler (1999). Building on the single-cell results from monkeys, research using brain-imaging techniques is making progress in identifying the neural underpinnings of cross-modal integration in the human brain (Calvert, 2001).

5.4   Looking to the future

Multimodal research, as illustrated by the above exemplars, is vital as it demonstrates that communication in a wide number of taxa, including primates, is inherently multi-modal, at both a behavioural and neuronal level. Thus, focusing on communication in a single modality is likely to only tell part of the story and to overlook the complexity which is likely inherent in these multimodal systems (Partan and Marler, 1999). We thus strongly advocate that more researchers consider communication within a multimodal framework. A methodological framework for examining simultaneous actions in multiple modalities is provided by Forrester (2008), and this may be of use to researchers wishing to document and investigate the structure and occurrence of multimodal signals. We also want to emphasize that the role of olfaction in primate multimodal signals currently receives little attention, and future research should aim to address this.

Theories of language evolution could also benefit from more multimodal research (Slocombe, Waller and Liebal, 2011). First, by studying modalities side by side or examining the contribution of each modality to successful communication, scientists will use comparable methods and contexts for each modality. Such multimodal data will not be subject to the biases inherent in the current unimodal data set and can therefore be used by theorists to make valid comparisons across modalities. Data from multimodal studies may reveal a different pattern of results to the current unimodal research, on which most theories currently stand, and this may lead to modification of these theories.

Second, the multimodal research conducted to date highlights the continuity in multimodal communication across human and non-human primate species. This suggests that language may have evolved through an integrated combination of vocal, gestural and facial communication, rather than a unimodal system. If so, an acknowledgement of this could lead to the development of new theories of language evolution that avoid the traditional gestural/vocal/facial divide and instead focus on the evolution of a multimodal communication system.

In line with these suggestions, the next chapters of this book attempt to synthesize research done across modalities on four fundamental and important cognitive facets of communication. First, we will examine how signals are acquired through the lifespan (Chapter 6 Acquisition), before assessing the degree of flexibility primates exhibit in the usage, perception and sequencing of signals (Chapter 7 Flexibility). We then investigate the extent to which primates produce intentional signals (Chapter 8 Intentionality), before finally discussing the evidence for referential signals in primates (Chapter 9 Referentiality). We have chosen to focus on these features of communication as they have been treated as indicative of interesting cognitive mechanisms and complexity. In addition, these features all make important contributions to attempts to understand the evolution of human communication, including language.

Summary

A systematic review of the primate communication literature revealed that 95% of empirical studies consider only a single modality and that vocal, gestural and facial research has attracted differing theoretical and methodological approaches. While experimental methods tend to be applied to vocalizations of captive and wild monkeys, the majority of gesture research uses observational methods to investigate the production of gestures in captive great apes. Observational methods are also favoured to examine facial expressions in captive monkeys. These differences in approach and methodology, inherent in unimodal research, create several problems and currently prevent us from having a complete and objective understanding of primate communication. Unimodal research makes it very difficult to make valid comparisons across modalities and this causes real problems for theories of language evolution that currently rely on such comparisons. We propose that adopting an integrated multimodal approach to primate communication will address many of these issues and offers a new, more accurate level of understanding. Although multimodal signalling has been thoroughly studied in a number of non-primate species and there is a clear framework available for the classification and study of multimodal signals, it remains surprisingly absent in primate communication research. There are a few excellent examples of both observational and experimental approaches being applied to multimodal signalling in primates and we encourage more researchers to take their lead. We argue that integrated multimodal research is essential to avoid methodological discontinuities and to better understand the phylogenetic precursors to human language as part of a multimodal system.

Part III

Cognitive characteristics of primate communication

6 Acquisition

An understanding of how and when animals acquire communication can be useful to expose the cognition underlying the production and perception processes of communication. As a great many changes occur in early life, of particular interest is acquisition of communication from birth to reaching adulthood (during ontogeny or development). Acquisition and modification of communicative processes could still occur during adulthood, however, which may be similarly (or even more) indicative of interesting underlying processes. The extent to which communication is flexible in the adult, therefore, is also an important consideration. Chapter 7 focuses on flexibility in more detail, but the two topics are necessarily intertwined.

Scientists are interested in how and when communicative processes develop during an animal's lifetime for two main reasons. First, by studying the sequence in which early communication emerges, the extent to which communicative skills are reliant or connected to the emergence of other skills and cognitive processes can be determined. For example, if a communicative skill manifests in humans before Theory of Mind has been acquired, this demonstrates that Theory of Mind is not necessary for that specific communicative skill to be used. In this way, the components of complex communicative abilities can be disentangled (which may be difficult to do in the adult) and the manner in which they interact can be better understood.

Second, examining the ontogeny of communication can reveal the extent to which communication requires input from the social and physical environment, or is present from birth. A communicative skill that emerges slowly, for example, could be dependent on the specific environment encountered. Demonstrating how skills are related to different rearing environments, and whether new signals can be generated during an animal's lifetime, is crucial to understand the role of learning in communication, and (similar to temporal emergence) can elucidate the underlying psychological features of communication systems, individual differences and perhaps even function of the specific skill. Importantly, *production*, *use* and *comprehension* of a specific communicative skill could all exhibit different developmental trajectories, which could also be indicative of different cognitive components.

In general, although many primatologists have great interest in these questions (particularly scientists with a psychological leaning), the ontogeny of communication is relatively understudied. Interestingly, this is in stark contrast to human communication, which is regularly studied from a developmental perspective. Recently, there has been a greater empirical focus on the flexibility of communication in adult primates, but development before adulthood is not well understood, particularly in wild populations.

6.1   Ontogeny of communicative signals

Generally, primates seem to acquire communicative behaviour gradually in the early stages of life. Emergence of all communicative modalities seems to follow some form of developmental trajectory, sometimes very short (a matter of days after birth) and sometimes much longer (a number of years). As primate communication is rarely present in its fully fledged form at birth, it seems highly likely that communicative skills are dependent on a combination of cognitive, social and physical development in the individual, tied in with interaction with the social and physical environment. Communication often occurs within highly complex social interactions, so it is unsurprising that individuals require feedback from their environment before communicative skills fully manifest. Sensitive periods of development, however, are not unusual in the animal kingdom. The vast majority of animals exhibit clear ontogenetic stages prior to adulthood. The croaking gourami (a species of teleost fish), for example, develops multimodal agonistic displays consisting of vocalizations, fin movements and biting motions in increasing complexity prior to sexual maturity (Henglmüller and Ladichm, 1999). The developing complexity of these signals and their order of appearance in ontogeny corresponds to the order of appearance in aggressive encounters between adults, suggesting that social feedback is an important developmental impetus.

6.1.1   When does communication emerge?

Few studies are able to document the temporal sequence of communication development in primates precisely, largely due to methodological constraints. For example, it is difficult to get access to large sample sizes since most primates give birth to only one infant at a time, to follow several individuals over long periods of time and to observe young individuals (particularly in wild populations) because of the infant's dependence on the mother. However, there are some notable captive and wild studies that report data on emergence and change during early life. Using a combination of observational and experimental playback approaches, Seyfarth and Cheney (1986) documented the production, contextual use and comprehension of calls in wild vervet monkeys. The different components of vocal communication emerged at different points during an individual's first 4 years of life (Seyfarth and Cheney, 1986). Immature individuals first responded to the calls of others, and then started to produce the vocalization in appropriate contexts later. Similarly, Hauser (1989) documented the development of intergroup 'wrr' vocalizations in vervet monkeys, a signal used in the context of encounters with other social groups. Although infants used 'wrr' vocalizations in the appropriate social context at 10 months of age, the acoustic morphology of their calls was still different from adults' 'wrr' vocalizations, indicating that infant calls undergo changes in form and function.

De Marco and Visalberghi (2007) found that facial expressions were fully absent at birth in zoo-housed tufted capuchins. Specific expressions then emerged from 1 month of age onwards, starting with 'lipsmacking', followed by 'scalp-lifting', 'relaxed open mouth' ('playface'), 'silent bared-teeth display' 'open-mouth silent bared-teeth display' and finally the 'open-mouth threat face'. In contrast to the vocalization examples mentioned above, context of use did not differ between infants and adults for all but one expression. The 'silent bared-teeth' display was used exclusively in affiliative contexts in infants, but also in submissive contexts in adults. The authors make comparisons between this display and the human smile, which humans learn to use in increasingly broader contexts as they age.

Bard (2003) reported the emergence of facial expressions and vocalizations at a much earlier stage in laboratory-raised chimpanzees. First, 'fussing' and 'crying' were reported at 4–5 days from birth, 'greeting' vocalizations at 7 days, 'smiling' at 11 days, 'effort grunts' at 14 days and 'laughter' at 37 days. The difference between the two studies could be a species difference, in that chimpanzee communication develops earlier, but caution should be taken due to methodological differences between the two studies. Bard (2003) had a unique opportunity to document chimpanzee behaviour very closely within the nursery setting and may have captured emergence at the very earliest point, which may not be possible when observing interactions within a social group. Since the behaviours were exhibited during interaction with human caregivers and not conspecifics, it is unknown how they may be related to behaviours in conspecific interaction. Also, the correspondence between these infant chimpanzee behaviours and adult expressions is yet to be explored.

A recent longitudinal, systematic study documented and compared the emergence of gesture in four great ape species: gorillas, bonobos, chimpanzees and orangutans (Schneider, Call and Liebal, 2012a). The apes were observed during their first 20 months of life, and the onset, use and modality (tactile, auditory or visual) of gestures documented. Across all species, tactile and visual gestures were used first (initially appearing at about 10 months), but visual signals increased and tactile signalling reduced over time. Importantly, this is one of the few studies examining multiple species at the same time, and presents some very interesting data on species differences. Orangutans exhibited distinct differences to the other great apes, in that they showed the latest gestural onset, did not use auditory gestures at all, and used gestures more often in food related interactions. These differences may be due to their slower life history (Schneider, Call and Liebal, 2012a). Equally, however, these differences may be reflective of need, as orangutans socially interact less frequently, although, arguably, they can exhibit equally complex social organization (van Schaik, 1999). This study highlights the importance of comparing different primate species to understand how gestural communication may have evolved. Although observational studies of gesture are rarely conducted in primates other than apes (see Chapter 5), some studies of gesture in monkeys in naturalistic settings are starting to appear (for example, Laidre, 2008, 2011; see section 6.3.1 on generation of novel signals).

6.1.2   What can affect the emergence of communication in an individual?

The importance of the social environment for the development of behaviour has been emphasized by Mason (1960, p. 388) who concluded that 'among animals whose

socialization has been restricted, the cue function of many basic social responses is poorly established if not absent altogether'. To investigate the effect of the social environment on the development of behaviour, early scientists turned to social isolation paradigms, separating infants from their mothers and peers at birth and raising them in socially impoverished environments (Harlow and Harlow, 1966). These controversial studies, although ethically questionable, were nevertheless very important to demonstrate the role of experience and social factors on behaviour.

Less well known than the Harlow studies, some studies focused specifically on the development of communication. Miller, Caul and Mirsky (1967) studied the communicative abilities of socially isolated rhesus macaques in an experimental setting. In the 'cooperative-conditioning' paradigm, socially isolated and wild caught monkeys were compared in their ability to use the information present in the face of a conspecific. The monkeys were conditioned to avoid an aversive stimulus by pressing a lever when they saw a visual stimulus (flash). The monkeys were then placed in pairs, but the flash could only be seen by one monkey (the stimulus monkey) and the lever was only available to the other (the responder monkey). The facial expressions of the stimulus monkey could be seen by the responder. Wild caught monkeys were able to avoid the aversive stimulus by pressing the lever. As the only information available to the responder monkey was the face of the other animal, it seems reasonable to conclude that they were using the information present in the face (presumably, negative facial expressions) to predict the forthcoming aversive stimulus. Isolated monkeys, however, were not able to do this, and were similarly impaired in their production of facial expressions during the experiment. Berkson and Becker (1975) showed that facial expressions were morphologically preserved in infant crab-eating macaques who had been blinded at birth, but (concurring with the study by Miller, Caul and Mirsky, 1967), they occurred at much lower frequencies compared to sighted individuals.

The effect of isolation on vocalizations seems to show a different picture. Squirrel monkeys were raised by mothers who had been experimentally muted (Winter *et al.*, 1973), but acoustic characteristics of their 'peep' and 'cackle' calls were no different to infants exposed to species-specific vocalizations during rearing. Thus, duration, starting frequency, mid-frequency and end-frequency were identical. Whether these calls were produced in appropriate contexts, or responded to appropriately, however, was not examined. Given that wild vervet monkeys produce species-specific calls at birth, but require experience to respond appropriately to these same calls in others (Seyfarth and Cheney, 1986), it seems likely that isolation and cross-fostering experiments could affect detection, understanding and usage even if the structure of calls is preserved. However, Owren *et al.* (1992) caution that cross-fostering experiments between related species need to be vigilant of individual variation *within* species, especially as sample sizes tend to be low in these experiments. The authors cross-fostered rhesus macaques and Japanese macaques and examined the development of food calls. Variation in the food calls themselves, however, made the data difficult to interpret: it was difficult to assess whether variation was due to normal within species variation, or due to cross-fostering. Equally problematic was that food calls did not seem to differ between the two species to a great extent. Thus, whether development was affected by the rearing environment was difficult to determine. A later study focusing on a wider range of calls was more successful at detecting the effect of cross-fostering (Owren *et al.*, 1993). The authors similarly cross-fostered rhesus and Japanese macaques. Fostered Japanese macaques developed species-typical vocal behaviour (predominately using 'coos' only), but rhesus macaques developed vocal behaviour intermediate between their foster species and their true species. Whereas rhesus macaques would typically produce a range of 'coos' and 'gruffs', when fostered they produced more 'coos' than would be expected. Thus, there is some evidence that the specific rate of production of vocalizations can be modified by social environment.

If development of the communicative system is dependent on the rearing environment, which aspects of the environment are important? Infants from social species usually find themselves in environments composed of conspecifics, and so it is by its very nature a social environment. Adults in particular (mothers even more so) seem to play an active role in the development of a growing animal's communication. For example, when infant vervet monkeys give predator alarm calls, adults respond more strongly when they produce the call in response to a genuine predator (e.g. eagles) as opposed to a non-predator (e.g. storks) (Seyfarth and Cheney, 1986). These observations cannot rule out the possibility that the adult also saw the predator, of course, but nevertheless, this pattern could still reinforce the behaviour in the infant. Interestingly, when infants hear an experimental playback of an alarm call, they are more likely to respond appropriately if they look first at an adult. This could function as a form of **social referencing** (the use of another's response to evaluate a situation and to inform one's own response), which some studies have shown primates are capable of. For example, Russell, Bard and Adamson (1997) demonstrated that infant chimpanzees (between 14–41 months old) respond differentially to novel objects depending on the behaviour of a human caregiver, who either reacted with a happy or fearful expression to this object, suggesting that chimpanzees can use such information and alter their behaviour accordingly already at this very young age.

Similarly, the mother seems to play an important role in the development of gestures in wild orangutans. Bard (1992) observed wild orangutan interactions between mother and offspring during feeding, and suggested that the mother plays a particularly active role in shaping these food request gestures. Gestures were present in younger individuals, but then exhibited refinement in how they were used in the first few years of life. The infants made solicitation gestures towards both the food and the mother (described as goal-directed, intentional behaviours), and the mother encouraged these behaviours by responding positively (i.e. sharing) when the infant coordinated their behaviour with her own.

The effect of actively responding to the communicative behaviour on infants was tested by examining the difference in communicative and expressive behaviours in two groups of laboratory-reared chimpanzees (Bard, 2003). One group ('responsive care') received more intensive daily interaction with human caregivers and conspecific peer groups, designed to encourage species-typical behaviour. The other group ('standard care') received standard laboratory care. The 'responsive care group' exhibited facial expressions and vocalizations earlier than the 'standard care group', indicating that the quantity and quality of social interaction influences the emergence of infant chimpanzees' communicative behaviour.

Interestingly, Møller, Harlow and Mitchell (1968) found that rhesus macaques raised by mothers who have been socially isolated at birth (and therefore received no mothering themselves) produced species-typical dominance displays ('crook tail') but at different frequencies to other animals. In sum, they displayed less dominant behaviour, and so, if they were part of a group, would likely be low ranking. The implications of this study are particularly interesting, as they suggest that the communicative system is built not only on the presence of conspecifics (including the mother), but that the mother needs social experience herself in order to provide the appropriate learning opportunities for her own offspring. In this way, communicative systems are (in a sense) cumulative.

Very little is known, however, about *how* mothers play a role in the development of their offspring's communicative repertoire (see section 6.2 for a discussion on the mechanisms underlying the signal acquisition). In great apes, there is little evidence that infant bonobos and chimpanzees learn their gestures by imitating their mothers (Schneider, Call and Liebal, 2012b), since shared gestural repertoires were more prevalent within age groups than within mother–infant dyads. These findings imply that if gestures are learnt through some sort of imitation process at all, it is more likely to be from peers rather than the mother. However, an alternative explanation for this finding could be that infants use different gestures than their mothers, because different social contexts and therefore different gestures are relevant at this stage of life. There is, however, some evidence that captive chimpanzees acquire certain attention-getting sounds from their mothers, so it is possible that some form of social learning process takes place (Taglialatela *et al.*, 2012; see section 6.2.2).

This section demonstrated the very important role of the social environment for the development of the communicative system. At the same time, studies can differ enormously in social setting (social groups, pair-housed animals, singly housed animals, presence of offspring), but also their physical setting (wild, captive zoo populations, captive laboratory populations). The possible influence of one or several of these factors may complicate not only comparisons between modalities, but also between species or different studies.

6.1.3   What does change in relation to life stage mean?

Changes in communication over the lifetime of an individual may not always (or only) reflect development in the sense that a final, adult form is being honed. Instead, changes over the lifetime may reflect different needs at different life stages.

In several ape species, the repertoire of gestures increases as infants develop, and then decreases again in adult individuals (Call and Tomasello, 2007). This probably reflects the importance of the play context for younger individuals in comparison to adults, since many ape gestures are used in this context. As a result, young apes use a wider range of gestures than adult individuals. In addition to such changes in repertoire sizes, some gestures seem to be used for different functions depending on the age of the individual. For example, while young siamangs use the 'throwback head' gesture to engage in play, the same gesture is used by adult females to initiate a copulation with the male (Liebal *et al.*, 2004c).

Figure 6.1 Examples for 'muzzle–muzzle contact' in crested macaques (a) and chimpanzees (b). This behaviour is described in many species, but different labels have been used across studies and species, e.g. 'approach face' for orangutans (Liebal, Pika and Tomasello, 2006) or 'peering' in bonobos (Stevens *et al.*, 2005) and gorillas (Luef and Liebal, 2012; Pika, Liebal and Tomasello, 2003) (Photos: Jérôme Micheletta (a), Simone Pika (b)).

Similarly, several primate species including mandrills, drills and olive baboons, as well as orangutans and chimpanzees, perform an interesting 'muzzle–muzzle contact' behaviour that varies in use between young and old individuals (see Figure 6.1 for some examples). 'Muzzle–muzzle contacts' occur specifically when one individual is chewing food, and seem to function as olfactory communication in the sense that individuals can smell the conspecific's mouth and so determine what they are eating, and (importantly) what they are consuming safely (Laidre, 2009). Laidre demonstrated the potential function of this behaviour in some Old World Monkeys by showing that 'muzzle–muzzle contacts' were usually followed by the actor consuming the same food as being consumed by the recipient, and that they occurred most often when animals were exposed to novel foods. Crucially, 'muzzle–muzzle contacts' were more often displayed by younger individuals than older individuals, and usually towards more experienced, older individuals. Thus, animals can use this method of communication at various life stages, but the benefit to younger individuals that are relatively food naïve is clearly greater.

Furthermore, a variant of the play facial signal ('playface' or 'relaxed open-mouth display') exhibits age-related changes in frequency of use in gelada (Palagi and Mancini, 2011). Both adult and immature individuals produce a standard 'playface' and a 'full playface', which includes an additional element of upper teeth exposure. The 'full playface', however, is used infrequently by immature individuals, in contrast to adults, who use it more regularly than the 'playface'. The 'full playface' is probably a blended display with elements of the standard 'playface' and the 'bared-teeth' display (Thierry *et al.*, 1989). It thus has multiple functions – signalling play as well as appeasement and affiliation. In gorillas, a similar 'full playface' display is used most frequently during intense, ambiguous play and may help in elongating intense play bouts that have the potential to terminate quickly (Waller and Cherry, 2012) (Figure 6.2). Such age-related use in gelada could, therefore, reflect that adults have a greater need to signal affiliative intentions during play (Palagi and Mancini, 2011). Importantly,

Figure 6.2 Variants of the 'playface' in gorillas. 'Playface' with upper teeth covered (a) and 'full playface' with upper teeth exposed (b) (from Waller and Cherry, 2012) (Photo: Jérôme Micheletta).

immature individuals selectively used the 'full playface' when playing with adults (interactions which may also require extra confirmation of non-aggressive intentions), so the age-related changes are unlikely to reflect an increase in cognitive competence underlying production of the display. Rather, these age-related changes seem to reflect a difference in need depending on life stage. The two explanations need not be mutually exclusive, of course (and indeed may complement each other), but a differential function at various life stages is not often considered in studies of development.

For vocalizations, there is evidence that at least some age-related differences in certain acoustic parameters can be explained by maturational changes. For example, Hammerschmidt and colleagues found that a variety of acoustical parameters of 'coo' calls in rhesus macaques varied as a function of their increasing body weight (Hammerschmidt *et al.*, 2000). However, some parameters unrelated to physical growth and body size were found to differ in De Brazza's monkeys of different ages. De Brazza's monkeys of various ages produce 'on' vocalizations in mainly affiliative contexts, but juveniles show a higher degree of acoustic variability in the structure of this call type compared to adults: juveniles produce 'on' units that are longer in duration compared to those of adults (Bouchet, Blois-Heulin and Lemasson, 2012). Bouchet and colleagues suggested that this difference between juvenile and adult monkeys might indicate a more limited vocal control in younger individuals and/or that juveniles still have to learn some more subtle temporal patterns of these calls.

Age not only changes various aspects of the acoustic structure of calls, but also the *usage* of vocalizations (Seyfarth and Cheney, 1986). Chimpanzee infants produce 'pant-grunt' vocalizations from a few days old, but they do not direct them to others until 2 months of age (Laporte and Zuberbühler, 2011). Whilst very young infants direct 'pant-grunts' at any group member, with increasing age, the youngsters' use of this signal becomes more similar to that of adults, who produce 'pant-grunts' as a greeting vocalization to higher-ranking individuals. While the frequency of 'pant-grunt' production fluctuates until they reach adolescence, the specificity in use is already increasing in juvenile individuals. These studies indicate that although primates produce certain call types from an early age, their frequency of use and context specificity change over time.

In regard to the comprehension of vocalizations, several studies indicate that primates have to learn to respond appropriately to calls of other group members. For instance, Fischer and colleagues (2000) demonstrated that infant baboons' ability to discriminate between two types of barks, 'contact barks' and 'alarm barks', increased with age. By the age of 6 months, they also showed the appropriate, adult-like response and thus strongly responded to 'alarm barks', while ignoring 'contact barks' (Fischer, Cheney and Seyfarth, 2000).

6.2   Mechanisms underlying ontogeny of communicative signals

Primate communication (or at the very least appropriate *use* of communication) requires a period of development with exposure to appropriate environmental stimuli. The mechanisms that underlie this process of acquisition, however, are not necessarily well understood, and not easy to study. Adding to the problem is that the more in-depth hypotheses of acquisition tend to focus on a single modality only, usually gestures. Three major mechanisms have been suggested and are introduced in the following section.

6.2.1   Ontogenetic ritualization

This theory attempts to explain the development of gestures and borrows principles from theories of signal evolution over phylogenetic time.

The original theory of signal evolution over phylogenetic time by Niko Tinbergen (1952) proposed that most animal displays become fixed in a species' repertoire through a process of 'ritualization' over generations. The theory is that animals exhibit movements and behaviours in readiness for specific actions (e.g. opening the mouth before biting), which eventually become honest signals of their 'intention' to perform that specific action (i.e. biting). If receivers respond appropriately before the specific action is performed (i.e. run away before being bitten), they are likely to have an advantage over others who do not respond, and thus are more successful. In this way, 'ritualized intention movements' are proposed to evolve. Tomasello (1996, 2008) is one of the main proponents of a similar theory explaining signal development through ontogeny: ontogenetic ritualization is a process proposed to shape signals that are 'abbreviations of full-fledged social actions' (Tomasello, 2008). For example, chimpanzees use an 'arm-raise' gesture to initiate play. Tomasello suggests that this gesture is learned through the following process:

1. An immature animal approaches another, raises an arm to play-hit the other, hits the other and play commences.
2. The pattern is repeated, and the recipient learns to anticipate the consequent play interaction from the initial arm-raise.
3. The actor learns to anticipate the anticipation of the recipient, and so raises the arm, monitors the recipient and waits for a reaction. Thus, the actor expects that the arm-raise initiates play.

Therefore, proponents suggest that intentional gestures are shaped through ontogeny by individuals anticipating and responding to each other's behaviour, through a relatively simple process of learning and reinforcement. Idiosyncratic gestures are argued to be good evidence of ontogenetic ritualization, as individuals must have developed them through a process other than imitation (as otherwise they would be similar to the gestures of others). Such idiosyncratic gestures are commonly reported in gesture research. For example, Pika, Liebal and Tomasello (2005) refer to three gestures that were repeatedly observed in just a single individual and thus interpreted as idiosyncratic. However, others have argued that idiosyncratic gestures are an artefact of small total observation time, and in reality may not exist at all (Hobaiter and Byrne, 2011a). Hobaiter and Byrne (2011a) found that individual repertoire size correlates with sampling time, suggesting that uncommon gestures may appear to be idiosyncratic in small samples.

Interestingly, Tomasello (2008) distinguishes the signals shaped through ontogenetic ritualization from attention-getting gestures, which serve to attract the attention of others, usually through auditory or tactile means (see Chapter 8 for further discussion). As attention-getters seem to be used flexibly in many different contexts, and not tied to a specific social activity (such as play), they are not thought to be ritualized from action in the same way. Rather, the sender learns them through learning about the response of the receiver. The sender learns that the attention of another is caught through use of a particular action or gesture. Thus, Tomasello (2008) argues that these gestures are more complex and novel in the animal kingdom, as they are necessarily tied up with an understanding of the other's intention and attention. He suggests that attention-getters may even be referential, in the sense that the sender is requesting (and thus referring to) the attention of another individual (Tomasello, 2008). However, this suggestion might be controversial, since referentiality is still hotly debated, both in terms of how to define and detect it (see Chapter 9 for a more in-depth discussion of referentiality).

6.2.2　Imitation

An alternative account of acquisition suggests that communicative signals are acquired through some sort of social learning process, such as imitation. Such a perspective generates interest from those interested in culture, as learning via imitation could give

Figure 6.3 Neonatal imitation. An infant rhesus macaque imitates the tongue protrusion of a human demonstrator (from Gross, 2006).

rise to cultural differences in the communication used in different populations. In which case, perhaps infants acquire their signals by observing them in others.

There is some evidence that attention-getting sounds such as 'extended grunts' or 'raspberries' in captive chimpanzees are socially learned, since chimpanzees reared by their biological mothers were more likely to produce attention-getting sounds in comparison to nursery-reared chimpanzees. More importantly, the offspring was more likely to use the same type of attention-getting signal type as their mothers (Taglialatela *et al.*, 2012). A recent study suggested that wild orangutans invent new calls that spread through social learning (Wich *et al.*, 2012). Thus, as opposed to existing studies that refer to geographic variation in the structure of certain call types in primates, this study found geographic variation in the presence or absence of particular call types across different populations that could not be explained by genetic variation. Although the absence of a particular behaviour is always difficult to show given different observation times and study group demographics across sites, the authors argue that their findings suggest the existence of 'call cultures' in wild orangutans.

Imitation of basic facial movements such as mouth opening and tongue protrusion has been demonstrated in neonatal chimpanzees (Bard, 2007; Myowa-Yamakoshi *et al.*, 2004) and neonatal rhesus macaques (Ferrari *et al.*, 2006; see Figure 6.3). In addition, delayed imitation of 'lipsmacking' facial expressions has been shown in neonatal rhesus macaques (Paukner, Ferrari and Suomi, 2011). Neurophysiological findings suggest that this facial imitation could result from the activation of mirror neurons (Gallese *et al.*, 1996). Mirror neurons are a class of neurons which become active both when actions (such as grasping) are performed, and also when the same action is observed in another individual (see also Chapter 3, **Box 3.2**). Hence, the neuron 'mirrors' the observed action. A recent study suggested that these neurons are active from the monkeys' very first days of life (Ferrari *et al.*, 2012) and thus could be driving this early neonatal facial imitation. Interestingly, however, although there is evidence of mirror neuron activation for facial movements (such as mouth opening) (Ferrari *et al.*, 2003) and hand movements associated with action (Gallese *et al.*, 1996), this has not been demonstrated (as yet) for communicative manual gestures or vocalizations.

It is important to emphasize, however, that although rhesus macaques possess a mirror neuron system that links the observation and execution of manual or oral movements, there is very little evidence that mature individuals are capable of imitating others' actions (Visalberghi and Fragaszy, 2002). This seems paradoxical given the evidence for neonatal imitation in these monkeys (Ferrari *et al.*, 2006). However, in both humans and nonhuman primates, neonatal imitation seems to disappear after several weeks (Fontaine, 1984; Myowa, 1996; Myowa-Yamakoshi *et al.*, 2004). Therefore, it seems unlikely that imitation is a major mechanism underlying the acquisition of signals in primates. This is supported by some studies on great apes that show that there is little overlap between gesture repertoires of mothers and their offspring or even between the repertoires of all individuals within one group, which would be indicative of imitation playing a major role in gesture acquisition (Schneider, Call and Liebal, 2012b; Tomasello *et al.*, 1997). Altogether, although the mechanisms of the acquisition of signals are not well understood (yet), it is important to emphasize that 'the primary function of mirror neurons cannot be action imitation' (Rizzolatti and Craighero, 2004, p. 172).

6.2.3   Genetic transmission

Finally, another account of acquisition is that some communicative signals derive entirely from a 'hard-wired' biological process of inheritance. Tomasello (2008, p. 8) argues that facial expressions, vocalizations and some gestures fall into this camp, and are 'almost totally genetically fixed and based on almost no learning'. Evidence from social isolation and cross-fostering studies do support this view in part, as species-typical vocalizations and facial expressions are produced with the same form regardless of environmental input (Geissmann, 1984; Owren *et al.*, 1992, 1993). However, the appropriate use and response to these signals is still affected by development, so although the structure of such signals relies less on shaping from the social environment, they nevertheless require triggers to function correctly. Thus, although a signal is passed on by genes to the next generation, this does not necessarily mean that its structure and use remains unchanged over the lifetime. The precise mechanisms underlying this particular process of development are, however, rarely discussed.

6.2.4   Evidence for the three mechanisms

Few studies have attempted to test between these competing explanations for the acquisition of communication and they might not be mutually exclusive. Genty *et al.* (2009) examined and compared the gestural repertoires of three captive groups and one wild group of gorillas, with the explicit goal of testing between these theories. Interestingly, the authors found no evidence to suggest that ontogenetic ritualization, cultural transmission *or* biological inheritance were more or less important in influencing the gestural repertoire of the study groups. Ontogenetic ritualization suggests that gestures should resemble an intention movement that could precede an action, or resemble actions that could bring about the presumed goal. Although such gestures were identified, these gestures were not used any more flexibly than gestures that did not fit this criterion and they were also not used in an intentional manner more so than any other gesture type. Thus, it is unclear if this process is a key developmental factor in the acquisition of flexible signalling.

In a later study in chimpanzees, however, Hobaiter and Byrne (2011a) reported no evidence of gestures that resembled an action that could have been a precursor. The extent to which ritualized gestures *should* resemble the original action, however, is questionable (Liebal and Call, 2012). Hobaiter and Byrne (2011a) examined very specific parameters (such as whether the fingers were curled or straight in begging gestures versus begging actions), which could conceivably change during the process of ontogenetic ritualization. This again highlights the importance of an approach that identifies structural properties of gestural signals to be able to identify morphological changes – in addition to possible changes in function – over an individual's lifetime.

In gorillas, there were very few gestures found commonly in one population, and absent in another (without a clear environmental cause), which rules out a cultural transmission explanation. Instead, Genty *et al.* (2009) suggest that the distribution of gestures is more consistent with a species-typical, universal repertoire. In short, the authors conclude that the gestural repertoire they observed is not easy to divide into those that are innate, species-typical displays, and those that are learned, intentional and flexible. Many processes could be at work, but the gestures are still firmly rooted in a species-typical repertoire, which can be employed flexibly and intentionally. In fact, Hobaiter and Byrne (2011a) also suggest that there may even be a suite of 24 gestures common across various species of great ape, thus pointing to a family-typical repertoire.

One characteristic of most (if not all) of these studies is the unwavering focus on one modality at a time, most often gestures. Where a different modality of communication is found within the focus of a unimodal study, efforts are made to widen the definition of the original modality to include the signal. Facial movements, for example, are often termed 'facial gestures' if they are caught in the net of a gestural study. Scientists clearly find this helpful methodologically, and meaningful theoretically. But driving a wedge between these different communicative channels might not be the best way to understand how they are acquired. Primate communication is not unimodal, since gestures are used in combination with facial expressions, vocalizations and whole body actions. Development, must, therefore, also occur in combination on some level.

6.3   Can primates generate and modify communicative signals as adults?

The ability to generate, modify and increase the number of signals in a communicative repertoire is one of the key facets of human language. Changes that occur in adulthood (after the repertoire should be fully formed) may therefore be even more revealing of complexity, than those that occur during development. Hence, scientists often consider such characteristics highly indicative of complex communication. In contrast, an inability to generate novel communicative signals implies that the animal has a fixed and closed repertoire, where the possibility for complexity is typically regarded as more limited. Chapter 7 focuses on flexibility of use and function, which may be indicative of even further complexity, but in this section the focus is purely on the structure of signal production.

Examining the variability that can be generated with existing signals, within adulthood, allows us to assess the control primates have over the structure of their signals. If for instance a primate is capable of modifying the structure of a vocalization as a result of exposure to conspecific signals, this demonstrates some flexibility within the system and the ability for vocal learning. In the literature this is distinguished from ***vocal plasticity***, which refers to the ability to generate novel vocalizations. However, given that many primates have highly graded vocal repertoires (e.g. Marler, 1976), it is sometimes very difficult to distinguish between generation of a novel signal and modification of an existing signal. This is particularly problematic given our lack of knowledge about how conspecifics perceive different signals. In humans, for example, several human phonemes are on a graded continuum (e.g. ba–pa: these phonemes only differ in terms of voice onset timing), yet we perceive the continuum as two discrete categories of phoneme (either ba or pa, but never something in between; Liberman *et al.*, 1957). In nonhuman primate species, there is evidence that macaque monkeys have the ability to perceive graded conspecific calls categorically (Fischer, 1998). Unfortunately, however, very little is known about how the vocal repertoire of most other primate species is perceived by conspecifics. If the majority of primates do perceive graded conspecific calls categorically, then generating variability in existing signals may be sufficient to create a novel signal from the receiver's perspective. However, in the absence of this data, we currently have to assume that modifying the structure of an existing signal is distinct from generating novel signals. Similar problems are encountered when partitioning facial expressions and gestures that can often appear highly graded (Parr *et al.*, 2005a; Roberts *et al.*, 2012b).

6.3.1   Generation of novel signals

The basic vocal repertoire of most primate species appears to be largely fixed, but there is some evidence which shows that novel vocalizations and sounds can be produced. First, population-specific vocalizations have been identified in orangutans, with seven call types present at some study sites, but absent at others (Hardus *et al.*, 2009a). The authors suggest that these vocalizations may be local 'cultural' innovations; however, to conclude that this is indeed the case relies on the absence of these calls in all other populations. Such a conclusion is difficult given differing observation time and research foci across sites.

More robust evidence for the production of novel sounds comes from captive chimpanzees populations, which often produce 'raspberry' sounds that have never been reported in the wild (Hopkins, Taglialatela and Leavens, 2007). These sounds seem to function as attention-getting signals and the authors argue that they represent novel acoustic signals. Marshall and colleagues (1999) reported that captive chimpanzees learnt to produce a novel variant of a 'pant-hoot' vocalization that incorporated a 'raspberry'-like sound 'made by blowing air through pursed lips', after a male who

Figure 6.4 'Eye-covering' gesture invented by mandrills. (Photo: Mark Laidre).

Figure 6.5 'Hand-clasp' behaviour in chimpanzees. (Photo: Daniel Haun).

produced this variant was introduced into the group (Marshall, Wrangham and Arcadi, 1999). In addition, a female orangutan has been reported to spontaneously copy human whistling and is now proficient in producing this novel sound (Wich *et al.*, 2009). The orangutan's whistles do not seem to have a particular communicative function, but she does imitate the duration and number of whistles produced by human models. This study shows that novel sounds can be acquired in great apes and are likely spread through imitative processes. It is important to note, however, that the sound innovations discussed here do not engage the larynx, which is required to define a primate sound as a vocalization. One documented example of a novel vocalization that is presumed to engage the larynx is the 'extended grunt' that chimpanzees have been observed to produce in captivity when begging for food from humans (Hopkins, Taglialatela and Leavens, 2007).

There is also some evidence of novel gesture production. For example, a unique arm movement gesture covering the eyes has been documented for only one out of several groups of mandrills. Laidre (2011) analysed responses to this gesture in comparison to controls, and concluded that it functions to communicate unwillingness to interact (Figure 6.4). Thus, the gesture must have been generated within this group, and has since been transmitted and maintained amongst other group members. Similarly, some gestures have been found to be unique to certain communities of chimpanzees (e.g. the 'hand-clasp' gesture during grooming), suggesting that it has been acquired through novel generation, and subsequent cultural transmission within the group (Bonnie and de Waal, 2006; McGrew and Tutin, 1978; van Leeuwen *et al.*, 2012) (Figure 6.5). However, this gesture is not unique to one population only, and so if novel generation has occurred simultaneously, generation must still be constrained by certain parameters. Furthermore, it is not clear whether the 'hand clasp' should be considered a gesture, since its communicative function is not entirely clear.

To date, there has been no similar documented evidence of novel facial expression generation (although the 'raspberry' sounds include facial movement), but there are several reports of chimpanzees and bonobos producing seemingly experimental idiosyncratic facial movements, often termed 'funny faces' (de Waal, 1988). Whether these facial movements are communicative and/or subject to cultural transmission is unknown.

A problem affecting identification of novel signals is how to define genuine novelty. For example, a signal may or may not appear to be novel depending on how the communicative repertoire has been parsed previously. Likewise, a behaviour may indeed be a novel generation, but whether this also has a communicative function may not be clear. Importantly, as with many other features of communication discussed in this book, 'novelty' seems to be defined differently within each specific modality.

6.3.2   Modification of signals in response to the social environment

There is some evidence that the structure of primate vocalizations can be modified according to changes in the social environment (see also Fischer, 2003). Pygmy marmosets have been shown to change the structure of their calls in response to long-term changes to their social environment. For example, after placing two unfamiliar populations of pygmy marmosets with acoustically distinct 'trill' calls in the same acoustic environment, the two populations modified the acoustic parameters of these contact calls making them acoustically more similar to each other (Elowson and Snowdon, 1994). A similar convergence of call structure was observed in the 'trill' calls of newly paired pygmy marmosets (Snowdon and Elowson, 1999).

In order to investigate how large-scale social change and exposure to a different 'culture' of calling affects the structure of chimpanzee calls, an ongoing project is examining what happens to the structure of 'rough grunts' produced by chimpanzees in response to food, when two groups of chimpanzees with different call variants for different types of food are integrated (Slocombe, unpublished data). Call structure before integration and a year after integration will be compared. Social data on the extent individual group members mix with individuals from the other group and thus have been exposed to different call variants in response to certain food types will enable us to make predictions about the individuals whose call structure should have shifted and the individuals whose call structure should have remained stable.

Contact with humans and exposure to human culture are examples of a special social environment that might have a far-reaching impact on the communicative behaviours of primates (Tomasello and Call, 2004; but see Bering, 2004). The extent to which the human environment shapes the communicative skills of primates, however, may differ between different settings. For example, nonhuman primates in zoos and laboratories live with other conspecifics, while contact with humans is limited to certain times of a day and to particular contexts, such as feeding, cleaning, or research (see Chapter 8, **Box 8.3** on interactions of nonhuman primates with humans when requesting food). In sanctuaries and occasionally also in zoos, primates are raised by humans and are later integrated into groups with other conspecifics. Intensive contact of nonhuman primates with humans is therefore mostly limited to certain life stages, particularly infancy, when humans act as substitutes for their biological mothers. Finally, some apes are raised in a human environment (sometimes like a human child) with intensive contact to humans that is usually not limited to early life stages. This is particularly the case for apes that are being taught some form of linguistic communication, e.g. the use of American Sign Language or lexigrams (see Chapter 10, Table 10.1, for an overview of studies).

There are very few studies that investigate the impact of human contact on the communicative behaviour of primates across these different settings, but there is some evidence that the intense and long-lasting exposure to the social environment of humans has far-reaching impact on a variety of cognitive skills of primates, both in the physical and social domain (Russell *et al.*, 2011). These 'enculturated great apes' – more specifically chimpanzees and bonobos that are often part of ape 'language' projects – performed significantly better than great apes in standard laboratory settings in tasks that include the comprehension of gestures (object-choice, see Chapter 4, section 4.2.4), the production of gestures to indicate the location where food is hidden to an ignorant human, and the production of gestures depending on the attentional state of the human, reflecting the cognitive skills of a 2.5-year-old child (Russell *et al.*, 2011). Most interestingly, however, nursery-reared individuals did not differ from individuals that were raised by their biological mothers suggesting that this initial intensive contact with humans is not sufficient for the development of more sophisticated cognitive skills in nonhuman primates.

Finally, there is some evidence that exposure to human culture and language specifically has an impact on the vocal communication of great apes, for example, the distinct 'peep' calls produced by Kanzi, a 'language'-trained bonobo (see Chapter 10, Table 10.1). During interactions with humans, the 'peeps' Kanzi produces in different semantic contexts are distinct 'peep' variants, both in terms of temporal and spectral structure. This indicates that Kanzi can modulate the structure of his vocalizations (Taglialatela, Savage-Rumbaugh and Baker, 2003). Unfortunately, this study did not attempt to compare these 'peep' variants to a representative sample of 'peeps' from other bonobos, so it is difficult to know if these 'peep' variants are a unique product of the social context in which Kanzi has been raised, or simply part of the species-typical repertoire that Kanzi uses in a context-specific manner.

6.3.3 Modification of signals in response to the social context

Signal structure can also be modified in a more dynamic manner, in response to much more temporary and short-term social changes. Whether this represents modification in the sense that hints at complex cognition such as intentionality is unclear. Palagi and Mancini (2011) have demonstrated that the 'playface' of geladas can change in form and function depending on the context, which may not demonstrate voluntary modification per se, but instead use of signals appropriate to the context. Nevertheless, such modification in relation to social factors may still help us to understand how signals are shaped. For example, Bouchet and colleagues (2012) showed that the degree of variability and individual distinctiveness of different vocalizations in red-capped mangabeys is highest for 'contact' and 'threat calls' that are used for interactions within a social group as opposed to long-distance vocalizations such as 'loud calls', but also 'alarm calls' (Bouchet *et al.*, 2012). The authors suggest that the acoustic variability of these calls that differs depending on the social context in which they are used highlights the important role of such social factors for the evolution of the mangabeys' vocal repertoire.

Flexible adjustment of the structure of individually distinct vocalizations has also been found in female Diana monkeys (Candiotti, Zuberbühler and Lemasson, 2012a). The fundamental frequency contours of the main social call of this species varied systematically between individuals in a sufficiently stable manner to convey individual identity; however, individuals were also able to temporarily alter the acoustic structure of their calls. As a consequence, they either converged or diverged their calls with those of other individuals depending on physical factors (e.g. low visibility resulted in increased divergence of calls to enable individual recognition) and social context. Thus, calls that were exchanged between females in affiliative or neutral contexts were more similar than non-exchanged calls due to females matching the frequency contour of their own call with that of the preceding call of another female during vocal interactions (Candiotti, Zuberbühler and Lemasson, 2012a).

Chimpanzees sometimes join in with another individual's 'pant-hoot' call and perform a so-called 'pant-hoot' chorus. Males are able to flexibly modify the duration of various elements of this call to increase the likelihood of another male joining in to form a chorus (Fidurek, Schel and Slocombe, 2012). When males call together in a chorus, it has also been found that 'pant hoots' produced in a chorus with a particular partner are acoustically more similar than to 'pant hoots' produced in choruses with others, suggesting that males may actively adjust the acoustics of their calls to match their partners' calls (Mitani and Brandt, 1994). A further study suggests that this acoustic accommodation is more likely to happen with strongly affiliated individuals, indicating that 'pant-hoot' choruses may have a social bonding function (Mitani and Gros-Louis, 1998).

A remarkable new finding indicates that wild orangutans use a leaf tool, which they hold to their mouth to modify their 'kiss-squeak' sounds in situations of significant danger (Hardus *et al.*, 2009b). The use of this tool lowers the frequency of the call and as the frequency of these calls reliably varies with body size, the leaf-call combination signals a larger body size than a call alone. Hardus and colleagues (2009b) conclude that such functional deception may deter predators from attacking.

In the facial domain, rhesus macaque mothers appear to augment their facial expressions depending on the intended receiver (Ferrari *et al.*, 2009). Thus, mothers make exaggerated and augmented facial expressions whilst communicating with their infants in the first month of life. First, mothers produce 'lipsmacking' whilst physically manipulating the infant to establish mutual gaze. Second, if the mother and infant are physically separated, the mother moves her face close to the infant before producing the 'lipsmacking' and sometimes produces the facial display in combination with bobbing her head to attract the gaze of the infant. These exaggerated 'lipsmacking' displays in combination with sustained mutual gaze are specific to communication with very young infants and have never been observed between adults (Ferrari *et al.*, 2009). A similar modification of communicative patterns by adults when interacting with infants is documented for female rhesus macaques that alter the acoustic spectrum of 'girney' vocalizations when they direct them towards infants (Whitham, Gerald and Maestripieri, 2007), as well as for gorillas that repeat gestures more frequently when they interact with infants, but not individuals of other age classes (Luef and Liebal, 2012).

There is some preliminary evidence that the form of ape gestures may vary according to the social context (Liebal, Bressem and Müller, 2010). Orangutans use a 'slap' gesture in two distinct social contexts and the structure of this gesture is different across these contexts: in agonistic contexts the palm of the hand is directed downwards and the whole arm is moved, while in play contexts the palm of the hand is upwards and only the lower arm is moved. However, in the absence of data on how conspecifics perceive the two variants of this gesture it is impossible to say if they are two distinct context-specific gestures that carry quite different meanings, or if this example shows that modifications to the structure of gestures can be made in response to different social contexts.

There is stronger evidence that the structure of facial expressions is subtly modified according to social context. Van Hoof (1973) observed three different variants of the 'silent bared-teeth' facial expression in chimpanzees: the 'horizontal bared-teeth display'; the 'vertical bared-teeth display' and the 'open-mouth bared-teeth display'. The variant produced is influenced by the age of the receiver in relation to the producer; with the 'vertical bared-teeth display' given more by older individuals towards younger group members and the 'horizontal bared-teeth display' given by younger individuals to older group members. The 'open-mouth bared-teeth display' and 'vertical bared-teeth display' were commonly produced in affiliative contexts, whereas the 'horizontal bared-teeth display' was more strongly associated with submission. Again, as with the gestures, it is difficult to determine in the absence of data on the perception of these signals whether these are distinct context-specific facial displays, or evidence of flexible modification of signals in response to social context.

6.3.4   Modification of signals in response to the group

The ability to modify an existing signal to converge with a group-specific 'norm' indicates not only that the production of the signal has a degree of flexibility, but also that specific learning processes must play a role in acquisition. Signals that can be modified, conventionalized and socially transmitted have the potential to increase the complexity of the communicative system as a whole, and could reflect the manifestation of culture. Such characteristics are key features of language (see Chapter 5, **Box 5.1**).

In humans, local dialects (modifying the acoustic structure of a common word) seem to play an important role in social bonding within communities and the establishment of group identity (Cohen, 2012; Cohen and Haun, 2013). Given that many primates are territorial group-living animals, it may be advantageous to converge on group-specific communicative signals, which quickly identify an individual as a group member or a stranger. Evidence for such group-specific variants has been found in chimpanzee vocalizations. Early work mapped out group-specific acoustic differences in the production of 'pant-hoot' calls, the chimpanzees' species-specific long-distance call (Mitani *et al.*, 1992). This seminal work compared the calls of geographically distant communities of chimpanzees and thus it was very difficult to exclude the effects of differential genetics and anatomical factors (e.g. body size) across the two populations. It was also difficult to exclude the effect of differences in sound propagation through the two different habitats (Mitani, Hunley and Murdoch, 1999). Subsequent work aimed to control for these potentially confounding factors. Group-specific 'pant-hoot' calls were found in two captive groups of chimpanzees, where the acoustic environments of the two enclosures were very similar and genetics were unlikely to play a role given the largely unrelated and genetically diverse nature of the two groups (Marshall, Wrangham and Arcadi, 1999). More evidence for group-specific 'pant hoots' comes from a more recent study conducted with the chimpanzees in three neighbouring communities in Tai Forest, Ivory Coast (Crockford *et al.* 2004). The habitat of these bordering communities was relatively homogenous, making it unlikely that these differences between 'pant hoots' across groups reflect differential sound propagation properties of the environment. Equally, Crockford and colleagues tested whether genetic relatedness between individuals within and between communities could account for the observed group-specific calling pattern and found no evidence to support the hypothesis. The calls of a fourth community, located 70 km away, overlapped in structure with 'pant hoots' of the three neighbouring communities, indicating that maximal differences occurred between neighbours. Taken together, these studies seem to indicate that chimpanzees are capable of converging on a group-specific 'pant-hoot' variant, thus modifying their existing 'pant-hoot' call to conform with the group-specific signature.

In regard to call perception, chimpanzees also seem sensitive to these subtle acoustic differences in the 'pant hoots' of different groups. In a playback study, 'pant hoots' of group members, neighbours and strangers were played to chimpanzees in Tai Forest (Herbinger *et al.*, 2009). Differential vocal, gestural and locomotor responses to the 'pant-hoot' calls of these three categories of individual were observed, with the strongest response elicited by neighbour calls. Chimpanzees may actively converge on a group-specific call that differs maximally from those of neighbours (Crockford *et al.*, 2004), which could aid the quick recognition of hostile territory invasion by neighbouring males.

Group-specific gestures are documented in great apes in a variety of contexts (e.g. 'hand-clasp grooming hold' and 'leaf clipping' in chimpanzees, Whiten *et al.*, 1999; 'offer food with arm' in orangutans, Liebal, Pika and Tomasello, 2006) and there is also some evidence for such group-specific behaviours in monkeys (Perry, 2011; Perry and Manson, 2003). However, other studies have only found limited evidence of culturally transmitted gestures (e.g. Genty *et al.*, 2009; Tomasello *et al.*, 1997). The existence or absence of a particular gesture in a population has been used as evidence of the pivotal role learning plays in the development of gestures and for cultural variation in signal production (Whiten *et al.*, 1999). This literature concerns the existence or absence of a particular gesture, however, and no study has yet examined group-specific variants of the same gesture in a parallel way to vocalizations. This may also be due to the difficulty in conducting fine-grained analysis on manual gestures. Unlike vocalizations, where very subtle acoustic differences can be measured with modern acoustic analysis software, similar techniques are not yet established for gestures. Equally, evidence for group-specific variation in facial expressions has not been documented, although there have been anecdotal observations of expressions unique to certain populations (for chimpanzees: van Hooff, 1973). Again, this may be partly due to a historical lack of established methods for identifying and quantifying individual or group variation in facial expressions. With the recent development of anatomically based observational tools to record primate facial expression production systematically (Caeiro *et al.*, 2013; Parr *et al.*, 2010; Vick *et al.*, 2007; Waller *et al.*, 2012; see Chapter 4, section 4.3) this avenue of research may be easier to pursue.

Importantly, we know very little about the specific mechanisms underlying convergence on a group-specific variant of a signal. It is not clear if individuals converge on the signal of a dominant individual, whether the group-specific variant remains stable over time across changes in dominance hierarchy and generations, how long this process takes, and whether input during ontogeny may play a role in the process.

6.3.5 Modification of signals in response to physical factors

It is well established that propagation of sound through different primate habitats can be remarkably different (Waser and Brown, 1986; see also Maciej, Fischer and Hammerschmidt, 2011), so it could be adaptive to modify signals accordingly. Morton (1975) proposed the *Acoustic Adaptation Hypothesis*, which suggests that animal signals should evolve to maximize transmission through the acoustic environment they work in. Support for this idea has been forthcoming in a number of species including primates (e.g. Brown, Gomez and Waser, 1995). Therefore, while the influence of the acoustic environment on the evolution of vocal signal structure has been explored, the sender's flexible modification of signal structure to accommodate dynamic changes in habitat, and therefore improve transmission of their signals to the receivers, has been investigated far less. In one recent study, however, baboons were shown to alter the temporal structure of their 'grunt' vocalizations according to their local environment (Ey *et al.*, 2009). Baboons gave longer 'grunts' whilst travelling through closed forest habitat (where 'grunts' propagated less efficiently) in comparison to open habitat. Thus, baboons seem to be flexibly adjusting their call structure to adapt to local environmental conditions.

Marler (1965) suggested that the modality of communication should also be influenced by habitat features. There is preliminary evidence that chimpanzees' reliance on vocal and visual modalities is flexibly altered depending on their environment. In the Budongo Forest in Uganda, the chimpanzees of the Sonso community live in a selectively logged area of the forest that is characterized by low visibility levels. In this community, preliminary data indicates that vocalizations are produced at a higher rate than visual gestures (Slocombe and Zuberbühler, 2010). In contrast, two captive groups of chimpanzees, kept in open enclosures, are reported to produce approximately twice as many gestures as vocalizations (Pollick and de Waal, 2007). There are obviously other confounding factors in this fairly crude comparison of wild and captive populations, such as group size, group composition and space, so comparisons should be made cautiously. In captivity, the limited space may require individuals to communicate differently in order to avoid conflict and to reconcile quickly with past opponents. In the future, a comparison of relative reliance on visual and vocal modalities in wild populations inhabiting environments with differing levels of visibility would be very valuable.

6.4   What could a multimodal approach bring to this topic?

A dominant theme of the research reviewed in this chapter has been the focus on single modalities. In this section, we discuss the potential and highlight the advantages a more integrated multimodal approach could bring and how this may challenge our current understanding of the topic.

Multimodal signals may have a different developmental trajectory to unimodal signals and so exploring the ontogeny of multimodal signals alongside unimodal signals could change our understanding of the mechanisms involved in signal acquisition. For instance, unimodal signals may develop first, with multimodal combinations only emerging when other key social cognition skills have developed, or when there has been sufficient time and exposure to learn multimodal signals. On the other hand, holistic, multimodal signals may emerge first and may then be pruned to form effective unimodal signals. Alternatively, they could emerge together and follow very similar developmental pathways. Multimodal signals may not only emerge at a different point in development compared to unimodal signals, but they may be grounded on different cognitive mechanisms. For instance, unimodal signals may be part of a genetically specified species-typical repertoire that emerges during maturation, yet combining signals to form multimodal signals could rely on learning mechanisms. Thus, individuals could require exposure to the signalling behaviour of others or experience in interacting with others in order to acquire them. Equally, the developmental trajectory of the comprehension of multimodal signals with different (emergent) meanings compared to the unimodal components could reveal if these combinations are part of a genetically specified repertoire that automatically elicit certain responses, or whether the comprehension of the multimodal combination emerges later, with repeated exposure to the signal providing the opportunity to learn its meaning.

Currently, our understanding of the development and underlying cognitive mechanisms in primates is limited to unimodal signal production. Thus, the communication system of a species is likely to be characterized as a 'closed genetically specified' or an 'open learned' system on the basis of unimodal signals alone. Whilst multimodal signals may follow a similar developmental pathway and support such classifications, they may show a different pattern of emergence and suggest the involvement of different cognitive mechanisms. Thus, in a hypothetical system where the unimodal signals are genetically specified, but the multimodal signal combinations are learnt, the current unimodal approach to primate communication would overlook such learnt elements and thus provide an incomplete understanding of the system. This could be avoided with the use of an integrated multimodal approach to primate communication.

Summary

Primate communication does not emerge fully formed in the newly born infant. Therefore, acquisition of communication requires a period of development and input from appropriate (usually social) stimuli. While form and structure of communicative signals sometimes seem fairly robust in response to varied and/or impoverished environments, appropriate use of signals is less so. Such reliance on a developmental period implies that some form of learning must play a role in signal acquisition, which is also supported by evidence that signals have the potential for modification in the adult, or for novel signals to be generated from scratch. Different mechanisms have been suggested to underlie signal acquisition, including genetic transmission, ontogenetic ritualization and imitation. Gestures have received considerable research attention in this regard, as they are frequently reported to exhibit more flexibility than other modalities, which is assumed to be associated with learning mechanisms as opposed to genetically determined maturation processes. Whilst this has been challenged by recent research, it seems possible for great apes to create some new gestures, something not evident in either facial or vocal signals. There is, however, considerable evidence that vocalizations can be modified in both structure and use in response to social and ecological factors, while less is known in regard to facial expressions. There is still comparatively little data available on the developmental trajectories of the different modalities. However, it seems that acquisition mechanisms are not mutually exclusive and that several mechanisms are likely to underlie signal acquisition within and between modalities and across different life stages of an individual. Adopting a multimodal approach to this topic may improve our understanding of the developmental trajectories of both unimodal and multimodal signals and the mechanisms that underlie signal acquisition.

7 Flexibility

The previous chapter examined the generation of new signals, the modification of the structure of existing signals and acquisition of signals through development. This chapter will build on this previous discussion of ontogenetic flexibility and examine the extent to which existing signals can be used in different ways, combined into sequences and comprehended flexibly.

7.1 Why is it interesting to investigate flexibility?

Many very effective communicative displays in the animal kingdom are involuntary, inflexible, reflexive responses to specific stimuli. For instance, in response to a rapidly advancing predator, a moth will rapidly open its bottom wings to reveal two large, intimidating eye spots. This display is successful in deterring predators, but in the presence of the correct category of stimuli (any rapidly advancing object in this case) the moth will invariably give this response. Such automatic, reflexive, stimulus-response behaviour is very effective in many instances, but it is unlikely to require complex cognition and thus it tells us very little about the mental capacities of the animal producing the display. For instance, human pupils tend to dilate in response to attractive potential mates and this in turn makes the producer of this display appear more attractive: it is an effective display, but one not under the conscious or voluntary control of the producer. If looking for cognitively complex communication, such displays are not particularly informative. In contrast, in a flexible communication system where there is not a one-to-one correspondence between the stimulus and response, it is likely that more complex cognition is required to operate the system. Although it is difficult to elucidate the cognitive mechanisms underlying successful communication, attribution of cognitive processes can sometimes be the most parsimonious explanation for behaviour. Thus, where multiple signals and responses can be produced, it may be more parsimonious to assume that cognitive processes underlie the system, rather than a large number of automated responses. For these reasons, flexibility is often seen as a key hallmark of cognitive complexity in a communicative system.

7.2   What do we mean by flexibility?

Flexibility in communication can be evident in many different ways, including the ability of individuals to modify and add to existing repertoires through ontogeny, as reviewed in the last chapter. We are going to review three additional commonly researched aspects of flexibility in this chapter. First, we will discuss the usage of signals over different contexts. The extent to which signals are produced invariably and rigidly in response to a specific stimulus and are thus considered 'context bound' (Tomasello, 2008) will be compared to the extent that signals are produced flexibly across different contexts and thus seem less bound by strict stimulus-response associations. Second, as such flexible signal production also requires flexible signal perception from receivers, who have to consider contextual information in addition to the signal in order to infer the meaning of the signal (Smith, 1977), receiver flexibility will then be considered. Finally, we will discuss the combination of signals into sequences. The ability to produce meaningful sequences of signals requires flexibility on the part of both the signaller and receiver and can offer the potential for generativity within a communication system

7.2.1   Flexibility in usage

Great ape gestures have been found to be flexibly produced across many different contexts. Gestures of captive gorillas (Pika, Liebal and Tomasello, 2003), bonobos (Pika, Liebal and Tomasello, 2005), orangutans (Liebal, Pika and Tomasello, 2006) and chimpanzees (Tomasello *et al.*, 1997) are all characterized by the use of several different gestures in a single context and the use of a single gesture in multiple contexts. For instance, bonobos used 50% of their gestures in two or more contexts and multiple gestures were observed in each of the eight defined contexts (Pika, Liebal and Tomasello, 2005), while gorillas used over 75% of gestures in two or more contexts (Pika, Liebal and Tomasello, 2003). This is supported by a more recent study that also included data from a wild population and found a high number of gestures used across several contexts (Genty *et al.*, 2009). Genty and colleagues also examined the instrumental function of 10 common gestures and found that gestures were multipurpose, with no gesture having a single, simple function. Such a **means-ends dissociation** between the signal and context is a hallmark of flexible signal use (but see **Box 7.1** for the important differentiation between flexibility on an individual versus group level).

It is not just gestures that have been observed to be flexibly produced. Liebal *et al.* (2004c) directly compared the use of gestures and facial expressions across contexts in siamangs. Although considerable flexibility across all different signals was evident (71% of a total of 33 signals were observed in three or more contexts), the facial expressions were produced in a significantly higher number of contexts than tactile gestures or visual gestures, indicating facial expression was the most flexibly deployed type of signal.

A very different pattern was found in captive chimpanzees and bonobos (Pollick and de Waal, 2007). This study also directly compared the flexibility of signal use across

---

Box 7.1 Differentiating between flexibility on an individual and group level

One important methodological issue with studies of flexible signal production is whether flexibility is being demonstrated on a group or individual level. From a cognitive point of view, flexibility in terms of being able to produce signals across a variety of contexts needs to be demonstrated at an individual level. Group analysis of several individuals who each use a signal in a highly context-specific manner may result in a claim of flexible signal production, yet this flexibility will not have been shown at the crucial individual level. Some studies fail to present any data at the individual level and thus it is very difficult to know if flexibility on the individual level has been demonstrated.

Flexible production of signal across contexts on an individual and a group level

Individual	Signal produced in these contexts?			Flexibly produced in 2+ contexts?
	Context A	Context B	Context C	
1	Yes	Yes	Yes	Yes
2	Yes	Yes	No	Yes
3	Yes	No	Yes	Yes
Group	Yes	Yes	Yes	Yes

Flexible production of signal across contexts on a group level, without any evidence for flexibility on an individual level

Individual	Signal produced in these contexts?			Flexibly produced in 2+ contexts?
	Context A	Context B	Context C	
1	Yes	No	No	No
2	No	Yes	No	No
3	No	No	Yes	No
Group	Yes	Yes	Yes	Yes

---

modalities in seven different contexts. The authors combined vocalizations and facial expressions (18 facial/vocal signals) and compared them to gestures (31). Gestures were reported to be significantly less associated with a 'typical' context than vocal/facial signals. In an analysis of eight gestures and five facial/vocal signals, gestures were not found to be significantly associated with a particular context as opposed to facial/vocal signals that were tied to a specific context.

As the study by Pollick and de Waal (2007) suggests, facial expressions seem to be heavily context bound and inflexibly deployed. Parr *et al.* (2005a) support this notion with the finding that chimpanzees produced different facial expressions depending on the behavioural context: because facial expressions were not equally distributed across the 30 behavioural contexts. This does not mean, however, that there was no flexibility in signal production. There was considerable variation in the degree of flexibility of the usage of individual signals. For example, while the 'relaxed open-mouth' face was highly context-specific and recorded in play contexts 83% of the time, the 'tense face' was only recorded in its most typical context 17% of the time and thus seems to be a much more flexibly used facial expression. It is difficult, however, to directly compare these results to the findings of the reviewed gestural research, as Parr *et al.* (2005a) discriminated 25 different facial expressions and considered 30 contexts compared to just seven or eight contexts usually referred to in gesture studies (Call and Tomasello, 2007; Pollick and de Waal, 2007).

One outstanding question concerning the flexible use of gestures and facial expressions across contexts is whether the underlying drive or message behind the signal remains constant. For instance 'silent bared-teeth' facial displays in chimpanzees are thought to function to signal appeasement and affiliative intent to others (van Hooff, 1972; Waller and Dunbar, 2005). Thus, there may be a variety of contexts in which an individual may want to convey such a message, however, if the message remains constant, but is simply needed in multiple contexts, is this really flexible usage of a signal? Equally, chimpanzees use an 'extend arm' gesture towards another to request things, be that meat in a food-sharing context or coalitionary support in an agonistic context. Again, the gesture may simply convey the producer's desire for action from the receiver and this same message of request is required in a variety of contexts.

Interestingly, due to the long-standing link between emotion and facial expressions (Darwin, 1872), more effort seems to be made to find links between motivational states and facial signal production, than is made with gestures (see also Chapter 8, **Box 8.1**). Parr *et al.* (2005a) identify seven factors that account for 80% of the variation in facial expression production and whilst some of these are behavioural contexts, such as play, some are characterized in terms of motivational factors such as nervousness, fear and distress. The question remains of whether a similar picture would emerge if similar efforts were made to determine the motivational states underlying gesture production.

Primate vocalizations, in contrast to great ape gestures, are typically characterized as being context bound, with strong one-to-one mappings between certain signals and eliciting stimuli. Pollick and de Waal's (2007) analysis of chimpanzee and bonobo vocal/facial signals supports this notion. Much of the vocal work to date has concentrated on evolutionarily urgent contexts, particularly predator defence. In these contexts there may be a greater evolutionary selection pressure for unambiguous signals with a close mapping between signal and eliciting stimulus. Thus, on hearing an alarm call, the receivers can react immediately with an appropriate anti-predator response, rather than having to integrate information from the signal with other contextual information before responding adaptively. A recent study examined the acoustic variability of red-capped mangabeys' vocal repertoires across different contexts and found that the acoustic variability was indeed lowest for alarm calls and variability was more pronounced in their contact calls, which are emitted in a range of evolutionarily less urgent contexts (Bouchet *et al.*, 2012).

The work on evolutionarily urgent signals has also revealed many vocalizations to function referentially. This is discussed in more detail in Chapter 9, but one of the defining characteristics of a functionally referential call is that they are given reliably in response to a specific eliciting stimulus (context-specific signal production). Thus, by definition, a functionally referential call cannot be produced flexibly across contexts. The influential work on referential vocalizations (Seyfarth, Cheney and Marler, 1980a, b) has shaped the common approach to vocal research. Many vocal researchers therefore search for context-specific call production in primates. This has two consequences. First, research tends to be more concentrated on signals that occur in contexts where unambiguous and context-specific signals are likely to be produced (generally evolutionarily urgent contexts where such signals are likely to have direct fitness benefits). Second, if researchers fail to find context-specific calling the investigation may simply be abandoned or not published, rather than being presented as evidence for flexible use of a communicative signal across contexts, which is often reported for production of gestures (Pollick and de Waal, 2007). This bias towards searching for context-specific vocalizations rather than flexible call usage is evident from the emphasis of Crockford and Boesch's (2003) analysis of chimpanzee 'bark' calls. 'Barks' were recorded in six different contexts and 'barks' given in response to aggression, travel and hearing members of the same or neighbouring community were not acoustically distinguishable. The authors comment that a 'generic bark' seems to be given in these contexts, yet flexible production across contexts is not mentioned and the paper instead focuses on the two contexts in which a distinct context-specific 'bark' is produced. Our prior assumptions about a modality seem to constrain the focus of our investigations, analysis and presentation of data (Slocombe, Waller and Liebal, 2011).

It could be that primates do produce vocalizations flexibly, particularly across socially more relaxed and thus less urgent contexts, in a similar way to that documented for gestures, but research effort has simply not focused on trying to find this kind of behaviour. In particular if vocal signals produced in the same relaxed social contexts that have revealed flexible gesture production in apes are examined, comparable attributes may be found. For instance, many primates vocalize during play and affiliative interactions and these signals may be more flexibly produced, but research examining this possibility is currently lacking. Although one recent study investigated the vocal production in a more relaxed social setting, it also focused on context-specificity rather than flexible use and suggested that laughter in chimpanzees varied acoustically as a function of whether the play partner was also laughing (Davila-Ross et al., 2011).

The different methodologies available for the analysis of the different modalities also seem to contribute to the current lack of flexible vocal production in contrast to gesture and to some extent facial expressions. Modern acoustic analysis software allows researchers to examine the structure of vocalizations in an extremely detailed and fine-grained manner and this enables researchers to identify very subtle differences in the structure of signals given in different contexts, just as it enables them to identify subtle modifications to the structure of existing signals (as reviewed in

Chapter 6). In addition, playback experiments allow researchers to test if these subtle acoustic differences elicit differential responses from listeners, giving strength to the interpretation of these subtle structural differences representing psychologically important differences for conspecifics. In contrast, the structure of gestures is currently much more challenging to analyse in the absence of any agreed standardized and objective coding scheme for the characterization of these signals (but see Roberts *et al.*, 2012b, for a recent suggestion for a structure-based coding system). As a consequence, the level of analysis possible on the structure of these visual signals is much cruder. In contrast, detailed objective coding of facial expressions in a number of primate species is now possible with the development of FACS for chimpanzees (Vick *et al.*, 2007), macaques (Parr *et al.*, 2010), gibbons (Waller *et al.*, 2012) and orangutans (Caeiro *et al.*, 2013). Although detecting the production of subtly different facial expressions is possible, it is much more challenging to test receiver responses to subtle differences in visual signal structure (gestures and facial expressions) in an analogous way to playbacks of vocalizations. Therefore it seems natural that the identification of context-specific signals is easier in modalities where finer-grained analysis of signals given in different contexts and testing of receiver responses to the signals are possible. For instance, wild chimpanzees produce 'pant-hoot' vocalizations in a wide variety of contexts including travelling, arriving at feeding trees, feeding, resting, displaying and replying to other communities and to other parties within community. On the surface this seems like a signal that is produced very flexibly in a wide variety of contexts, but due to the fine-grained analysis that is possible on acoustic signals we know that the acoustic structure of 'pant hoots' given in at least some of these contexts are distinct context-specific signals (Notman and Rendall, 2005, Uhlenbrock, 1995). If the same fine-grained analysis of gesture were possible, would we find a similar pattern? Would an 'extended arm' gesture, given in agonistic and food-sharing contexts be found to be subtly different in each of these contexts? Preliminary work on the structural properties of great ape gestures indicates that this may be the case (Liebal, Bressem and Müller, 2010).

The common characterization of vocalizations and facial expressions as context-bound inflexible signals in contrast to flexibly produced gestures may be in part an artefact of the methodological issues and inherent researcher biases discussed above. More integrated multimodal research would help us avoid these confounding factors and develop a more accurate picture of the degree of flexibility primates exhibit with signals in different modalities. It is also important to remember that many other factors influence the flexibility of signal use, including the social structure of a species. The finding that the flexibility of specific facial expression usage across monkey species was mediated by the social dominance structure of the species led Preuschoft and van Hooff (1995) to propose the *Power Asymmetry Hypothesis of Motivational Emancipation* (see Chapter 1, section 1.3.1, and Chapter 10, section 10.3). This hypothesis posits that the degree of flexibility in the production of specific communicative signals is at least partly determined by the social structure of the species, with more flexibility possible in species characterized by more relaxed dominance structures (Figure 7.1).

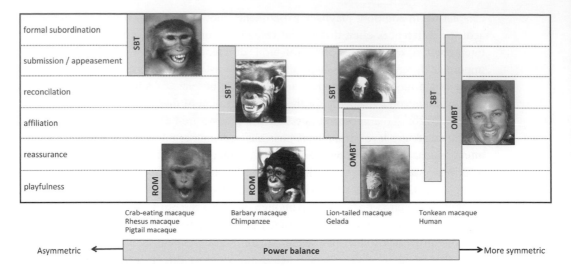

Figure 7.1 Use of facial expressions across different social contexts varies as a function of primates' social structure. More despotic (asymmetric) species are shown on the left, more egalitarian (symmetric) species are shown on the right. SBT = silent bared-teeth face; ROM = relaxed open-mouth face; OMBT = open-mouth bared-teeth face (modified after Preuschoft, 2004) (Photos: Seth Dobson, Lisa Parr).

7.2.2   Flexibility in the receiver

The receiver in any primate society is bombarded with a multitude of signals in different modalities. The ability to respond in an adaptive manner to these signals is vital to success in a social primate group. It is very difficult to disentangle the mechanism driving responses to communicative signals in primates: some signals such as colouration, olfactory signals and, some argue, vocal signals (Owren and Rendall, 2001) may have a direct physiological effect on the receiver, which then largely determines the response of the receiver. In such cases, the receiver responses are tightly tied to the perception of the signal and are viewed as inflexible, reflexive responses that the receiver has little volitional control over. In other cases it is argued that the receiver has to integrate the signal with information from the surrounding context or existing social knowledge in order to extract the message from the signal (Smith, 1977) or that the receiver forms some kind of mental representation based on the signal received and responses are therefore likely under more cognitive control (Zuberbühler, Cheney and Seyfarth, 1999). In both of these two latter cases there is much more scope for flexible, cognitively interesting behaviour from the receiver.

Cheney and Seyfarth (2005) make a convincing case for flexibility in receiver responses to vocal signals. Receivers seem to have an open-ended ability to learn novel sound-meaning associations, and learning occurs throughout the lifetime of the animal, across many different contexts. Tomasello (2008) argues, however, that this flexibility to form novel signal-meaning pairings relies on generic learning mechanisms, not communication-specific cognitive abilities, which are evident in signal production.

Regardless of the origin of the mechanism driving signal processing in the receiver, primates do not just respond in an automatic, reflexive way to specific acoustic signals. For example, baboons use their social knowledge to infer who a call is directed towards (Engh *et al.*, 2006) and thus only respond to a small proportion of the myriad of calls they hear within their social group each day. Research on Campbell's monkeys showed that the identity of the vocalizing individual has an impact on the frequency of responses from other individuals of the group. Thus, aged females are more likely to elicit a vocal response than younger females, indicating that these monkeys pay special attention to the vocalizations of the elderly (Lemasson, Gandon and Hausberger, 2010a).

Primates seem to process sounds on some kind of representational level, allowing them to respond flexibly to signals. For example, vervet monkeys use two acoustically distinct calls, 'chutters' and 'wrrs' in encounters with other groups. In an habituation experiment, Cheney and Seyfarth (1988) demonstrated that after repeated exposure to 'wrr' calls, vervet monkeys also ignored 'chutter' vocalizations and concluded that they habituated to these signals based on their similar meanings, despite their different acoustic properties. In contrast, habituation did not transfer across calls that denoted different predatory dangers. Similarly, Zuberbühler and colleagues (1999) showed that Diana monkeys treat shrieks from an eagle and a group member 'eagle alarm call' (two acoustically different signals) as functionally similar signals, and thus habituation to one signal transfers to the other signal, despite the marked differences in their acoustic structure. Together, these studies demonstrate that listeners are not just passive receivers of information. In most cases they need to actively integrate the information obtained from sounds with their existing knowledge about their world and other contextual sources of information before being able to make an adaptive response.

Receivers of gestural signals equally have to show flexibility in order to produce an appropriate reaction. Due to flexible gesture production across contexts, receivers have to rely heavily on contextual information to infer the meaning of the signal in order to produce an appropriate response. This means a simple, automatic signal–response pairing would not lead to adaptive responses across contexts and thus greater flexibility in the receiver is required for effective communication.

Facial expressions seem to allow receivers to predict the future behaviour of the signaller (e.g. Waller and Micheletta, 2013). Based on this prediction, primates may exhibit flexibility in their response to others depending on the facial expression they see. For instance, long-tailed macaques showed a stronger gaze-following reaction to a human experimenter shifting their gaze if it was accompanied by a socially meaningful facial expression (e.g. 'bared-teeth expression'; Goossens *et al.*, 2008).

There is little evidence, however, for flexibility in perception of as opposed to response to these signals. It is, however, exceedingly difficult to determine the mechanism driving receiver responses, because it is difficult to disentangle the perception of a signal and the response to a signal and therefore most studies only look at the overt behavioural response. Since very few studies examine receiver perception of signals, neurophysiological studies are very important for identifying how signals are processed at a neuronal level. For instance, recent work has shown that in rhesus macaques' various brain regions including the superior temporal sulcus and the auditory cortex contain neurons that respond to the integration of visual and auditory signals, such as dynamic vocalizing monkey faces (Ghazanfar *et al.*, 2005; Ghazanfar and Schroeder, 2006). On a behavioural level, Lisa Parr has also succeeded in examining perception of signals in chimpanzees by using a match-to-sample paradigm, where the chimpanzee is presented with a stimulus and is then asked to choose the matching stimulus from a choice of two (Parr, 2004) (see Chapter 4, section 4.3.4). Using this paradigm, Parr has been able to determine the modality that the chimpanzees perceive to be most salient in a multimodal representation of a signal. Having first established that chimpanzees can use visual or auditory samples alone to successfully match the corresponding facial expression for that signal, incongruent sample stimuli were then presented. For instance, the chimpanzees were presented with an incongruent multimodal video stimulus as the sample (e.g. dynamic 'screaming face' and 'pant-hoot' vocalization) and then presented with a 'scream face' or a 'hooting face': Through the use of this technique, Parr (2004) found that long-distance 'pant-hoot' calls are more readily matched in the auditory modality, and 'screams' by visual. The differential salience of the auditory and visual components of a multimodal signal provide valuable information about how the chimpanzees perceive these communicative signals.

7.2.3   Flexibility in combining signals into sequences

A key facet of human language is 'duality of patterning', which is the ability to combine meaningless phonemes into words, which in turn are ordered and structured according to syntactical rules into an infinite number of meaningful utterances (see de Boer, Sandler and Kirby, 2012). If we were constrained to only producing single signals, our communication system would be very restricted. Thus, the extent to which animals have the flexibility to move beyond single signal production to combine signals into sequences is highly indicative of the potential for complexity in the system. In this section we are not going to consider the simultaneous combination of signals into multimodal signals that were reviewed in Chapter 5, but concentrate on the production of multiple signals in consecutive sequences.

Gesture sequences are produced in a number of primate species and are commonly defined as multiple consecutive gestures produced by one individual towards the same goal, directed at the same recipient (Liebal, Pika and Tomasello, 2004c; Tomasello *et al.*, 1994). Some studies specify this further and define a time interval in which one gesture has to follow the previous one, e.g. 1 or 5 s (Hobaiter and Byrne, 2011b; Liebal *et al.*, 2004a). Captive chimpanzees produce approximately a third of their gestures as part of a sequence (Liebal *et al.*, 2004a). Sequences ranged in length from 2 to 39 gestures, however, almost two-thirds of sequences consisted of two gestures. Repetitions of the same gesture were common and accounted for almost 40% of observed sequences. There was no evidence that gesture sequences were more likely to occur than single gestures in any one context. The authors suggest that sequences occurred most often after the failure of the first gesture to elicit a response in the receiver and that gesture sequences in chimpanzees are ad hoc responses to unresponsive recipients rather than planned sequences aimed to increase signal efficacy. A very different view of gesture sequences is given by a recent study on gorilla gesture sequences (Genty and Byrne, 2010). Although gesture sequences in gorillas occurred at roughly similar rates (31%) and the majority of sequences represented combinations of two gestures, there was no evidence that gorillas produced sequences as a response to an unresponsive receiver. Indeed, gesture sequences of gorillas were not more successful than single gestures in eliciting responses from receivers. As in chimpanzees, there was no evidence that gesture sequences conveyed different meanings from their constituent components and the authors highlight that this should discourage any interpretation of gorilla gesture sequences as syntactical. In terms of function, Genty and Byrne (2010) suggest that sequences of gestures that were most commonly observed in the play context, allow for subtle, ongoing adjustments to be made to social interactions. This suggestion is supported by Tanner's (2004) largely qualitative analysis of dyadic gesture exchanges and sequences in two captive gorillas. However, a study on gesture sequences in orangutans found that they do not take the recipient's behaviour into account and continue to gesture regardless of whether there is a response or not, suggesting that these sequences may be simply the result of high arousal in some specific contexts, such as play (Tempelmann and Liebal, 2012).

The apparent discrepancy in empirical support for great ape gestures being a response to failed communication may have been resolved by a recent study on wild chimpanzee gesture sequences. Hobaiter and Byrne (2011b) distinguished between gesture *bouts* consisting of multiple gestures and response-waiting and gesture *sequences* of rapid-fire gesturing without pauses. They found that gesture bouts were commonly produced in response to a recipient's failure to produce an appropriate response and largely consisted of repetitions of the same gesture. In contrast, the rapid-fire gesture sequences often consisted of different gestures and were mainly produced by young individuals. Hobaiter and Byrne (2011b) argue that through development, chimpanzees shift from relying on initially long and largely redundant sequences of rapid-fire variable gestures to selecting more effective single iterative gestures as adults.

Sequences or combinations of signals are not always unimodal. Multimodal combinations consisting of gestures and vocal or facial signals made up 8% of the signals recorded for bonobos and 22% of the signals recorded for chimpanzees (Pollick and de Waal, 2007). Interestingly, only in the bonobos, where these multimodal signals were relatively rare, were receivers significantly more likely to respond to these multimodal signals as opposed to single modality signals. In chimpanzees, response rates to multimodal combinations and single signals were very similar. This seeming lack of efficacy for the more complex signals mirrors the data from gorillas indicating that gesture sequences were no more likely to elicit a response than single signals (Genty and Byrne, 2010) and wild chimpanzees where gesture sequences were less successful in eliciting appropriate responses than single gestures (Hobaiter and Byrne, 2011b). More detailed analysis of the types of combinations made by bonobos that are successful in increasing the efficacy of their signals is needed.

Gestures and facial expressions are sometimes also combined in simultaneous combinations, as has been shown for siamangs (Liebal, Pika and Tomasello, 2004c). However, as with the chimpanzees, these multimodal combinations were not more

Figure 7.2 A blended facial display. The 'stretch-pout-whimper'. The two parent types are at the far left ('pout') and far right ('bared-teeth' display). The 'stretch-pout-whimper' is the fourth from the left. The second and third images reflect grading between the 'pout' and the 'stretch-pout-whimper' (from Parr, Cohen and de Waal, 2005).

effective in eliciting a response from the receiver than single modality signals. While facial expressions are often reported as one of several parts in multimodal sequences, they are not commonly investigated as unimodal sequences containing facial expressions only. Most research focuses on the occurrence of prototypical facial displays but not on the more subtle transitions between expressions. Parr *et al.* (2005a) describe the occurrence of several 'blended displays' in chimpanzees, which share characteristics with two or more prototypical expressions (Figure 7.2). These blended facial expressions are not only evident as one expression changes to another in a sequence, but also occur independently as single signals. For instance, a 'stretch-pout-whimper' shares characteristics with both the 'pout face' and 'bared-teeth display' and occurs in the transition between these two expressions as part of a sequence but also alone. It is evident from this research that sequences of facial expressions do occur, but the contexts or the efficacy of sequences compared to single facial signals in eliciting responses from receivers have not been investigated.

Considerably more research effort has concentrated on combinations or sequences of primate vocalizations. This may in part be because the onset and offset of vocalizations is more clearly distinguishable in contrast to delimiting and defining the boundaries of facial expressions and gestures. Primates always have some kind of facial expression when the face is visible and so pinpointing the start of a new facial expression is difficult. Similarly, some gestures may not have clearly defined boundaries. Therefore, sequences of vocal signals should be easier to detect, both for conspecific receivers and human researchers.

Perhaps one of the most striking examples of sequences of calls comes from the primates who produce elaborate choruses, duets or songs. Gibbons for instance produce songs, often as duets with their partner, which seem to perform a territorial function. Males often produce a number of different songs, each composed of varying combinations of notes (Mitani, 1988). Male agile gibbon songs have been analysed in some detail and 13 different 'notes' have been identified. These notes or call elements are not just randomly sequenced within a song. Instead, there are regularities in the order of individual elements in songs. Playback experiments have shown that violating these regularities by broadcasting a song with rearranged elements elicit some subtly different responses from listening gibbons compared to normal songs, indicating that gibbons are sensitive to these regularities (Mitani and Marler, 1989).

More recent work with white-handed gibbons has shown that songs produced in response to predators differ probabilistically in their composition (e.g. certain notes are more likely or more numerous in one song type compared to the other) compared to territorial duet songs (Clarke, Reichard and Zuberbühler, 2006). Anecdotal evidence indicates that group members can distinguish these predator and territorial songs, but more systematic data is required to test this thoroughly. Certain sequences of calls can also be given to different kinds of predatory threat. For example, predator model experiments have revealed that black-and-white colobus monkeys produce specific call combinations in response to specific predators (Schel, Tranquilli and Zuberbühler, 2009).

Although much more limited in scope than the gibbons, there is also evidence that great apes produce simple sequences of calls. Bonobos produce sequences of up to five different food-associated call types in response to different value foods. In a similar way to the gibbons, the composition of these sequences varied systematically with the context. The occurrence and frequency of each of these five call types within a calling sequence was modulated by the value of the food present (Clay and Zuberbühler, 2009); however, the temporal order of these different call types within the sequence was not investigated.

One study has revealed that call combinations are a common aspect of chimpanzee vocal communication, with just under half the vocalizations produced by the chimpanzees of Tai Forest being produced in combination with other call types or drumming to form a sequence (Crockford and Boesch, 2005). The combination of 'bark calls' and drumming has already been shown to be important in increasing the context specificity of some 'bark calls' in this community (Crockford and Boesch, 2003). Most of the recorded call combinations were reliably produced in a certain order, mirroring similar findings in cotton-top tamarins ('chirps' always precede 'whistles'; Cleveland and Snowdon, 1982) and titi monkeys ('chirrups' always precede 'pants'; Robinson, 1979). Some call combinations, however, demonstrated more flexibility in the ordering of the component calls. Whether listeners are sensitive to call order and whether this is important in changing the message conveyed to listeners is yet to be tested.

The evidence for call sequences so far has focused on the structure, order or composition of sequences and their use across contexts. The regularity in the order of notes within a gibbon song, or the rules determining call order in simple combinations in chimpanzees show that primates are capable of sequencing calls according to basic rules. This is sometimes referred to as phonological syntax (or just phonology) (Marler, 1977), where the individual component sounds do not convey meaning. None of these studies has, however, provided any evidence that vocal sequences mirror any of the key properties of **lexical syntax** (Marler 1977). Lexical syntax refers to a sequence of meaningful signals that are grouped according to certain rules to form a meaningful overall structure (Marler 1977). Robinson (1984) tried to elaborate on and operationalize this definition, indicating that lexical syntax could be represented by one call acting as a contextual modifier for another, or sequences where the meaning of the combination is the sum of the constituent parts. Equally important is the search for semantically combinatorial syntax in nonhuman species, which is characterized by the combination of two or more signals that mean something when given in isolation, but come together to mean something different when produced as a combination (Hurford, 2011).

The first study to claim that calls could modify the meaning of another call given in a sequence focused on the alarm-calling system of forest-dwelling Campbell's monkeys. Zuberbühler (2002) found that male Campbell's monkeys emitted loud 'boom' calls in addition to acoustically distinct 'eagle' and 'leopard alarm calls'. In a playback experiment, Zuberbühler created artificial sequences designed to test the effect of the 'boom' call on a subsequent alarm call. The test animals were Diana monkeys who commonly form sympatric groups with Campbell's monkeys and have been shown to understand their alarm-calling system (Zuberbühler, 2000a). In response to alarm calls alone, Diana monkeys responded with their own alarm call appropriate to the predator type. In response to the 'boom + alarm call', Diana monkeys gave significantly fewer alarm calls. Zuberbühler concluded that the 'boom' call acted as a modifier to the subsequent alarm call and signalled that the threat was not immediate. One complication with this study is that Campbell's monkeys are not reported to naturally produce this sequence with any regularity, so the inhibition of the Diana monkey response could have been driven by the perception of a novel combination they did not know how to interpret.

More recent systematic research on Campbell's monkeys' call production has unveiled further complexity and flexibility in this vocal system (Ouattara, Lemasson and Zuberbühler, 2009a; Candiotti, Zuberbühler and Lemasson, 2012b). The inclusion of 'boom' calls into sequences with specific alarm calls (Zuberbühler 2002) is not the only way in which the meaning of calls can be modified. Males can transform a loud 'hok' call that is given specifically to eagles into a 'hok-oo' call that acts as a more general alarm call, indicating arboreal disturbance. Similarly, a loud 'krak' call that is given in response to leopards can be transformed into a more generic alarm call with the addition of the 'oo' element to form a 'krak-oo' call. The flexible modification of the meaning of these calls by the addition of this 'oo' element is argued to be functionally similar to **suffixation** in human language (Ouattara, Lemasson and Zuberbühler, 2009a). In addition, the potential for this species to produce more varied and complex call combinations has been highlighted by Ouattara and colleagues (2009b). They observed that male Campbell's monkeys produce six acoustically distinct loud calls. Sequences of these calls are produced in a highly context-specific manner and can be combined to potentially convey new messages. For instance, a 'boom' sequence is given to elicit movement of the group towards the male, whilst a 'krakoo' sequence is given as a general alarm. The concatenation of these calls into a 'boom + krakoo' sequence is produced in response to falling trees or branches. The use of different predator-specific sequences also allows for the level of urgency to be labelled (different sequence produced if the eagle has been sighted or merely heard) in addition to predator type. If playback experiments reveal that these call sequences are meaningful to listeners, this calling system will represent an excellent example of how combining existing calls in a repertoire can increase the complexity of information that can be communicated.

One of the most compelling pieces of evidence for call combinations that alter the meaning of the message provided to listeners comes from the work of Kate Arnold with West African putty-nosed monkeys. Male putty-nosed monkeys produce two loud calls ('hacks' and 'pyows') that can act as alarm calls. Unlike the functionally referential alarm calls documented for other West African guenons, such as Diana monkeys (Zuberbühler, 2000b) and Campbell's monkeys (Zuberbühler, 2001), putty-nosed monkeys do not reliably label eagles and leopards with a particular alarm call (Arnold and Zuberbühler, 2006). They are more likely to produce 'hack' sequences in response to eagles and 'pyow' sequences towards leopards, but these call sequences are also given to a number of other stimuli. Whilst the meaning of the 'pyow' sequences and to some extent 'hack' sequences does not seem to be very specific, males also combine these calls into a 'pyow-hack' sequence, which seems to convey a distinct message for listeners (Arnold, Pohlner and Zuberbühler, 2011). This sequence is used to initiate group travel and carefully controlled playback experiments have shown that broadcasting this combination reliably elicits travel responses in listeners, which is not a reaction either constituent call elicits on its own (Arnold and Zuberbühler, 2006). Further experiments demonstrated that the 'let's go message' conveyed by the 'pyow-hack' sequence was contingent on the sequence of the calls, not the acoustic structure of the calls themselves (Arnold and Zuberbühler, 2008). Whilst this combination is not likely to be compositional (e.g. the meaning of individual calls contributes to the overall meaning of the combination), this study illustrates how combining calls can enable a greater flexibility in terms of the functional usage of calls: by combining calls, the number of messages that can be conveyed with a closed repertoire of calls increases, providing the potential for a more complex communication system (Arnold and Zuberbühler, 2012).

7.3   What could a multimodal approach bring to this topic?

Whilst some studies reviewed in this chapter have notably examined signals from different modalities together, in this section we highlight further advantages that may arise from the wider application of a multimodal approach to this topic. In particular we think that our understanding of flexible use of signals across contexts and combination of signals into sequences would benefit from a multimodal approach.

In order to study a specific modality, researchers often, perhaps inadvertently, degrade multimodal signals to focus on one unimodal component in isolation. When this happens we are not studying the composite signal as it is perceived by the receiver and thus we may be drawing incorrect conclusions about its function and in this case its flexibility. A signal that is identified as flexible when studied at a unimodal level may be much more context specific, if other signals that it is naturally combined with are taken into consideration. For example, although the macaque 'bared-teeth' facial expression is given over a wide number of contexts, indicating flexible usage, these composite multimodal signals can be highly context specific if the vocalizations and postures it is often combined with are considered. Therefore, at the level of the composite signal, there can be a one-to-one correspondence between a context and a multimodal signal, even if the unimodal components appear in more diverse contexts. Thus, it is likely that a multimodal approach may reveal more context specificity in signals that have been characterized as flexibly produced by unimodal research.

In terms of flexible combination of signals into sequences, researchers applying a unimodal approach may fail to identify salient signal sequences by only considering signals from a single modality. For example, if a male chimpanzee produces a 'soft hoo' call to encourage an oestrus female to follow him and then produces a 'branch shaking' gesture, a signal sequence has occurred. Yet both the unimodal vocal researcher and the unimodal gesture researcher would miss this sequence. A multimodal approach is thus likely to allow us to identify more signal sequences that a unimodal approach would overlook.

Summary

This chapter has shown that primate gestures, facial expressions and vocalizations are used, perceived and combined with varying degrees of flexibility. It is clear that there are aspects of flexibility in all three modalities, indicating the potential for complexity: flexible communication systems that lack a one-to-one correspondence between single stimuli and responses are more likely to require complex cognitive skills. Different kinds of flexibility were considered, including flexibility in the perception as well as production of signals with special consideration of signals sequences. Unfortunately, comparably little research is dedicated to the flexible use of facial expressions, partly because they are commonly assumed to lack the potential for substantial flexibility and instead are driven by specific emotional states. While gestures are often referred to as highly flexible in regard to their usage, many vocalizations are characterized as highly context specific. This chapter, however, argued that these findings are at least partly the result of the theoretical and methodological approaches that differ fundamentally between modalities. Furthermore, while there is substantial evidence for flexibility in the receiver of vocal signals, this topic has so far been largely neglected by gestural researchers, perhaps due to methodological difficulties in examining receiver responses experimentally. Finally, while there is little evidence that primate gesture sequences represent meaningful combinations, vocal sequences can convey different messages depending on the type and order of calls combined. This demonstrates that combining elements of a closed repertoire can increase the number of messages that can be communicated, thus resulting in a more flexible, more complex communicative system. We highlight that there is a risk that unimodal investigations into signal flexibility may result in inaccurate characterizations of signals and argue that a more accurate understanding of signal flexibility could be obtained through the use of a multimodal approach.

8   Intentionality

Intentionality is a key feature of human language. Much attention has therefore been dedicated to the communication of nonhuman primates to assess whether they also use their signals intentionally. The purpose of this chapter is to introduce the criteria scientists have used to identify intentional communication in primates and review the extent to which these criteria have been met for each of the different modalities. We then evaluate the validity of these criteria and discuss how different studies define and identify intentional communication inconsistently. Based on this, we propose a tentative set of criteria that may be most appropriate for identifying intentional communication across modalities and for multimodal signals.

8.1   What Is Intentional communication?

The question of what intentional behaviour actually is, and the extent to which the concept is relevant to communication, is hotly debated, and could easily constitute an entire book on its own. In philosophy, intentionality is often seen as the crucial characteristic that differentiates mental from non-mental states (Brentano, 1874 [1973]). Unlike non-mental states, mental states – including communicative acts – are 'about' certain objects and/or states of affairs in the world (Benga, 2005). Because of this, intentionality is often simply described as 'aboutness' (Dennett, 1983, 1987). For a mental state to be about the world is for it to have an intentional content, where this content is a way of representing the object or state to which the mental state is directed. A large philosophical literature has been devoted to considering what it is to have content, and to the varieties of content that minds might have (e.g. Cussins, 1992; Dennett, 1987; McDowell, 1994).

Often, intentionality is used to refer specifically to states of mind that are about beliefs, desires and goals. A concept of intentionality then becomes explicitly connected to a concept of *Theory of Mind* (**ToM**), which is a term used to suggest that an individual has a belief (theory) about the mind of another. While the first level or order in a hierarchy of intentionality refers to an individual's own goal-directed intentions without the consideration of the other's mental states, ToM involves the understanding of the other's intentions and thus at least second order intentionality (Dennett, 1983).

When linking intentionality to communication, some strict definitions emphasize that 'true' communication takes place only if the signaller and the recipient take into account each other's state of mind and intentions (Grice, 1957). If this criterion is applied to primate communication, however, it would be impossible to consider any animal signal communicative without evidence that the species is capable of second order intentionality, and understanding other minds (Seyfarth and Cheney, 1993). Thus, under this strict definition only humans and potentially chimpanzees could be considered to be 'truly' communicating. As discussed above, there are, however, more basic levels of intentionality and the extent to which communication in primates is influenced by any level of intentionality is both an interesting topic in itself and vital for our understanding of how 'true' communication (Grice 1957) may have arisen in humans.

In primate communication research, therefore, the focus is mostly on first order intentionality. The term *intentional communication* is used to describe purposeful, goal-directed behaviour, and in an even narrower sense, controlled voluntary actions (Benga, 2005). This means that the sender is acting in an intentional, goal-directed manner by means of voluntary control over the production of a particular signal. It does not automatically imply that the individual receiving the signal *understands* that this signal is an intentional act of communication (Genty *et al.*, 2009). Therefore, it is important to consider that the production and comprehension of signals can often involve different processes, and should be treated as different issues (Moore, in press). The focus of this chapter, however, is primarily on the *production* of signals that are used intentionally in the sense of goal-directed, voluntary and purposeful behaviours.

8.2   Why is the intentional nature of communicative acts of interest?

From a biological perspective, communicative signals are behavioural patterns or physical characteristics that have evolved under very specific selection pressures to influence the behaviour of other organisms of their own or other species (see Chapter 1, section 1.1.1 for understanding communication at different levels). As a result, all individuals of one species share those communicative behaviours that are often used for specific functions, often in very specific contexts. As opposed to biologists, however, psychologists focus more on the proximate explanations for behaviour and thus the mechanisms underlying primate communication. Psychologists in particular, therefore, are interested in distinguishing between mechanisms driving signal production and in identifying instances where cognitive processes, rather than physiological states, might be driving signal production. This distinction between intentional signals as opposed to affective, reflexive signals that are not under voluntary control has been a driving force for the interest in this topic for many years, since from a cognitive perspective the intentional production of signals implies voluntary control and the associated potential for more flexible and complex use of a signal. Although interest in intentional signals seems warranted, as **Box 8.1** points out, the division between intentional and emotional signals is not a particularly helpful one.

The intentional production of signals has also received much interest due to its prominence as a defining feature of human communication. Researchers interested in

Box 8.1 Intentional versus emotional signals: a useful distinction?

Characterizing signals as intentional implies that the production of such signals is under the voluntary control of the signalling individual, and suggests that cognitive processes are involved in signal production. Such behaviours are often contrasted with behaviours that express internal affective states, which are assumed to be involuntarily produced behaviours: the result of physiological processes that are immune to cognitive influence. This classical dichotomy between intentional and emotional signals is reflected in the divide between intentional gestures and involuntary, emotionally driven vocal and facial signals (e.g. Tomasello, 2008). We argue that this implicit (and sometimes explicit) dichotomy and its application to primate communication is unhelpful for four main reasons:

1. The dichotomy is false, since signals could be both under voluntary control and still be associated with emotional correlates. Virtually all intentional human communication can have a considerable emotional component, which affects the structure of our signals (e.g. prosody of speech, or speed and intensity of speech-accompanying gestures). Voluntary cognitive control and emotion are not mutually exclusive influences on signal production, but might represent two potential underlying mechanisms that both affect signal production.
2. By pitting cognitive and emotional influences on signal production against each other, rather than viewing them as complementary processes, we are likely clouding our understanding of signal function. As we have argued throughout the book, deconstructing the individual components of multicomponent and multimodal communication can often fail to capture the full complexity. Likewise, by ignoring the fact that cognitive and emotional processes might be involved simultaneously we could miss important evolutionary functions, such as any potential social bonding afforded by the emotional aspect of the exchange.
3. 'Emotionally driven' is not a useful explanatory term for signal production and is often used in the absence of any relevant physiological data showing how emotion may be mediating signal production. Indeed, when and how to identify emotional processes in nonhuman animal signals is a matter of fierce debate (e.g. Waller and Micheletta, 2013).
4. The application of this dichotomy to primate signals of different modalities is based on an incomplete and biased data set. This chapter discusses this in detail, but there is now evidence available that indicates, contrary to the typical categorization, that vocalizations and facial expressions may be used intentionally. Equally, more research is needed into the emotional aspects of gesture production. For example, the intensity of a gesture might be modulated by the arousal level of the producer.

language evolution, and particularly the proponents of a gestural origin, have therefore often asserted that language must have originated from an intentional communicative system, since acquiring control over the production of signals is an essential prerequisite for adding new components to an existing communicative repertoire (Corballis, 2003; Rizzolatti and Arbib, 1998; Tomasello, 2008). They have thus focused on examining the phylogenetic origins of this aspect of human communication, by searching for evidence of intentional communication in our primate cousins. Seminal research into intentional non-linguistic communication was first conducted on prelinguistic children, focusing primarily on their gesture use (Bates *et al.*, 1979; Bates, Camaioni and Volterra, 1975). This important body of work laid the foundations for intentional communication to be examined in our closest living relatives and the methods and criteria established in the infant work have been adapted to primate research. Intentionality in the production of signals in primates has been highly unimodal to date and has focused heavily around the study of gestural communication of great apes, particularly chimpanzees, in both wild and captive settings (Plooij, 1978, 1979; Tomasello *et al.*, 1985, 1989, 1997). The focus on gesture in this body of research may be because of the original work on children also centred on their gestural communication, which means that some of the criteria for identifying intentional communication are only applicable to signals produced in the visual domain. This chapter will explore the extent to which these findings are influenced by the nature of the criteria used to identify intentional communication and critically evaluate the different criteria used.

8.3   How can intentional communication be identified?

Taking the human's perspective observing an interaction between two nonhuman primates, it is rather difficult to reliably identify features that characterize a given signal as the product of intentional, voluntary behaviour. To give an example, imagine observing an interaction between two chimpanzees. Chimpanzee A spins around and around on the spot. Suddenly another chimpanzee B approaches, and the two start to play. From the observer's perspective, the difficult issue here is to decide whether A was performing the spins for reasons unrelated to play-initiation (e.g. solitary play) and B was merely reacting to the behaviour of A or whether A intended to use her spins to initiate a playful interaction with B. Thus, from a cognitive perspective, it is not trivial to differentiate the production of intentional, goal-directed signals and unintentional signals, as both can be highly effective in eliciting responses from receivers and it is not possible to behave without communicating something to nearby individuals (Watzlawick, Bavelas and Jackson, 1967). However, only if A was performing the spins in an intended and directional way to initiate play, would this signal count as an intentional act of communication (MacKay, 1972).

Therefore, with the aim of identifying acts of intentional communication, a variety of criteria have been used by different researchers (e.g. Gómez, 1990; Tomasello *et al.*, 1985). Leavens and colleagues (Leavens, 2004; Leavens, Russell and Hopkins, 2005) offer a set of criteria to identify intentional communication that are directly derived from

> **Box 8.2** Criteria used to identify intentional acts of communication in nonhuman primates, based on Leavens (2004) and Leavens, Russell and Hopkins (2005)
>
> 1. Social use: sender's sensitivity to the presence (or absence) of other individuals, since intentional communication requires an audience
> 2. Visual-orienting behaviour or gaze alternation: sender visually monitors the recipient and looks back and forth between the social partner and a distant object or event
> 3. Influence of attentional state: sender adjusts its signals depending on whether the recipient is visually attending or not
> 4. Attention-getting behaviours: sender uses particular signals to attract the recipient's attention if the recipient is not visually attending
> 5. Persistence: sender repeats signals in case of failure of its initial communicative attempts
> 6. Elaboration: sender uses multiple different signals in case of failure of initial communicative attempts

those suggested by Bates and colleagues (Bates, Camioni and Volterra, 1975; Bates *et al.*, 1979) in their original work on prelinguistic communication in children. The aim of the current chapter is to discuss and integrate the evidence from the different modalities with special reference to the criteria suggested by Leavens and colleagues (see **Box 8.2**).

8.3.1   Social use

8.3.1.1   Presence of an audience

By definition, the successful transfer of information, be it in the vocal/auditory, facial/visual or tactile domain, requires an audience of at least one or more individuals. For some signals, however, the audience may not always be present at the same time the signal is produced. For example, olfactory signals can persist in the environment and thus be perceived even once the signalling individual has moved on. In contrast, for more momentary signals such as gestures, facial expressions and vocalizations, the potential recipient needs to be present as the signal is produced in order for the behaviour to be communicative. However, the distance between signaller and receivers can vary depending on the modality and the habitat (see Chapter 1, Table 1.1). The sensitivity to the presence of this audience, resulting in a signal only being used when another is present and able to perceive the signal, is called an *audience effect* (Rogers and Kaplan, 2000).

Several studies indicate that primates' vocalizing behaviour is sensitive to the presence of an audience in several different contexts, such as predator detection and feeding. For example, when encountering a predator, vervet monkeys rarely produce alarm calls when alone (Cheney and Seyfarth, 1990). Similarly, when discovering a tiger model, wild male Thomas langurs only produced alarm calls when they were in a group, but not when they were solitary (Wich and Sterck, 2003). In interactions with humans, there is also ample evidence that chimpanzees, gorillas and orangutans only vocalize if the human they are trying to beg food from is present (Hostetter, Cantero and Hopkins, 2001; Poss *et al.*, 2006). In contrast to these studies showing that vocal signal production is sensitive to the presence of an audience, another study found that the rate of food call production in captive tamarins was not mediated by visual contact with their partner (Roush and Snowdon, 2000). Whilst at first glance this finding may indicate that vocal signals do not always meet the criteria of social use, this study highlights the importance of defining the audience appropriately in terms of the signal modality being produced. In this case a partner was always present and in auditory contact and it was just the visual access that was manipulated. Auditory signals do not require visual contact to be accurately perceived, so as long as group members are in auditory contact, an appropriate audience is present.

For the use of facial expressions, very little is known about whether nonhuman primates consider the presence of an audience. Humans produce at least some of their facial expressions when they are alone, and such solitary faces are sometimes considered pure expressions of emotions, since they are not constrained by any social demands (Ekman, 1984; Ekman, Davidson and Friesen, 1990). Others suggest a more cognitive explanation for such solitary facial expressions in humans, since they might reflect the imagination of previous interactions (Fridlund, 1991). From the little that is known about nonhuman primates, it is difficult to draw any consistent conclusions. One study presents observations of a female gorilla repeatedly hiding her 'playface' by covering it with her hand (Tanner and Byrne, 1993). These observations seem to indicate that she had no control over the production of the facial expression itself, but used the hand instead to cover her face. To our knowledge, however, there are no studies that specifically address the question whether the production of certain facial expressions is mediated by the presence or absence of an audience in monkeys or apes. Thus, arguments and assumptions that are commonly made asserting that nonhuman primates lack the capability to control their facial expressions are premature, given the paucity of evidence available.

Gestures, like facial expressions, have yet to be systematically investigated for social use in terms of how the presence of a conspecific audience mediates signal production. A possible explanation for the lack of studies in this regard could be that gestures are less well defined in terms of their structural properties, which makes it more difficult to distinguish them from general body movements in the absence of a social interaction. In fact, many researchers use the presence of a recipient as an a priori criterion for identifying gestures, raising the possibility that to date researchers may have overlooked gestures that are produced in the absence of audiences. Substantive evidence for social use of food-begging gestures is available from studies conducted in experimental settings where great apes or monkeys are interacting with a human experimenter. There is now good evidence that selective production of gestures in the presence, not the absence, of a human who has access to food occurs not only in great apes (Call and Tomasello, 1994; Hostetter, Cantero and Hopkins, 2001;

Leavens, Hopkins and Bard, 1996; Poss *et al.*, 2006), but also several monkey species such as squirrel monkeys, capuchin monkeys and rhesus macaques (Anderson, Kuwahata and Fujita, 2007; Anderson *et al.*, 2010; Blaschke and Ettlinger, 1987; Hattori, Kano and Tomonaga, 2010; Mitchell and Anderson, 1997), although monkeys were less likely to consider whether the human was actually looking at them or not (see section 8.3.2).

In summary, there is good evidence that primate vocal production is mediated by the presence of conspecific listeners, whilst there is a lack of corresponding studies that investigate the effect of an audience for gestures and facial expressions. Experimental studies where primates beg for food from humans demonstrate that gestures are also only produced in the presence of an appropriate audience.

One could argue that the presence of an audience triggers communicative behaviours through low-level mechanisms such as increased arousal as a result of social facilitation. Therefore the fact that signals are only used in the presence of an audience may not tell very much about the degree of control over the production of a vocalization, facial expression or gesture. Therefore, a more conservative test of control over production of signals might be to investigate whether the composition of the audience has an impact on the sender's behaviour. If nonhuman primates adjust their communication to the composition of the receiving audience, this could indicate that signal production is not merely based on social facilitation, but may involve more complex cognitive processes and a higher degree of flexible use representing better evidence for intentional acts of communication.

8.3.1.2 Composition of audience

Moving beyond signal production being mediated by the mere presence or absence of an audience, we now examine the extent to which signal production is affected by *who* is present as a potential receiver. Thus, depending on the composition of the audience, the sender could potentially either vary the rate or structure of signals, or could even choose not to produce a signal in the presence of particular individuals, indicating that the sender has control over the production of these signals. The majority of these features are studied in regard to the vocal modality, and to a much lesser extent in facial expressions or gestures.

In regard to adjustment of the rate of vocalizations, female vervet monkeys produce alarm calls more often when they are with their offspring than when they are with unrelated juveniles (Cheney and Seyfarth, 1985), suggesting that monkeys vary the rate of alarm calling depending on who is present. Similarly, chimpanzees who had seen a snake model were more likely to increase their alarm-calling behaviour in response to the arrival of an individual if that individual was an important social partner (i.e. 'friend'), irrespective of their rank (Schel, Townsend *et al.*, in press). Rhesus macaques vary the rate of food-associated calls in a way that females call more in the presence of kin than if unrelated individuals are present (Hauser and Marler, 1993), and chimpanzee males are more likely to produce food-associated calls if an important social partner is nearby (Fedurek and Slocombe, 2013; Slocombe *et al.*, 2010b). Experimental work has confirmed that male chimpanzees seem to direct their food-associated calls at specific individuals and when lone males were feeding silently, they were significantly more likely to start producing calls in response to playbacks that simulated the arrival of individuals who were both higher-ranking and important social partners (Schel, Machanda *et al.*, in press). In the context of sexual behaviour, female chimpanzees varied the production of their copulation calls by calling more while they were mating with high-ranking males, and by suppressing their calls if high-ranking females were nearby (Townsend, Deschner and Zuberbühler, 2008; Townsend and Zuberbühler, 2009). Male chimpanzees, 'pant-hoot' production is mediated by the presence of alliance partners, but not oestrus females, suggesting that they call to particular members of their respective group (Mitani and Nishida, 1993).

There is also evidence that primates do not only adjust their calling rate, but also the structure of their vocalizations in regard to the composition of the audience. For example, in aggressive within-group encounters in chimpanzees, victims alter the temporal and spectral structure of their 'screams' as a function of the severity of the attack (Slocombe and Zuberbühler, 2007). This may not seem very surprising at the first glance, since the more serious an attack is, the more the victim might scream, with the call structure simply reflecting the caller's level of arousal. However, in specific circumstances, the victim also produced 'screams' that exaggerated the severity of the attack. Such exaggeration only occurred if they were facing severe aggression and when there was at least one individual nearby who matched or surpassed the aggressor in rank. In other words, the acoustic structure of their calls changed in cases where there was a third party present who could effectively help them, but only when they most needed the support.

Overall there is good evidence that call production is mediated by subtle social factors, since both the rate and structure of calls can be modified depending on the composition of the audience, indicating a potentially sophisticated degree of control. However, it could be argued that the presence of different individuals causes different levels of arousal in the caller. Thus, the extent to which these audience effects are indicative of cognitive control or arousal mechanisms is still difficult to determine.

In contrast to vocalizations, whether primates consider the composition of the audience when using visual signals such as gestures or facial expressions has not yet been studied in as much detail. For vocal signals, researchers define the audience as all individuals in auditory contact of the caller by estimating the transmission distance of each call and recording the presence of individuals within this range. Determining the audience for visual signals is more difficult, as it depends on the visual attention of the potential recipients. Thus, research in the visual domain has so far focused on whether certain characteristics of a dyadic partner (e.g. age class, sex, rank) mediate signal production. There are some studies that show that particular facial expressions or gestures are directed to specific individuals but not to others. For example, female stump-tail macaques direct the 'bared-teeth face' almost exclusively to higher-ranking individuals (Maestripieri, 1996b). However, this example does not necessarily reflect sensitivity to the composition of the audience, since the social function of this facial expression is to communicate submissive behaviour and thus it is *always* used by lower-ranking towards higher-ranking individuals. There is also research indicating that the use of facial expressions can be more flexibly mediated by whom the signaller is interacting with. For example, macaque mothers have also been observed to produce exaggerated forms of common facial signals, such as 'lipsmacking', selectively towards their young infants in the first 3 weeks of life (Ferrari *et al.*, 2009).

Evidence of such selective use of gestures during interactions with certain conspecific individuals is currently lacking, however, interesting findings are available from experimental human–chimpanzee interactions. During such interactions, great apes selectively gesture towards an experimenter they have learnt is cooperative rather than competitive. For example, chimpanzees point towards a cooperative human (who will share food with them) to indicate where food is hidden, but will withhold any pointing gestures or even point to the wrong location when interacting with a competitive human (who will use the ape's pointing gesture to find the hidden food and then take it for himself) (Woodruff and Premack, 1979).

To summarize, while the sensitivity to the presence of an audience could be mediated by low-level processes such as social facilitation, the flexible adjustment of the communicative behaviour according to the composition of an audience may represent stronger evidence that the corresponding signals are voluntarily produced and thus intentionally used. Audience effects have been mostly studied in regard to vocalizations and the majority of studies show that monkeys and great apes only call if an audience is present and that they can vary their calling rate and the structure of their vocalizations as a function of the composition of the audience. Recent findings indicate that the production of certain facial expressions is also mediated by the identity of a dyadic partner, although empirical tests of the more basic presence/absence audience effect are still needed. In contrast, studies that address the mediation of gesture production as a function of conspecific dyadic partner identity or wider audience composition are lacking. Audience effects for gestures have, instead, been studied in interactions with human experimenters and these studies have clearly shown sensitivity to the presence and absence of a human with regard to gesture production. All modalities therefore have some evidence indicating that they meet the criteria of social use. Whether meeting a single criterion is sufficient to warrant the use of that signal as 'intentional' is a matter that will be debated further in section 8.5.

8.3.2 Visual-orienting behaviour and gaze alternation

Gazing at others can signal the intention to communicate or – in the case of gaze aversion – not to communicate (Emery, 2000; Gómez, 1996) Thus, eye gaze is a communicative signal by itself, but more importantly for the purpose of this chapter, it can also accompany other communicative behaviours. The initiation and maintenance of eye contact is particularly used in research into gestures to assess whether a certain behaviour is directed at another individual. 'Audience checking' in the form of looking at the recipient before the production of a signal has been observed in wild chimpanzee gesture production (Hobaiter and Byrne, 2011a). Most recently, a similar behaviour has also been documented in the vocal domain, with wild chimpanzees visually monitoring their audience before they produce 'waa bark' or ' alarm huu' call bouts in the presence of a python model (Schel, Townsend *et al.*, in press).

Eye gaze and its relation to signalling has been studied in more detail within the context of great apes or monkeys interacting with humans, since eye gaze is easier to identify in such experimental settings. Unlike the studies with wild populations mentioned above that rely on head direction to infer gaze direction, eye contact and exact gaze direction can be identified with greater ease by human experimenters who are directly interacting with the primates. For example, chimpanzees look at the face of the human experimenter while pointing (Leavens, Hopkins and Bard, 1996) and a human-raised gorilla used eye contact as a prerequisite for a communicative interaction with a human (Gómez, 1996). More recently, a study that used a non-invasive eyetracking technique examined whether the way that great apes look at eyes and faces reflects species-specific strategies used in facial communication (Kano, Call and Tomonaga, 2012). It was found that overall, great apes showed similar patterns to humans when scanning the eyes of other apes and humans; however, humans differed from great apes in their prolonged looks into the eyes, regardless of the eye colour of the presented faces. For monkeys' interactions with humans, findings are more inconsistent since rhesus macaques and human-reared capuchin monkeys gaze at the human experimenter while pointing (Blaschke and Ettlinger, 1987; Mitchell and Anderson, 1997), while spider monkeys do not (O'Connell, 1994). At this point, it is not clear whether there are consistent differences between monkey species or whether differing rearing histories or testing procedures are more likely explanations for the different findings of these studies.

However, there are several problems related to reliably identifying eye gaze in communicative interactions. First, eye gaze is often very difficult to measure given the dark sclera of nonhuman primates (see Chapter 2, section 2.3.1), particularly when observing them in their natural habitats, which may be characterized by low light levels. Second, gazing at others might not always represent a reliable criterion to identify the intentionality of an interaction, since some species avoid directly staring at others (Kaplan and Rogers, 2002). Third, eye contact might be important for interactions of individuals in close proximity that involve gestures or facial expressions, but it might not be a useful criterion for some longer range vocalizations, since the target audience may be recipients who are out of sight of the signaller.

Signals accompanied by gaze alternation are used by some scientists to identify intentional acts of communication. Gaze alternation occurs when the sender repeatedly alternates his gaze between the recipient and another individual or a distant object or location (Leavens and Hopkins, 1998; Tomasello *et al.*, 1994). It is different from eye gaze alone, since the alternation of gaze may convey not only the signaller's intent to initiate a communicative interaction: by repeatedly looking at the recipient and a particular object, the signaller may also direct the attention of the recipient to this specific target. The relation between intentionality and signals accompanied by gaze alternation may not, however, be quite as straightforward as this. A more simple explanation for this behaviour is that the signaller has two competing goals of interest (the receiver and an external object/event) and the gaze alternation may simply represent oscillation between these two objects of interest.

It is mostly research into gestures that refers to gaze alternation as a criterion to identify intentional communication between conspecifics (Plooij, 1984; Tomasello *et al.*, 1994), with the exception of a few studies that concern vocal communication of chimpanzees in their interactions with humans (Leavens and Hopkins, 1998; Leavens, Russell and Hopkins, 2010). However, most of these studies do not actually *measure* how frequently gaze alternation co-occurs with the corresponding communicative behaviours, but use it as the a priori definition for a signal to be considered being intentionally used. One exception is the study by Leavens and Hopkins (1998), who found that – despite differences between gesture types – 80% of chimpanzees' gestures used to beg for food from a human were accompanied by gaze alternations between the human and the banana. Interestingly, this study also considered vocalizations and found that 75% of these vocalizations were also accompanied by gaze alternation. Further evidence that supports the occurrence of gaze alternation during vocal production comes from a recent study on wild chimpanzee alarm-calling behaviour. In this study, 75% of chimpanzees presented with a python model performed gaze alternation between the snake and a recipient, whilst producing alarm calls and gaze alternations were significantly more likely to be observed within calling bouts than outwith these communicative periods (Schel, Townsend *et al.*, in press).

In summary, we know that chimpanzees, at least, produce gestural and vocal signals accompanied by gaze alternation in food-begging contexts with humans. The question still remains, however, as to whether gaze alternation is a reliable marker of intentionality, or whether simpler explanations for this behaviour are more appropriate. In addition, whilst this particular criterion can be successfully applied across different modalities, it can only be usefully applied to referential or triadic communicative contexts. This severely limits its applicability to primate communication as a whole, as the majority of primate signals do not refer to a particular object or an event in the environment (see Chapter 9). Thus, if gaze alternation were considered to be an essential prerequisite to defining a certain communicative behaviour as intentional, the majority of primate signals could not be considered as intentional signals, simply because triadic interactions are relatively rare.

8.3.3   Influence of the recipient's attentional state

Whilst previous sections of this chapter reviewed evidence for the social and directed use of signals in terms of a receiver being present and who the receiver was, the next sections consider adjustment of signal use depending on receiver behaviour.

The term attentional state refers to the state of a recipient immediately before or while perceiving a communicative behaviour and researchers commonly discriminate between individuals who are visually attending and not attending. If signals are only produced when recipients can perceive them, this implies that the signaller has a goal-directed intention to communicate with the receiver. The same logic applies to the more basic 'social use' criterion, where the first step towards an intended receiver being able to receive the signal is their being present. Whereas low-level explanations for basic audience effects may bring into question the role of cognitive processes and intentionality, understanding the visual attention of your receiver is a much more sophisticated skill that requires cognitive processes such as visual perspective taking. Thus, this criterion has been argued to be a more resilient marker of intentional communication. However, even here, simpler processes such as learning the context for using a visual signal needs to include being able to see another individual's face or eyes, are still plausible alternatives. To perceive a facial expression or visual gesture, it is important that the recipient is visually attending, while vocalizations and gestures with tactile or auditory components can be perceived regardless of the recipient's attentional state. Thus, this criterion usefully applies only to visual signals; however, facial expression researchers have not used this criterion to date.

Several primate species including monkeys, gibbons and great apes have been shown to use their visual gestures in interactions with conspecifics *only* if their social partner is visually attending (for an overview, see Call and Tomasello, 2007). For interactions with humans, there is considerable evidence that several species of great apes and monkeys adjust their communicative behaviour to the attentional orientation of a human experimenter, since they gesture more if the human is oriented towards them (capuchin monkeys: Hattori, Kuroshima and Fujita, 2010; squirrel monkeys: Anderson *et al.*, 2010; mangabeys: Maille *et al.*, 2012; chimpanzees: Hostetter, Cantero and Hopkins, 2001; orangutans: Poss *et al.*, 2006). However, in experiments with chimpanzees that involved more complex situations with two human experimenters with differing attentional states and body orientations, they did not seem to show sensitivity to the attentional state of the human (Povinelli and Eddy, 1996; Theall and Povinelli, 1999). A possible explanation is that the orientation of the face and thus the attentional state of the human might communicate different information compared to the orientation of the body: while the face reveals whether the human can see the chimpanzee's begging gesture, the body orientation actually informs the ape whether the human is able to give any food at all (Kaminski, Call and Tomasello, 2004). This is supported by research showing that if the orientation of the face and the body is varied across conditions, but the human is potentially able to give food regardless of body orientation, then all great apes use the orientation of the face independent from the orientation of the body to decide when they should beg for food and thus use visual gestures (Tempelmann, Kaminski and Liebal, 2011).

All the above studies of nonhuman primates interacting with humans are based on the paradigm that the orientation of the face and body of a human experimenter changes across conditions, but the location of the monkey or great ape – sitting opposite to the human – remains stable throughout all conditions. However, when great apes are given the opportunity to choose their position in relation to the orientation of a human experimenter, they preferentially walk in front of the human to beg for food rather than using attention-getting gestures such as 'clapping' to attract the human's attention (Liebal *et al.*, 2004b). This touches the very important question whether nonhuman primates actually manipulate the attentional state of others and the next section will address this issue.

8.3.4   Attention-getting behaviours

Attention-getters are proposed to be signals that function to attract the attention of the recipient. They are referred to as strong indicators of an intention to communicate with the recipient, since the sender directs the recipient's attention to a subsequent action or even further communicative signals to make sure that these communicative attempts are successfully perceived. Humans have a large range of attention-getters that enable us to directly convey our intention to communicate, which we then follow with a range of signals once the receiver's attention is captured. Some researchers argue that such signals – that make an individual's communicative intentions explicit to its audience – were vital for the evolution of more complex communication (Scott-Phillips *et al.*, 2012).

In nonhuman primates, vocalizations as well as particular types of gestures (auditory and tactile) have the potential to serve as attention-getting behaviours, since they can be perceived regardless of whether the recipient is visually attending or not. Tomasello (2008) suggests that the main function of attention-getters is to get others to look at the self. Once the recipient's attention is captured, they can perceive further, potentially involuntary, behaviours of the sender. It is thus argued that attention-getters do not convey a specific meaning and are used across a wide variety of functional contexts, most often when the recipient is not attending (Tomasello, 2008; Tomasello *et al.*, 1994). For example, young chimpanzees use gestures such as 'throw stuff', 'poke at' or 'ground slap' to attract the attention of an individual that is not attending and then start play (Tomasello, Gust and Frost, 1989). Another example that is used as supporting evidence for attention-getting signals is the observation by Nishida (1980) reporting that male chimpanzees use a 'leaf-clipping' signal to attract the visual attention of the female to their erect penis, which then initiates copulation. These two examples demonstrate some potential problems in defining a behaviour as an attention-getter. First, to rule out the alternative explanation that attention-getting behaviours indeed have a specific function (e.g. to solicit play), it is essential to demonstrate that they are used across multiple contexts. Second, in the case of a highly context-specific gesture, such as 'leaf clipping', which is almost exclusively used by adult males to solicit copulations, it is impossible to know whether the 'leaf clipping' or the erect penis elicits copulatory behaviour. To investigate this issue, the responses of females to 'leaf clipping' when the penis is visually obscured need to be examined. Thus, if 'leaf clipping' is a true attention-getter in the absence of the further visual signal of the penis, the female should not approach and present for copulation. Unfortunately, to date, studies have not systematically addressed these issues.

Furthermore, results in regard to the use of potential attention-getting behaviours are somewhat inconsistent. In their interactions with humans, there is evidence that some signals of great apes may function as attention-getting behaviours. For example, orangutans, gorillas and chimpanzees use auditory gestures in food-begging interactions with humans more when the human is facing away compared to situations when the human is facing them (Hostetter, Cantero and Hopkins, 2001; Poss *et al.*, 2006). Similarly, chimpanzees vocalize more if the human is not attending (Hopkins, Taglia-latela and Leavens, 2007). These studies support the conclusions from the above section that great apes are sensitive to the visual attention of a human recipient and can produce the appropriate signal depending on whether the human is attending or not. In contrast, other studies found that vocalizations are used independently of the attentional state of the experimenter (Theall and Povinelli, 1999). Furthermore, some studies focusing on

Table 8.1 Attention-getting signals. Evidence that would be required to establish that attention-getting signals exist in primates. One individual would need to give the same attention-getting signal combined with different subsequent visual signals across two or more contexts.

Attention-getter	followed by	Visual signal	Context
A	+	C	1
A	+	D	1
A	+	E	2
B	+	C	1
B	+	D	1
B	+	E	2

conspecific interactions indicate that auditory and tactile gestures were used regardless of the attentional state of the recipient (Liebal *et al.*, 2004a; Tomasello *et al.*, 1997). In other words, they are not used more often when a recipient was *not* attending compared to when he was attending. Therefore, it was suggested that the aim of such gestures is to trigger others into action, but not necessarily to 'call their attention' in the first place; the fact that they also serve to capture attention may be a by-product (Liebal and Call, 2012).

A related, but more rigorous approach to the use of attention-getting behaviours is to investigate whether nonhuman primates use attention-getters in a kind of premeditated manner by using a auditory/tactile and/or a particularly effective signal first to attract the recipient's attention to the following, visual signal. Thus, producing an initial signal in order to enable a subsequent communicative signal to be successfully received is a strong indicator of an intention to communicate with the receiver (see also Chapter 7, section 7.2.3 on signal combinations). Table 8.1 illustrates the kind of data that would provide convincing evidence for the existence of attention-getters, and shows that a particular attention-getter needs to occur across different contexts and can precede different visual signals. However, research investigating gesture sequences, which considers gestures of several modalities (auditory/tactile, visual), found no evidence that nonhuman primates use attention-getters to attract the recipient's attention to a following signal in the manner suggested in Table 8.1 (Liebal *et al.*, 2004a; Tempelmann and Liebal, 2012). This supports the idea that the gestures that have been previously identified as 'attention-getters' may simply be signals that have a clear function (e.g. initiate play) and that attracting the attention of the receiver is a by-product due to their physical properties.

An alternative explanation for these inconsistent findings in regard to the use of both gestures and vocalizations as attention-getting behaviours is that the experimental setting might influence the communicative behaviour of the apes in these studies (see **Box 8.3** on interactions of apes with humans). As described in the previous section (8.3.3), great apes are less likely to use attention-getting behaviours if they have the choice to position themselves in relation to a human's body orientation. Alternatively, if they cannot reposition themselves, they are more likely to use both auditory gestures and vocalizations. Therefore, great apes seem to use different strategies depending on

> **Box 8.3** Is communication in interactions of apes begging for food from humans special?
>
> For several markers of intentional communication and other aspects of communication discussed in this book (e.g. pointing), research has produced divergent results when examining conspecific interactions and interactions where apes are begging for food from a human. Generally more sophisticated communicative performance is seen with humans. This may be partly due to the superior control this experimental set-up affords compared to observational research with conspecifics. Thus, researchers have better opportunities to accurately test and detect interesting features of communication in ape–human interactions. Begging from an individual who readily shares high-value food, but whom you cannot make physical contact with, also represents a highly unusual scenario that would not occur in ape–ape interactions. Thus, it is possible that captive apes exposed to this novel scenario develop novel behaviours to best exploit it. Thus, although ape–human food-begging paradigms are useful for revealing what apes are capable of doing, they may not be representative of strategies that apes naturally employ in conspecific interactions.

the set-up of the situation, but if they can move freely they prefer to move into the visual field of the potential recipient and produce a visual gesture, rather than using auditory or tactile gestures or vocalizations to attract the human's attention (Liebal *et al.*, 2004a)

8.3.5 Persistence and elaboration

One important way to determine the intention or goal of a signaller is to examine cases where a communication attempt is unsuccessful. When a signal is produced, there are three different potential outcomes, regardless of the signal's modality: first, there is an appropriate response from the recipient; second, the recipient reacts but not in the appropriate way; and third, the recipient does not react at all. This section specifically concerns the two latter outcomes – they are similar inasmuch that the sender was not able to achieve its goal. Instances in which a recipient does not react are very interesting in the framework of intentional communication, since they reveal how flexibly primates can react in such situations. The behaviour of the signaller after such a communicative failure gives an insight into whether their signal was produced intentionally to achieve a specific goal and we would expect to see the signaller do one of two things: they may show *persistence* and repeat the signal after an initial failure, or they may *elaborate* their communicative signals, by changing the type or intensity of the signal produced, in order to increase the chances of successful communication. Both of these behaviours could indicate that the signaller is producing signals intentionally in order to achieve a specific goal.

8.3.5.1 Persistence

The sender's behaviour in response to an initial failure of its communicative attempts has been mostly examined in the examination of gesture sequences (see Chapter 7).

This is based on the assumption that sequences or signal combinations are likely to emerge as a response to a lack of the recipient's reaction to the sender's initial signal.

In conspecific interactions, chimpanzees have been shown in both captivity and the wild to persist in producing the same gesture, after the first gesture failed to get a response from the recipient (Hobaiter and Byrne 2011b; Liebal *et al.*, 2004a). Thus, while chimpanzees responded to three-quarters of single gestures, they responded significantly less to the initial gesture of a sequence (46%). This seems to indicate that our closest living relative is capable of producing gestures in a goal-directed, intentional manner and persists in its communicative attempts. The question remains whether it is appropriate to argue that such signal sequences are evidence for persistence in case of failure, since the chimpanzees also continued to gesture in almost 50% of all sequences despite the recipient's response to the initial signal (Liebal *et al.*, 2004a). However, this relatively high proportion of responses might be exaggerated, since this study did not investigate whether the recipient responded *appropriately*. Thus, it did not differentiate between different types of response (e.g. change of attentional state, initiation of an action, production of gesture or leaving) and therefore did not discriminate whether the sender's goals were met by the recipient's response (see Cartmill and Byrne, 2010, suggesting an approach to this methodological problem). In contrast to chimpanzees, examination of gorilla and orangutan gesture sequences produced during conspecific interactions has revealed that they do not repeat gestures in response to a failed communication attempt (Genty and Byrne, 2010; Tempelmann and Liebal, 2012). The lack of persistence shown in these species cannot be explained by the lack of opportunities to respond to communication failures with unresponsive recipients, as in captive orangutans, recipients respond to less than 50% of single gestures given. Altogether, the little we know indicates that the strategy of persistence is not a widespread phenomenon in great apes' gestural interactions with conspecifics.

In the vocal domain, persistence has not been widely studied. This may be partly because vocalizations tend to be transmitted to a number of recipients (all in auditory contact) and therefore it is difficult for the human observer to record the response of all recipients to identify cases where responses were not forthcoming. Even in cases where this is feasible, it is possible that the signaller is targeting a particular recipient in the audience with a vocal signal and showing persistence in the face of a failure of the intended individual recipient to respond appropriately. In these cases analysis of a group response to a vocal signal will not be the appropriate unit of measurement to capture this behaviour. Whilst primates themselves may be adept at understanding who vocalizations are targeted at (Engh *et al.*, 2006), the task is much more difficult for human observers to perform objectively and reliably, making systematic investigation of vocal persistence a real challenge. The few studies that have risen to the challenge examined group-level responses to alarm calls and have found good evidence for persistence. Male Thomas langurs persisted in producing 'predator alarm calls' to other group members until each individual in the group responded with at least one 'alarm call' (Wich and de Vries, 2006). Vocalizations were repeated until the appropriate response from all recipients was given, showing clear evidence for persistence. Similarly, chimpanzees persist in producing 'alarm calls' in the presence of a snake model and when they stop it is significantly more likely than expected by chance that all other group members are safe, in that they are physically out of danger (10 m away or up a tree) or they are aware of the ambush predator (they have approached and visually scanned the snake area; Schel, Townsend *et al.*, in press). This indicates that the goal of the caller is to warn others of the danger and he is more likely to stop calling once this goal is achieved.

In contrast to gestures and vocalizations, no study has yet examined persistence in facial expressions. Once again we can see that this question has been given differential amount of attention and attracted different approaches across modalities. As highlighted above, differences in methods used to identify and code the behaviours of interest, not only between modalities but also within modalities, exist and are likely to impact our understanding of signaller persistence and subsequently intentionality.

8.3.5.2 Elaboration

It is possible to argue that elaboration is not qualitatively different from persistence in terms of measuring intentionality, since both strategies indicate that signals are used in a goal-directed manner. However, elaboration may be a stronger marker of intentionality than persistence, as it may be less subject to lower-level alternative explanations. Consider the following case: a chimpanzee receives aggression from another and subsequently approaches a dominant individual with a 'bared-teeth' facial expression. The dominant individual looks at the signaller, but initially does not respond. The signaller continues to produce the facial expression and after 10 s the dominant starts to groom the signaller. At this point, the signaller stops producing the 'bared-teeth' expression. On the surface, this case seems analogous to the behaviour of apes, who persist in the face of initially unresponsive recipients and continue to repeat the same gesture until the desired goal is met. However, due to the traditional assumptions about the non-intentional nature of facial expressions, it is easy for us to suggest a different lower-level explanation for this behaviour. The production of the 'bared-teeth' expression may be driven by fear or arousal and it may be this emotional state that persists, until affiliative contact occurs and changes the emotional state and corresponding facial expression of the signaller. Many would follow this line of reasoning and argue that persistence of a facial expression is not under voluntary control and is instead driven by emotions (see **Box 8.1**). Emotions or arousal could be playing a similar role in gesture production, thus persistence does not necessarily imply intentional processes. In contrast, although similar arousal-based alternative explanations for elaboration are possible, it is perhaps less plausible that heightened arousal or a particular emotion will drive the production of a variety of different signals in a short time period. Thus, elaboration and sequences of *different* signals as opposed to the repetition of the *same* signal might represent a more suitable criterion, since arousal is a less plausible (although still possible) alternative explanation for longer, varied signal sequences.

Despite the potential of elaboration as a marker of intentionality, there is very little evidence of elaboration of gesture signals in the face of an unresponsive recipient in conspecific interactions, while research assessing elaboration in vocal or facial signals has not been conducted. Most gesture sequences are repetitions of the same gesture and even if different gesture types are combined there is little evidence that these sequences are elicited by unresponsive recipients, or that these elaborated sequences are more successful in obtaining an appropriate response from recipients than single gestures (Genty and Byrne, 2010; Liebal *et al.*, 2004a; Tempelmann and Liebal, 2012).

In contrast to conspecific communicative studies, research that has examined interactions of great apes with humans in a food-begging context has shown that both orangutans and chimpanzees elaborate their gesture use if their goals are not met (Cartmill and Byrne, 2007; Leavens, Russell and Hopkins, 2005). For example, orangutans adjusted their communicative behaviour when begging for food from a human depending on whether the human's response met their goal fully (whole banana), only partly (only half of the banana), or not at all (different undesirable food). Thus, they stopped requesting when they got the whole banana, they repeated the same gesture if they receive only half instead of the whole banana (persistence), and they switched to other gestures in cases when the human offered them a completely different food item than they requested (elaboration) (Cartmill and Byrne, 2007).

In contrast to studies of elaboration in children, research in primates to date has mostly focused on the use of another signal of the same or other modality in cases of the failure of communicative attempts. The augmentation of signals in terms of an increase in intensity has received very little attention. One reason to explain the lack of such a measure, at least in gesture research, is that intensity in general is rather difficult to measure, since there are hardly any studies on nonhuman primates that describe the structural properties of gestures that would allow conclusions in regard to a change in the size of a gesture, size of the gesture space or how fast a gesture is produced. However, a much more fine-grained analysis of how nonhuman primates vary their communicative means as a function of the recipient's behaviour is essential to grasp the intentional nature and complexity of primate communication.

To summarize, in interactions with conspecifics, chimpanzees have been shown to persist in their communicative attempts and continue to gesture when there is no response from the recipient, while comparable evidence for persistence in gorillas and orangutans was not found. If great apes continue to signal in the face of an unresponsive recipient, they mostly repeat the same gesture and are not more likely to switch to another gesture indicating that there is little evidence for elaborated gesture use. However, when begging for food from a human, both chimpanzees and orangutans demonstrate persistence and elaboration (see **Box 8.3**). In the vocal domain, persistence and elaboration have not received much research attention, but there is evidence for persistence in alarm calling in both apes and monkeys. These topics have yet to be tackled by facial expression researchers.

Most importantly, the combination of gestures, facial expressions and vocalizations in this context has been largely overlooked. By focusing on one modality only, we may be missing important information. For example, nonhuman primates might continue to use the same gesture repeatedly indicating persistence rather than elaboration, yet they may combine gestures with other signals. If analysis focuses on the composite multimodal signals, rather than the unimodal gesture signal, a different conclusion (e.g. elaboration rather than persistence) might be reached (Leavens *et al.*, 2004).

In addition to the criteria suggested by Leavens (2004) and Leavens, Russell and Hopkins (2005), there are some further criteria commonly used to define intentional communication: response-waiting and flexibility. Since they are used in many studies, particularly research on gestures, they will also be briefly evaluated here.

8.3.6   Response-waiting

Response-waiting refers to the situation where a sender performs a signal and then waits for the recipient's response (Tomasello *et al.*, 1985). After this waiting period, if no response is forthcoming, one can expect to see the production of further signals. Therefore this criterion is often coded alongside persistence and elaboration. Theoretically, response-waiting in itself may be indicative of intentional communication as it suggests that the signaller has an intended goal and is waiting to see if the signal is successful in achieving the goal. However, in the absence of persistence or elaboration it is difficult to see how researchers can reliably identify a period of response-waiting from any other activity: what behavioural markers are indicative of response-waiting if further signals are not produced? Even in cases where further signals are produced there are difficulties with consistency in operationalizing this criterion, for example, how much time needs to elapse before such a behaviour can be clearly defined as response-waiting. As with persistence and elaboration, this criterion can only be applied to a subset of cases, where there is no immediate receiver response. In conclusion, whilst this is a necessary component of persistence and elaboration, by itself it is difficult to meaningfully code and interpret.

8.3.7   Flexibility

Flexibility was originally introduced to differentiate phylogenetically ritualized, species-specific displays, such as fur colour, from ontogenetically ritualized signals that imply that the sender has some control over their production. Although there are many meanings of the term 'flexibility' (see Chapter 7), in relation to intentionality it typically refers to the production of a single signal across a variety of functional contexts and the production of several signals in the same functional context (means-ends dissociation; Bruner, 1981). This criterion has been exclusively applied to gesture research and it is argued to be indicative of voluntary control over signal production and therefore related to intentionality. This is often contrasted to context-specific signals that are assumed to be emotionally driven (see **Box 8.1**). There are several issues with this criterion. First, there is no reason why a highly context-specific signal cannot be given intentionally: humans have many signals that are highly specific to certain contexts or referents, yet we still produce them intentionally to communicate with others. Second, as Chapter 7 highlights, the same emotional or arousal state can occur across contexts and could therefore drive the production of the same signal in multiple contexts. Thus, whilst means-ends dissociation is important for identifying flexible usage of signals, it is not a suitable marker of intentionality.

8.4   How have intentionality criteria been applied in empirical studies?

The researchers that have undertaken the challenge to identify intentional primate communication have adopted very different approaches to this endeavour. This is illustrated in Table 8.2, which summarizes some studies that explicitly claim to investigate intentional communication in primates together with the criteria that have been used to identify intentional use of signals. These studies were selected since all of them specifically mentioned that their focus was on intentionally used signals. The majority of these studies are on gestures – produced in interactions with conspecifics or towards a human experimenter – as this is the dominant modality in which intentionality has been examined. It is important to emphasize that the list of studies in this table is neither exhaustive nor is the aim to criticize any of these studies. The point of Table 8.2 is to show that there is a lack of consensus across studies in the number and type of criteria used to identify intentional signalling. For example, none of the studies used *all* of the total of eight criteria, and none of them used even the six suggested by Leavens (2004) and Leavens, Hopkins and Bard (2005). There is also no consistency as to which criteria are chosen, or how many of the criteria a signal has to meet in order to be classified as intentional. Furthermore, studies differ substantially in regard to how they operationalize intentional communication. First, several studies measure how frequently a behaviour was characterized by one or more of these criteria ('x' in Table 8.2), but they differ considerably in regard to how many of them are actually considered. Second, other criteria serve as prerequisites for a behaviour to be included as intentional signal ('o' in Table 8.2). Again, studies differ substantially in regard to how many of these criteria are sufficient to define a behaviour as an intentional signal. Some of them require only one out of a set of criteria to be present, while others do not provide more details about how many of these criteria were required. However, as this chapter illustrates, reliance on single criteria is most likely not sufficient to identify intentional signalling with any confidence, as apparently sophisticated cognitive skills can also be explained by more low-level mechanisms, and because some criteria are more reliable than others to identify acts of intentional communication.

8.5   Validity of criteria to measure intentionality

This section aims to critically evaluate the validity of the different criteria introduced in the above sections and to identify the most robust criteria. We then consider how these criteria are currently being applied across different studies and discuss the implications of the varied use of criteria for our understanding of intentional primate signal production.

Throughout this chapter we have highlighted that although the behaviour identified by these criteria might be indicators of intentionality, there are also alternative explanations for the behaviours that suggest they may not necessarily be measuring intentionality. For instance, *social use* of signals, in terms of producing signals selectively in the presence of an audience, may be driven by heightened arousal levels caused by social

Table 8.2 Overview of some example studies in regard to the criteria used to identify intentional acts of communication. Criteria include those suggested by Leavens (2004), Leavens and colleagues (2005) in addition to some criteria commonly used by other groups of researchers. The letter 'x' indicates that the corresponding criterion was actually measured in this study (e.g. data presented in the results section), while 'o' means that this characteristic was used as a prerequisite criterion specified in the methods for selecting intentional signals for further analysis, but it was not explicitly measured (e.g. data not presented in the results section). 'o + x' indicates that the criterion was used both as a prerequisite to identify intentional signals initially and then explored quantitatively in more detail. For studies that used prerequisite criteria to identify intentional signals for further analysis (coded as 'o'), the last column summarizes how many of these prerequisite criteria had to be met for a behaviour to be included as an intentional signal. '?' indicates that the study did not provide sufficient details to determine how many of these criteria had to be met for a signal to be deemed intentional.

Studies	Modality: gesture (g), vocalization (v), Facial expression (f)	Partner: conspecific (c), human (h)	Social use	Gaze alternation	Attentional state	Attention-getters	Persistence	Elaboration	Flexible use	Response-waiting	No. of a priori criteria signals met to be categorized as intentional (only for studies with prerequisite definitions for intentional signals coded as 'o')
Anderson, Kuwahata and Fujita, 2007	g	h	x	x	x						n/a
Bard, 1992	g	c	x								n/a
Cartmill and Byrne, 2007	g	h					x	x			n/a
Cartmill and Byrne, 2010	g	c	o	o			o	o	o	o	?
Genty and Byrne 2010	g	c	o		o		x	x	o	o	?
Genty et al., 2009	g	c	o		x	x	o	o	x	o	?
Hobaiter and Byrne, 2011a	g	c	o		o + x	x	o	o	x	o	1/4
Hobaiter and Byrne, 2011b	g	c	o		o	x	o + x	x x x	x	o + x	1/4
Hopkins and Leavens, 1998	g, v	h		x							n/a
Hostetter, Cantero and Hopkins, 2001	g, v, f	h	x		x	x					n/a
Leavens and Hopkins, 1998	g, v	h		x							n/a
Leavens, Hopkins and Bard, 1996	g, v	h	x	x		x	x				n/a
Leavens, Russell and Hopkins, 2005	g, v	h	x	x			x	x			n/a
Leavens, Russell and Hopkins, 2010	g, v, f	h			x	x		x			n/a
Liebal, Call and Tomasello, 2004a	g, v, f	c	o		x	x	x	x	o		2/2
Liebal, Call, and Tomasello, 2004b	g, v, f	h	o		x	x		x			1/1
Liebal, Pika and Tomasello, 2004c	g, f	c	o		x		x	x	x	o	2/2

Table 8.2 (*cont.*)

Studies	Modality: gesture (g), vocalization (v), Facial expression (f)	Partner: conspecific (c), human (h)	Criteria according to Leavens (2004) and Leavens, Hopkins and Bard (2005)						Additional criteria		No. of a priori criteria signals met to be categorized as intentional (only for studies with prerequisite definitions for intentional signals coded as 'o')
			Social use	Gaze alternation	Attentional state	Attention-getters	Persistence	Elaboration	Flexible use	Response-waiting	
Liebal, Pika and Tomasello, 2006	g, f	c	o		x		x	x	x	o	2/2
Maille et al., 2012	g	h			x						n/a
Meunier, Prieur and Vauclair, 2013	g	h	x	x	x	x					n/a
Poss et al., 2006	g, v, f	h	x		x	x		x			n/a
Tempelmann and Liebal, 2012	g	c	o		x	x	x	x			1/1
Tomasello et al., 1985	g	c	o + x	o + x			x			o + x	1/3
Tomasello et al., 1994	g	c	o	o	x	x	x	x	x	o	?

facilitation. A much stronger indicator of intentional communication might be the demonstration that nonhuman primates are sensitive to the *composition* of their audience and adjust their communication accordingly. Whilst it can be argued that the presence of certain individuals can differentially affect arousal levels and subsequent signalling behaviour, considering *who* is present and adjusting communication in terms of *when* (rate) and *how* (structure) to produce a signal, this line of argument is less well grounded in physiological evidence. Signals accompanied by *gaze alternation* may indicate intentional production of the signal, but patterns of gaze alternation may simply be the product of competing foci of attention for the signaller, rather than a marker of intention. The sender's sensitivity to the *attentional state* of the recipient is a criterion often highlighted as evidence for more sophisticated cognitive skills, since it implies that the sender is capable of taking the recipient's visual perspective. As a consequence, the sender adjusts its communicative means and only uses visual signals if the recipient is visually attending. An alternative explanation for this behaviour is that the sender has simply learnt that the appropriate context for production of visual signals needs to include the face of the recipient. Just as vervet monkeys learn through development to narrow the set of stimuli that should elicit an 'eagle alarm call', similar processes could be occurring in gesture production, with individuals learning that a recipient's face needs to be present as a prerequisite for visual gesture production. *Persistence* in signalling in response to an unresponsive recipient has been identified as a marker of intentional, goal-directed communication, however, alternative explanations for this behaviour have been suggested. If signal production is driven by arousal or emotional state, it could be this internal state that persists until the recipient's behaviour changes the internal state that was driving signal production. Although a similar argument may be made for the production of different signals during *elaboration*, it seems less likely that arousal would drive production of several different signals in a short period of time. Thus, elaboration seems to be a stronger marker of intentional communication than persistence. *Response-waiting* is a necessary component of persistence and elaboration, but is not a good marker of intentionality if considered alone. *Flexibility*, as defined as means-ends dissociation, might indicate voluntary control over signal production; however, non-intentional processes (e.g. arousal) could also explain this pattern of behaviour, indicating this is not a particularly strong measure of intentionality. *Attention-getting behaviour*, as defined as the production of a signal with an auditory or tactile component to attract the attention of the recipient that is followed by a further visual signal, could be a strong indicator of intentional communication. Such manipulation of another's attentional state in order to improve the likelihood that subsequent signals are successfully received is indicative of a pro-active and strongly goal-directed communicative strategy. Although a complex series of learning discriminations could possibly explain such behaviour sequences on a lower level, these explanations seem less likely, since they lack parsimony.

In summary, nearly all of the criteria commonly used to define intentionally used signals are open to lower-level, less cognitive explanations. Thus, relying on single criterioa in isolation as markers of intentionality will not enable researchers to make a strong case for intentional communication. A more promising approach is to rely on a set of these criteria and look for signals that meet several of these criteria, thus producing convergent evidence for intentionality. Whilst different lower-level explanations will be possible for individual criteria, if convergent evidence from several criteria is obtained, it may be more parsimonious to attribute a single cognitive mechanism (e.g. intentionality) to the signaller, rather than a myriad of low-level learning and arousal-based explanations (Byrne and Bates, 2006).

We argue that the set of criteria with the highest validity should be used together in order to build the most convincing case for intentional communication. We propose these should be (1) producing a signal selectively to certain individuals in the audience, (2) producing signals in a manner that ensures the target recipient can perceive the signal by using visual signals only if the recipient is attending or manipulating the attention of the recipient using signals with an auditory or tactile component before producing a visual signal, and (3) persisting in and elaborating of signal use in the face of failure of communicative attempts.

The remaining criteria are argued to be weaker in regard to their suitability to capture intentional communication, because convincing alternative, lower-level explanations for the pattern of behaviour are available that do not require intentionality (e.g. flexibility, audience presence), they are poorly defined and/or difficult to operationalize (e.g. response-waiting) or are only applicable to a specific subset of communicative signals (gaze alternation).

8.6   What could a multimodal approach bring to this topic?

After evaluating the empirical evidence for intentionality in primate signals and investigating the biases and methodological differences in unimodal approaches to the topic, it seems premature to conclude that gestures are intentional and vocalizations and facial expressions are not. It seems that as human observers it is easiest to observe what looks like intentional signalling in great ape gestures, but without rigorous testing using comparable criteria in the other domains, where such behaviour may be less obvious and more difficult to measure, we should not fall into the trap of thinking that absence of evidence indicates absence of ability. In the future, if these valid multimodal criteria are used together to identify acts of intentional communication across modalities and species, we will finally be in a position to evaluate whether there are systematic differences in the extent to which the production of primate vocalizations, facial expressions, gestures and multimodal signals is intentional.

A multimodal approach will not only allow more valid comparisons between modalities, but also allow us to test whether multimodal combinations are produced intentionally. If multimodal signals exhibit more key hallmarks of human language, such as intentionality, than their unimodal components, this would indicate that multimodality may have been an important stepping stone for human language.

In addition, a more integrated analysis of signals from different modalities may reveal complexity that is overlooked by the unimodal approach. For instance, a gorilla may produce a facial signal that fails to elicit the appropriate response from the intended recipient and thus he may elaborate with gesture. Unimodal researchers interested in just facial expressions would conclude that no elaboration occurred, whereas the multimodal researcher would be able to detect this demonstration of signal elaboration. Equally, unimodal researchers may be misclassifying elaboration as persistence: if a gibbon produces a sequence of vocal signals when met with an unresponsive recipient, and he starts to augment his vocal signals with gestures, the vocal researcher would classify this as persistence, but miss the elaboration that would be evident to the multimodal researcher. To conclude, a multimodal approach would help us to assess intentionality in a comparable manner across modalities and multimodal signals, as well as ensuring we accurately detect and classify cases of persistence and elaboration.

Summary

This chapter has necessarily focused on gestural research in great apes, as this reflects the vast majority of available empirical work on intentionality in primate signals. It is clear, however, that systematically investigating this important cognitive facet of communication across modalities and in multimodal signals will help challenge our assumptions and understanding of what constitutes an intentional communicative act and the proximate mechanisms driving signal production across modalities. Therefore, the aim of this section was to identify a subset of criteria that can be usefully applied across modalities and multimodal signals.

We suggest that three criteria have relatively high validity and are applicable to a multimodal approach to primate communication. The first criterion is social use as evident in the sensitivity to the presence and critically the composition of an audience. Methodologically, it should be possible to apply this across modalities and to multimodal signals, although currently, its application to gestures may be limited to a smaller subset of salient, structurally distinct gestures that can be reliably identified independently of the social context. As more work is done on defining the structural properties of gestural movements, this subset can likely be expanded. Second, we suggest that persistence and elaboration should be examined across modalities and multimodal signals. The difficulties human observers can have identifying the intended recipient of vocal signals may make this challenging in the vocal domain, but we suggest two approaches that may help: (1) looking at group responses to signals likely directed to the whole group (e.g. 'alarm calls') and (2) systematically using eye gaze to try and determine the intended recipients of more social signals likely to be directed at specific individuals. Third, if researchers adopt an integrated multimodal approach, searching for attention-getters (signal with auditory/tactile component followed by a visual signal) would provide good evidence for intentional communication. Critically, these criteria should not be used in isolation, but instead be used together to provide convergent evidence of intentionality.

9 Referentiality

9.1 What is reference?

Humans are very adept at referring to events and objects in the external world, using both gestural and linguistic signals. A large number of human words are referential: they refer to specific entities in the world. In linguistic terms these words 'stand for' the external referent and the relation between the physical form of the word and the referent is arbitrary. In language there is usually a one-to-one mapping between a word and its culturally agreed referent(s). Although the referential specificity of words varies greatly (e.g. the number of referents associated with the sign can be small (e.g. banana cake) or large (e.g. food)), the referential meaning is stable and shared between speakers and listeners. In contrast to this, gestural reference, in the form of human pointing, has no one-to-one referential meaning and can only be successfully interpreted by integrating the signal with the shared common ground between signaller and receiver (Liebal *et al.*, 2009; Tomasello, 2008; Tomasello, Carpenter and Liszkowski, 2007). Points direct the attention of the receiver spatially to a location in the immediate perceptual environment, but the referent and meaning of the point can only be decoded by understanding the communicative intent of the signaller. Although this may seem cognitively complex, this ability emerges very early in human development. Humans usually comprehend pointing gestures before their first birthday and start to produce pointing gestures before any spoken words (Bates *et al.*, 1979; Liszkowski *et al.*, 2004; Tomasello, Carpenter and Liszkowski, 2007). Pointing is thought to scaffold the emergence of spoken language in infants. Referential signals in both the linguistic and gestural domains allow humans to direct the attention of others to specific external entities or events and share attention, feelings or thoughts about them, which is fundamental to the complexity of human communication. The extent to which these abilities are uniquely human or shared with other primate species has been a central theme of animal communication research in recent decades.

The operationalization of definitions of reference for animal communication has not been without controversy, and this may be partly due to an over-reliance on human linguistic terms (see Chapter 1, **Box 1.1**). In humans, both pointing and words are termed referential, yet as outlined above, there are important differences between them, particularly in terms of the cognition involved. The traditional split between gesture and vocal researchers has led to the investigation of pointing gestures in apes, building on our understanding of the processes underlying human pointing and the investigation of referential vocalizations in primates building on our understanding of referential words. Difficulties are then encountered when trying to come up with a holistic, cross-modal definition of how to identify 'referential' signalling in primates. Thus, commonly, primate gestures are termed 'referential' and primate vocalizations are termed 'functionally referential', with different definitions and criteria being applied to these two terms. In this chapter we provide traditionally accepted definitions of animal reference, but different researchers have questioned whether these definitions are either too broad or too narrow to adequately capture the behaviours of interest.

9.2   Why is reference interesting to investigate in primates?

Reference is a fundamental component of human communication (and language in particular) and as such many studies on primate communication have been motivated by a desire to search for similar behaviours in our closest living relatives. Such studies are using primate behaviour as a tool to further our understanding of the evolution of human behaviour (see Chapters 5 and 10 for more details). Reference within a communication system is, however, also interesting in its own right. The ability to communicate with others about events and objects in the environment enlarges the communicative repertoire of a species and the potential for communicating a larger number of messages. A referential communication system also enables individuals to move beyond purely dyadic communication and to engage in triadic interactions about external entities. In this manner, the power of a communicative system is potentially enhanced by referential abilities. Little research has been conducted with facial expressions or olfactory signals in regard to functional reference, thus this chapter necessarily focuses on vocal and gestural modalities.

9.3   Gestural reference: pointing

Pointing gestures are a fundamental part of human communication (Tomasello *et al.*, 2005). They emerge early in development and are used to direct and share attention, but also to inform others (Liszkowski *et al.*, 2004, 2006). In nonhuman primates, however, there seems to be a divide between their abilities to produce and comprehend pointing gestures. Therefore, the following sections refer to the production and comprehension of these gestures separately.

9.3.1   Production of pointing gestures

Human infants start to point at around 12 months of age and this behaviour builds on the infant's established abilities to follow gaze and engage in visual joint attention with a caregiver about an object (Carpenter, Nagell and Tomasello, 1998). Humans point imperatively to request action from other individuals (e.g. an infant will point to an out-of-reach toy to request the caregiver brings it to them) and **declaratively** in order to draw the attention of others to something of interest, or to simply point something out. This tendency to point declaratively for others illustrates our inherent motivation to share experiences with others (Tomasello and Carpenter, 2007). Pointing seems to be universal across human cultures, although there can be considerable variation in the physical form of this gesture, with the lips, the chin or the whole hand being used instead of the index finger in some cultures (Kita, 2003; Liszkowski *et al.*, 2012).

In contrast to the early emergence of this behaviour in human infants and the common and ubiquitous nature of these gestures in humans, pointing gestures have only been reported on a single occasion in a wild primate: one bonobo was observed to point twice at researchers hidden in undergrowth (Veà and Sabater-Pi, 1998). Despite extensive field research on other great apes and monkey species in their natural habitat, other observations of primates pointing for other conspecifics have not been forthcoming.

The absence of pointing in the behavioural repertoire of most wild primates is in contrast to the substantial evidence that great ape species spontaneously point for caregivers in captivity. Great apes will point with their index finger (Figure 9.1) or whole hand to indicate the location of desirable food to a human (Leavens and Hopkins 1999; Leavens, Hopkins and Bard, 1996; Leavens, Hopkins and Thomas, 2004). In a token exchange study, captive orangutans produced pointing gestures to conspecifics, and these **imperative** gestures were understood by the recipients who transferred more of the desired tokens to the signaller after receiving a pointing gesture (Pelé *et al.*, 2009). The emergence of this behaviour in an environment, which physically prevents them obtaining their desired goals, indicates that pointing is a tactical strategy employed to achieve a goal through manipulation of a human. Leavens, Hopkins and Bard (2005) argue that human infants share this 'referential problem space' with captive apes, due to

Figure 9.1 Production of a pointing gesture: A nursery-reared chimpanzee male points with his index finger to a bottle of juice placed on the ground, which is not shown in the picture (Photo: David Leavens; from Leavens and Hopkins, 1998).

both their physical inability to locomote effectively at a young age and the human tendency to restrain infants in papouses, slings or chairs. There seem to be commonalities, therefore, in the social and physical environment that can lead to the emergence of imperative pointing in humans and other great apes. Apes also seem sensitive to the presence of an audience when pointing (Leavens, Hopkins and Bard, 1996) and enculturated apes, such as the orangutan Chantek, are also sensitive to the importance of the eyes being open in the receiver (Call and Tomasello, 1994). In addition, ape pointing has the hallmarks of intentional communication (see Chapter 8), as chimpanzees are reported to combine pointing with gaze alternation between the caregiver and the desired object and show persistence until they achieve their goal (Leavens, Hopkins and Bard, 1996). Unlike prelinguistic children, chimpanzees do not point to absent entities to indicate that they want a desirable object that is usually in a particular location (Liszkowski et al., 2009). This suggests that displaced reference (referring to absent entities), an important characteristic of human language (Hockett, 1960), emerges during human ontogeny before language and may be unique to humans.

Although in the constraints of captivity apes readily point imperatively in order to request and obtain desirable objects (Leavens, Hopkins and Thomas, 2004; Pelé et al., 2009), there is little convincing evidence of spontaneous declarative pointing in apes. Declarative pointing is important in human development as a meta-analysis has revealed that the emergence of spoken language in infants is significantly correlated to the production of declarative points, but not imperative ones (Colonnesi et al., 2010). Tomasello, Carpenter and Liszkowski (2007) argue that human infants, as young as 12 months, engage in two variants of declarative pointing. First, expressive declaratives for sharing their attitude about an object with another and second, informative declaratives for providing others with useful information. Moore and Corkum (1994) suggested that early expressive declaratives are produced by infants to gain positive emotional reactions from caregivers and as such are not really referential (as the referent of the point is unimportant). This suggestion was refuted by carefully controlled experiments run by Liszkowski and colleagues (2004, 2007) that showed that infants were only satisfied if the caregiver shared attention with them about the object to which they pointed: if the adult simply responded with positive emotion to the infant, or attempted to share attention about a different object, the infant repeated their pointing gestures in an attempt to redirect their attention to the intended referent. In contrast, there is no compelling evidence that apes produce declarative points.

The lack of evidence for expressive declarative points in apes has come from a variety of different studies. To test whether apes are capable of producing informative declarative points, they were presented with situations where they can inform an experimenter about the location of a tool. Although orangutans (Call and Tomasello, 1994; Zimmerman et al., 2009) and bonobos (Zimmerman et al., 2009) will point to the location of a hidden tool so the experimenter can obtain food for the ape, the motivation behind the pointing still seems to be imperative (e.g. 'Get the tool to get the food for me'). This was explicitly tested by Bullinger et al. (2011) who gave chimpanzees the opportunity to point for tools that could be used to provide either the chimpanzee with a reward (selfish), or the experimenter with a reward (helpful). Whilst 2-year-old children pointed out both tools for the experimenter, regardless of who benefitted, chimpanzees only reliably pointed at the tool when it could be used to deliver a food reward to them. So whilst chimpanzees can point to items that only bring a reward indirectly, the motivation behind their gestures seems still to be imperative, rather than informative.

The imperative nature of ape pointing is also mirrored by their non-pointing gestures. Observational studies of captive orangutan gestures revealed that spontaneous gestures were produced to achieve one of six social outcomes that included movement of a social partner, initiation of affiliative interactions and sharing of objects (Cartmill and Byrne, 2010). Although 29 gestures were identified as having a single meaning, these meanings were all imperative.

9.3.2   Comprehension of human pointing gestures

It seems from the evidence reviewed in the previous section that although apes produce referential pointing gestures in captivity, the cognitive and motivational mechanisms that underlie this may be very different from production of declarative pointing in humans. This is supported by the surprising difficulty apes have in comprehending informative, declarative points from humans. Many studies have investigated this using a basic object-choice paradigm, where a helpful experimenter provides the ape with a pointing gesture towards one of two (or more) containers to indicate where food is hidden (Figure 9.2). A review by Miklósi and Soproni (2006) indicates that, overall, primates perform poorly on these tasks, rarely choosing the indicated container at above chance levels: although chimpanzees, orangutans, gorillas and capuchin monkeys have some success interpreting proximal points (finger is within 10 cm of the target; Figure 9.2a), rhesus macaques fail the same task and only enculturated apes succeed in interpreting distal pointing cues (finger more than 50 cm from the target; Figure 9.2b).

On many levels such poor performance of primates compared to human infants is unsurprising. Humans are interpreting the signals of members of their own species,

(a)   (b)

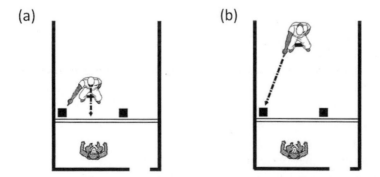

Figure 9.2 Comprehension of human pointing in an object-choice paradigm. The human produces a pointing gesture towards the container with food hidden inside. The primate then has to choose which of the two containers to search. Once a selection has been made, the experimenter shows the primate the contents of the container and if they have chosen the correct container, they receive the food that was hidden inside it. Different types of pointing gestures have been tested, including (a) the proximal point (finger 10 cm from object) and (b) the distal point (hand more than 50 cm from the object). (Figure adapted from Povinelli, Bering and Giambrone, 2000.)

whilst other primates have the much larger challenge of interpreting the signals of a different species. This requirement of cross-species interaction puts nonhuman primates at a clear disadvantage compared to humans. To address this issue, Tempelmann, Kaminski and Liebal (2013) tested bonobos, chimpanzees and orangutans in a setting where they had to use pointing gestures of another conspecific to find hidden food, but found no evidence that these great apes were using this information. These negative findings cannot be explained by the fact that great apes do not interpret pointing gestures as informative cues, since the great apes in this study were tested in a competitive setting.

The picture is further complicated by the excellent performance demonstrated on these tasks by dogs, dolphins and seals (Miklósi and Soproni, 2006). These species seem to overcome the cross-species barrier and in recent years considerable research effort has focused on examining reasons for these interesting species differences in performance. These include investigations into the effect of differences in methodology, particularly the spatial set-up of the equipment on performance (e.g. Mulcahy and Call, 2009), the competitive or cooperative nature of the species and thus the ability to understand the altruistic intent of the helpful experimenter (Tomasello, Carpenter and Liszkowski, 2007) and the effect of domestication on the ability to read social cues from humans (Hare et al., 2002). Whilst each of these factors can explain some of the variation between species, no unifying explanation has yet been found to explain the complete pattern of data across this diverse set of species. Although this book focuses on primate communication, this is a great example of the importance of applying a wider comparative approach to research questions and testing a wide variety of species.

9.4   Functionally referential signals

In the vocal realm considerable research effort has concentrated on trying to find the animal equivalent to referential words in human language. In contrast to pointing gestures, which have escaped rigorous definition, the use of the term 'functional reference' is associated with a set of clear criteria (as outlined below). This term also acknowledges that although the animals behave as if a signal provides them with information about an external event, the cognitive mechanisms underlying this may be quite different to those associated with human referential communication. Several researchers have recently questioned the utility of research into functional reference (e.g. Rendall, Owren and Ryan, 2009; Wheeler and Fischer, 2012) and reignited a debate on the wider characterization of communication that follows from this research as a process of information transfer. This debate, as outlined in Chapter 1 (see **Box 1.1**), highlights the need to move beyond surface behaviour and examine the cognitive processes underlying communication: something which is bypassed by the definition of functional reference. Whilst functional reference continues to be a controversial topic, its utility in furthering our understanding of animal communication has been defended by Seyfarth et al. (2010) and Scarantino (2010).

In order for a signalling behaviour to be termed 'functionally referential', the following well-established criteria have to be met.

9.4.1   Production criteria

The signal needs to be discrete and produced reliably in response to a specific event or stimulus (context specificity) (Evans, 1997). Eliciting stimuli should belong to a coherent category, although the broadness of this stimulus set can vary, just as it does for human utterances (e.g. 'animal' is elicited by a very large number of stimuli, whereas 'aardvark' is elicited by a much smaller number of stimuli, although both are referential). So the specificity of a call or the size of the stimulus set that elicits it does not preclude it functioning referentially. One potential problem with establishing context specificity is that sometimes insufficient efforts are made to document the range of contexts the calls are naturally produced in, therefore claims of highly specific calls can later be undermined by careful observations showing that a wider range of stimuli elicit the call.

9.4.2   Perception criteria

Signals should be 'context-independent' (Macedonia and Evans, 1993), in that the signal alone should be sufficient, in the absence of the eliciting stimulus, to allow listeners to respond in an appropriate way (the same way they would to the actual eliciting stimulus). It is critical that the receiver's understanding of the signal is tested and the most common method for doing this in the vocal domain is using playback experiments. As discussed in Chapter 4 (section 4.4.4), playback experiments allow researchers to examine the understanding the receiver has of the call alone, in the absence of other cues they may normally rely on to make the appropriate response (e.g. responses of others, visual cues, direct perception of the eliciting event). Although the production and perception criteria can be applied to any modality, the difficulty in testing receiver perception of visual signals (see Chapter 4) means that the focus of this section is necessarily on vocal research.

Functionally referential signals should allow listeners to extract information about external events in the environment, in addition to factors internal to the caller (e.g. species, motivational state, size, identity). Many signals across modalities, including olfactory signals, have different structures depending on intransient features of the signaller, such as gender, dominance and identity (Setchell *et al.*, 2010); however, functionally referential signals focus on consistent structural differences that occur more flexibly within an individual according to environmental events. However, the distinction between internal and external information easily leads to confusion on several levels. First, it is very difficult to accurately categorize all potential referents as external or internal, which is an issue we will revisit later in this chapter. Second, it is easy to confuse two issues, which should be kept separate (Scarantino, 2010): (1) how signals are produced, and (2) what information, if any, they carry. Signal production may be mediated by cognitive, affective and/or different physiological mechanisms, and although this is an important empirical question in its own right, the mechanism of production does not affect a signal's ability to function referentially. For example, alarm calls may be driven by the caller's internal state (e.g. level of fear), or a cognitive desire to inform others of danger, but as long as the caller reliably produces the same signal in this context and the listener can understand the likely external cause of the signal and therefore react appropriately, the signal is functioning referentially. It should also be highlighted that a signal is capable of providing listeners with information about both external events and internal states: signals need not to be either 'referential' or 'emotional'. This dichotomy is a false one and both human utterances and animal signals can provide receivers with varying amounts of information about external events and internal states simultaneously (Marler, Evans, and Hauser, 1992; Scarantino, 2010; Townsend and Manser, 2013).

Functionally referential signals, if evident as defined above, would provide evidence of continuity between human and animal communication. However, there are also some important differences between animal reference and human reference. First, whether calls evoke mental representations of eliciting events in the minds of receivers is unknown (Evans, 1997) and, if they do occur, the nature of the representation is unknown. For instance, in the case of vervet monkey alarm calls, it is unclear if a 'leopard alarm call' means 'big spotty cat' in a declarative manner or 'run up a tree' in an imperative manner (Baron-Cohen, 1992). Second, as already outlined, no assumptions are made about the intentions of the signaller in functionally referential communication. Unlike human communication where we are motivated to share information and inform one another about external entities, there is very little evidence that in the vocal domain signals are given by primates with the intention to inform others. However, recent studies on wild chimpanzee alarm call production may have begun to challenge this. First, alarm call production in response to a moving python model met several criteria for intentional signal production and crucially chimpanzees persisted in calling until all individuals in the immediate vicinity were safe from danger (aware of the ambush predator or physically distant from the snake; Schel, Townsend *et al.*, in press). This indicates the goal of the caller could have been to inform or warn others about the danger. It is, however, difficult to rule out the possibility that callers were motivated by a desire for proximity with others to gain reassurance in a dangerous situation. Second, a different study suggests that wild chimpanzees are more likely to produce alarm calls in response to a snake model in the presence of ignorant rather than knowledgeable group members (Crockford *et al.*, 2012). Whilst this seems to suggest that they may inform others about this potential danger, lower-level, more parsimonious explanations are available. In this case, once the confound of the caller's own knowledge about the snake model has been controlled for, the callers could have been responding to potentially inappropriate behaviour of ignorant individuals approaching the predator, rather than their knowledge state. Although promising, further experimental evidence is required before firm conclusions about whether chimpanzees call with the intent to inform can be made. Functionally referential signals thus facilitate our learning about behaviour patterns, which appear similar to those of humans, but further investigation is necessary to probe the cognitive mechanisms underlying their production and perception and how these relate to human mental processes.

Functionally referential signals are perhaps more likely to evolve in evolutionarily urgent contexts, where there is an adaptive advantage to the listener being able to extract precise information about an important external event and therefore act appropriately without having to witness the event first hand. This is because in contexts such as predator detection, unambiguous referential signals that facilitate rapid, adaptive responses in receivers have the potential to provide significant fitness benefits. In contrast, ambiguous signals that require the receiver to combine the call with contextual information before being able to make an appropriate response may lead to slower responses and thus might have negative fitness consequences. Unsurprisingly then, most research into functionally referential signalling in primates has concentrated on the evolutionary important contexts of predator encounters, food discovery, copulations and agonistic interactions.

9.4.3   Predator avoidance contexts

Researchers interested in functionally referential communication have dedicated considerable research effort to investigations of signalling in the context of predator defence. Visual signals (gestures and facial expressions) have not been systematically studied in predator avoidance contexts, mainly due to the difficulty in obtaining video footage of individuals in this context that is of sufficient quality to enable analysis of gesture or facial expression: animals tend to move fast and often hide in this context. Although functionally referential alarm vocalizations could represent a very efficient way of disseminating information about the presence of a predator to group members, who may be spread out over some distance, this does not preclude gestures and facial expressions also being an important part of communication about predators, at least in species that live in more open habitats. Indeed, we know that in other non-primate species multimodal predator alarms are produced. Grey squirrels produce calls in combination with a visual tail-flagging signal and ingenious experiments with robotic squirrels have revealed that although these two signals in isolation convey some important information to receivers, the combined multimodal signal elicits a significantly greater response (Partan, Larco and Owens, 2009). There are indications that visual information is also important in this context in primates, with putty-nosed monkeys moving towards the male caller after hearing him produce a non-specific alarm call, perhaps to gain visual contact with him (Arnold and Zuberbühler, 2012). Particularly in referential alarm systems that have low level of specificity, visual cues, such as gaze direction and gestures, may be crucial in helping receivers disambiguate the referent of the call. It is also possible when encountering a predator where the most adaptive response is cryptic behaviour and retreat, that visual rather than auditory signals would be most adaptive to reduce the chances of the predator eavesdropping on the signals and using them to accurately locate and hunt the signaller.

So whilst visual signals may be important in this context, we are lacking empirical studies that have examined them and the vast majority of studies have focused upon vocalizations. Just as the specificity of human referential utterances varies greatly from highly specific (e.g. my dog: single referent) to very general (animal: thousands of referents), primate calls also have varying degrees of specificity. In terms of alarm calls, the most specific systems function to refer to the type of predator, whilst the less specific systems refer to aerial as opposed to ground threats or primates have specific calles for aerial threats and other calls for a wide range of disturbances.

The seminal study of predator-specific functionally referential alarm calls was conducted with vervet monkeys in the savannahs of East Africa. Struhsaker (1967) documented the production of acoustically distinct calls produced by these monkeys when encountering pythons, eagles, leopards and humans. Robert Seyfarth and Dorothy Cheney, under the supervision of Peter Marler, then conducted playback experiments to test listener understanding of these calls. Seyfarth, Cheney and Marler (1980a) found that listeners responded to a playback of a group member's alarm call to a specific predator in the same adaptive way as they did to the actual presence of the predator (Figure 9.3). For example, when hearing a python alarm call they would take a bipedal stance and scan the ground for a snake, as they would when discovering a real python.

Since then, similar predator-specific alarm-calling systems have been found in a number of primate species: Zuberbühler, Noë and Seyfarth (1997) found that forest-dwelling West African Diana monkeys produced two distinct alarm calls reliably in response to leopards and eagles and that listeners responded as they would to the real predator, indicating these signals functioned referentially. A subsequent study using playbacks of predator sounds showed that the monkeys did not deviate from the labelling of the predator type, despite variation in the distance of the speaker (indicating

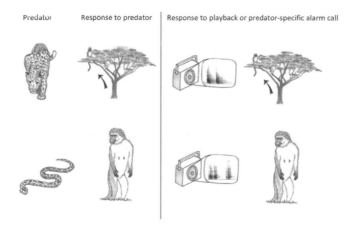

Predator   Response to predator │ Response to playback or predator-specific alarm call

Figure 9.3 Response of vervet monkeys to predator alarm calls. Vervet monkeys produce the same adaptive anti-predator responses to playbacks of predator-specific alarm calls as they do if they discover the actual predator themselves. In response to leopards, monkeys typically climb trees, and they produce the same adaptive response when a group member's 'leopard alarm call' is broadcast from a speaker. In response to pythons, monkeys will typically stand bipedally and scan the ground, and they produce the same adaptive response when a group member's 'snake alarm call' is broadcast from a speaker. As the call alone is sufficient to elicit these adaptive responses, these predator-specific alarm calls are considered functionally referential signals.

urgency of response) and the elevation of the speaker (above or below the group) (Zuberbühler, 2000b). Experiments then investigated whether listening monkeys extracted the referential meaning from these calls, or merely responded in a reflexive, automatic manner to the calls (e.g. responded to the low-level acoustic structure of the call). Zuberbühler, Cheney and Seyfarth (1999) played eagle shrieks to unhabituated Diana monkey groups and recorded their vocal responses. Before hearing this test stimulus, however, the monkeys were primed with one of three different sounds (Figure 9.4). These prime sounds were an eagle shriek (acoustically and semantically similar), a Diana monkey 'eagle alarm call' (semantically similar; acoustically different) or a Diana monkey 'leopard alarm call' (semantically and acoustically different). After being primed with both the eagle shriek and the 'eagle alarm call', calling rates to the test stimuli were low, as both priming stimuli, despite their acoustic differences, had already indicated the presence of an eagle. After being primed with the 'leopard alarm call', calling rates were high to the test stimuli as information about an eagle in the vicinity was novel. This shows that rather than just responding reflexively to low-level acoustic information in a call, listeners seem to process the meaning of the call, in terms of the event that elicited it (e.g. presence of eagle). Zuberbühler and colleagues suggest that these results indicate that receiver responses are likely mediated by mental representations of the referents of these calls. Similar results have been obtained for vervet monkeys and their perception of two different call types used in encounters between groups (Seyfarth and Cheney, 1986) and for food-associated calls in rhesus macaques (Hauser, 1998). These studies are important, but unfortunately rare examples of

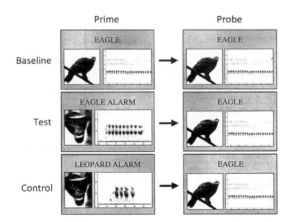

Figure 9.4 Illustration of the experimental design of one of the experiments of Zuberbühler, Cheney and Seyfarth (1999), used to examine whether monkeys processed the meaning of the calls they heard or simply responded reflexively to the acoustic features of the calls they heard. Diana monkey groups were tested on two stimuli separated by 5 minutes of silence. Stimulus pairs differed in similarity of the acoustic and conceptual features across conditions as follows: (a) baseline condition – both the acoustic and the conceptual features remain the same; (b) test condition – the acoustic features change but the conceptual features remain the same; (c) control condition – both the acoustic and the conceptual features change (Copyright © by the American Psychological Association – reproduced with permission; from Zuberbühler, Cheney and Seyfarth, 1999).

experiments that probe the cognitive underpinnings of communication, rather than focusing on surface behaviour.

Campbell's monkeys also produce functionally referential loud calls that seem to be specific to eagles and leopards. Playback experiments showed that their vocal response to these predators remained consistent, regardless of whether the eliciting event was a playback of predator sounds (e.g. eagle shrieks), another Campbell's monkey's alarm call (e.g. 'eagle alarm call') or a sympatric Diana monkey's 'alarm call' (Zuberbühler, 2001). This indicates that the monkeys hearing these stimuli understand that they all predict the presence of a certain predator (e.g. an eagle or a leopard) in the vicinity and therefore react with their own appropriate predator alarm call for their own group.

Ring-tailed lemurs produce 'rasps' or 'shrieks' in response to raptors and 'yaps' in response to carnivores. Playbacks of these call types elicit adaptive anti-predator responses from listeners (Macedonia, 1990): individuals climbed trees in response to carnivore alarm calls and moved down in trees and looked up in response to raptor calls. Although it is unclear whether these alarm calls referentially label the predator type or the likely direction of attack (terrestrial or aerial), experiments in captivity indicate that the alarm calls are not reflective of perceived threat or urgency of response (Pereira and Macedonia, 1991). Carnivore alarm calls given to a dog remained consistent across conditions where the dog either walked by calmly at a distance or charged at the lemurs from close range. Equally, lemurs did not deviate from production of raptor alarm calls when a model raptor was presented perched and stationary or swooping overhead. Thus, regardless of the severity of the threat, lemurs did not deviate from their functionally referential, predator-specific vocal responses.

Functionally referential alarm call systems that denote aerial or terrestrial threats have been identified in a number of primate species including moustached tamarins and black-and-white colobus monkeys (Kirchhof and Hammerschmidt, 2006; Schel, Candiotti and Zuberbühler, 2010; Schel, Tranquilli and Zuberbühler, 2009). Interestingly, the black-and-white colobus monkey alarm system distinguishes between aerial and terrestrial threats at the level of call sequence, not individual call type. Whilst eagles elicit long roaring sequences, terrestrial threats, such as leopards, elicit short roaring sequences. Playbacks of these roaring sequences indicate they are meaningful to listeners, eliciting looks in the direction danger can be expected from (up or down).

More recent research has also revealed a number of less specific alarm-calling systems in primates. Several species, including tufted capuchin monkeys, black-fronted titi monkeys, saddle-back tamarins, red-fronted lemurs and white sifakas are said to have a 'mixed' system, where a highly specific functionally referential aerial or eagle alarm call is produced, but a more general 'disturbance' alarm call is given to a wide variety of other threats and high arousal events including terrestrial predators and aggressive intergroup encounters (Cäsar *et al.*, 2012a, b; Fichtel and Kappeler, 2002; Kirchof and Hammerschmidt, 2006; Wheeler, 2010). For instance, black-fronted titi monkeys tend to produce a specific call type for dangers from the canopy or air (e.g. eagles, predatory capuchin monkeys) and a different call type for a range of terrestrial predators and non-predatory disturbances (Cäsar *et al.*, 2012b). Playback experiments showed that whilst the alarm calls originally given to a perched raptor model elicited looking up in recipients, the more general disturbance call originally given to a terrestrial predator elicited looks to the caller (Cäsar *et al.*, 2012a). White-faced capuchins have also been suggested to possess a mixed system (Digweed, Fedigan and Rendall, 2005), but playback experiments to systematically test listener responses to alarm calls are still needed. Putty-nosed monkeys have a probabilistic alarm call system where call sequences are produced to a range of stimuli, meaning that, for instance, calls associated with leopards are also given during intergroup encounters and call sequences associated with eagles are also given to trees falling (Arnold, Pohlner and Zuberbühler, 2008, 2011). The authors argue that this alarm system requires listeners to infer the likely meaning of the call by combining the call type with the context. So whilst not functionally referential, this alarm system shows some hallmarks of flexibility that are generally thought to be more characteristic of ape gestures (see Chapter 7). Finally, the ruffed lemur alarm call system seems to convey little information to listeners about the likely eliciting event and instead seems to reflect caller arousal. Regardless of the call type, receivers would look to the speaker, indicating they needed more information from the caller (e.g. gaze direction) before responding appropriately (Macedonia, 1990).

As this section highlights there is a broad range of specificity in the functionally referential alarm systems of different primate species. In evolutionary terms it seems several factors likely influence the specificity of an alarm-calling system and the type of information encoded (e.g. predator type, direction of danger, urgency of response, escape strategies) and these include the number and type of predatory threats a species faces and the number of different anti-predator responses the primates need to engage in, in order to survive (Macedonia and Evans, 1993)

9.4.4   Food discovery contexts

Much of the research investigating pointing gestures in captive apes has used food contexts to elicit pointing from apes to human caregivers (e.g. Leavens, Hopkins and Bard, 1996). Desirable out-of-reach food is definitely something that captive apes communicate about with humans in an imperative sense and, whether apes use points or other gestures, they can use gestures to refer to specific food items in their environment (Cartmill and Byrne, 2007). If human experimenters 'misunderstand' the referent of their gestures and provide them with an item that was not referred to, orangutans will elaborate and persist with their gesturing in order to repair the communication and thus reach their goal. Primates rarely actively share food (Jaeggi, Burkart and van Schaik, 2010), and so far no instances of referential gestures towards food have been reported in interactions between primates.

A considerable body of research indicates that many species of animal, including primates, produce specific vocalizations when finding and consuming food. Food-associated calls can function to reference the presence of food, in a similar way to alarm calls that reference the presence of an external danger. Whilst such calls are sometimes accompanied with specific facial expressions in some species, whether they can function as referential signals for the presence of food is yet to be tested. Putatively functionally referential food-associated calls produced by free-ranging toque macaques seem to reliably label the presence of large quantities of food, but there is no evidence that they have the potential to convey information about the type or quality of the food (Dittus, 1984). The acoustically distinct calls produced by toque macaques were elicited exclusively by abundant food sources in 98% of recorded cases, so the calls meet the perception criteria of functional reference (acoustic distinctiveness; stimulus specificity). The monkeys responded to these calls with rapid approach and their feeding bouts lasted significantly longer than comparable feeding bouts when calls were not produced. Although this observational evidence indicates the calls were meaningful to listeners, the case for regarding this call as a functionally referential one is weakened by the lack of playback experiments to test whether in the absence of other cues (e.g. olfactory, visual) the calls functioned to refer to the presence of plentiful food sources. More recent studies have used playback experiments to meet the perception criteria necessary to identify functionally referential calls. Tufted capuchin monkeys showed rapid and direct approaches to a speaker broadcasting food-associated calls compared to control stimuli (Di Bitetti, 2003). Similar findings were obtained for white-faced capuchins (Gros-Louis, 2004a), but as the author acknowledges these results remain ambiguous. Although these monkeys respond to the presence of an abundant food source with direct and rapid approach, this response is also elicited by many other stimuli. Therefore it is difficult to conclude that the responses of listeners are indicative of feeding. In contrast to these ambiguous responses, marmosets show higher frequencies of foraging and feeding behaviours in the 20 minutes after hearing a playback of food-associated calls, compared with two types of control stimuli (Kitzmann and Caine, 2009). This food-specific response in listeners parallels that of domestic chickens, where female chickens respond to male food calls with an anticipatory feeding stance: fixating downwards with the frontal binocular field (Evans and Evans, 1999). These food-specific responses to playbacks of food-associated calls show that these vocalizations unambiguously function to reference the presence of food.

It is proposed that several species of animal produce food-associated calls with greater degrees of referential specificity, where calls provide information about the nature of the food source, as well as its presence. Several species of New World monkeys are able to vary the rate of food-associated call production as a function of the quality or quantity of a food source. Golden lion tamarins, cotton-top tamarins and white-faced capuchins all produce food-associated calls at higher rates to high-quality or highly preferred foods (Benz, 1993; Benz, Leger and French, 1992; Elowson, Tannenbaum and Snowdon, 1991; Gros-Louis, 2004b; Roush and Snowdon, 2000). Whilst in these species the calls vary with preference, independently of quantity, red-bellied tamarins produce calls at a higher rate to both large quantities and higher quality food (Caine, Addington and Windfelder, 1995). These calling systems have the potential to provide listeners with information about the relative quality of a food source, but the absence of playback experiments to test this hypothesis leaves these results difficult to interpret.

Rhesus macaques produce five acoustically different types of food calls: 'warbles', 'harmonic arches' and 'chirps' to rare and highly desirable food, and 'grunts' and 'coos' to low preference foods (Hauser and Marler, 1993). Whilst the type of call produced varied reliably with the quality of food discovered, call rate varied with the hunger levels of the signaller. A playback experiment using an habituation–dishabituation technique illustrated that listeners attended to the meaning of the calls in terms of their referents, not to the different acoustic structures of the five calls (Hauser, 1998). This combination of observational and experimental research indicates that these food-associated calls do function to reference the quality of a food source to recipients.

Chimpanzees also produce functionally referential food-associated calls. For example, they commonly produce 'rough grunts' when approaching or consuming food and close examination of this graded call type has revealed that acoustically distinct variants of this call type are produced in response to foods of different values (Slocombe and Zuberbühler, 2006). As the value of the food source increases, 'rough grunts' become higher pitched and longer in duration. In a captive setting, where food type is the main determinant of food value (quantity, ripeness and quality of food are consistent), this calling system seems to allow specific acoustic variants to be produced to certain high-value food types. Indeed, chimpanzees at Edinburgh zoo were found to have distinct calls for bread, banana and mango and these calls remained consistent in structure across several feeding events. A playback experiment revealed that 'rough grunts' elicited by apples and bread were meaningful to listeners. After hearing a group member's 'rough grunts' to either bread (high value) or apples (low value), a young male chimpanzee exerted significantly more time and effort searching for the food type signalled by the calls (Slocombe and Zuberbühler, 2005a). However, this study has two main weaknesses. First, it is not clear if the listening chimpanzee was extracting information about the value of the food (high versus low), or the specific food type (bread versus apple). Second, it is based on the responses of a single individual whose behaviour may not be representative of chimpanzee behaviour in general. In order to address these points, more recent work has replicated and extended this study at a different site (Slocombe *et al.*, in prep). This study has first replicated the ability of listeners to distinguish between calls for high and low value food and to use this to guide their own search for food. Second, this study indicates that chimpanzees can distinguish between calls given to two high-quality food items, suggesting that this calling system has a very high level of specificity. These chimpanzee studies were the first to provide evidence for functional reference in any great ape species, but more recent work has revealed similar findings in bonobos.

Bonobos, like rhesus macaques, produce five different calls in response to food, but they do not have a simple one-to-one matching between a specific call type and the quality of the food encountered. Instead, bonobos produce sequences of calls, containing several different call types and it is the proportion of certain call types within those sequences that seems to indicate the quality of the food source (Clay and Zuberbühler, 2009). For instance, high-quality food seems to elicit sequences containing mostly 'barks' and 'peeps', whilst low-quality food is more likely to elicit sequences containing 'yelps' and 'grunts'. Listeners are able to use this probabilistic call system to infer the quality of food that elicited the calls and adjust their own foraging behaviour to take advantage of that information (Clay and Zuberbühler, 2011a). This playback study

showed that bonobos extract meaning at the level of call sequence, rather than responding directly to individual call types, and that these call sequences function referentially.

9.4.5   Social contexts

In contrast to communication about predators or food that are both clearly external
entities, the extent to which social signals can fulfil the criteria of 'functionally
referential' signals is unclear. Although receivers have been shown to extract useful
information about ongoing social events from social signals, the extent to which these
are truly external to the signaller is debated. It seems many social signals reference the
behaviour or motivational state of the signaller, and it is disputed if behavioural
referents can be counted as 'external' (Smith, 1981). It could be argued, however, that
signals that allow receivers to understand the nature of a social interaction are not
qualitatively different from signals that allow receivers to understand the nature of a
caller's interaction with a predator. Despite the contentious nature of the definition of
functionally referential signals in animals, we felt it was still important to review this
literature, as from a wider perspective it is still extremely helpful to understand the
conditions under which certain signals are produced and the extent to which receivers
can extract useful information about the eliciting event.

One social gesture that is suggested to function referentially is the chimpanzee
'directed scratch'. When grooming, members of the Ngogo community of chimpanzees regularly perform a noisy, rough scratch on a specific body part and Pika and
Mitani (2006) suggest that this is a referential gesture to indicate the body part they
wish the other to groom. Receivers of this signal groomed the signalled area more than
expected by chance and the authors argue that this shows receivers understand the
intentional meaning of the signal. However, in order for the gestures to be classified as
referential, receiver understanding of these gestures needs to be tested in an analogous
way to playback experiments. Otherwise alternative explanations for these results
cannot be ruled out. For instance, from the signaller's perspective, the 'directed
scratch' may simply be a signal given to maintain or re-engage a grooming partner
in the grooming session. The location of the scratching may be random, but the
receiver may be drawn to this spatial area through stimulus enhancement, thus making
it more likely when they start grooming again they will focus on the area they are
visually attending to.

In the vocal domain there is considerably more evidence for functionally referential
social signals, mainly in copulation and agonistic contexts. In many species, females
produce vocalizations when copulating and the extent to which these calls provide
listening individuals with useful information about the copulation event has been
investigated. It should be noted, however, that some would argue the information
extracted by listeners from these calls (e.g. identity, reproductive state of the female)
is not external and therefore these calls should not be considered functionally referential. The 'copulation calls' produced by yellow baboons are individually distinctive and
playback experiments have shown that males can discriminate between the calls of

different females (Semple, 2001). Calls in this species also vary in structure according to the reproductive state of the female and the rank of the male partner (Semple *et al.*, 2002). Although this systematic acoustic variation gives the potential for listening males to gain important information about the copulation, playback experiments are still needed to see if males are sensitive to these subtle acoustic changes and their corresponding contexts.

There is controversy over the extent to which Barbary macaque 'copulation calls' systematically vary in acoustic structure with the reproductive state of the female. A study comparing 'copulation calls' given at the start and middle of a female cycle, as determined by the size of the female's swelling, found consistent acoustic differences in the calls (Semple and McComb, 2000). Male macaques responded more strongly (more individuals looked towards the speaker and/or approached it) to the mid-cycle calls than the early cycle calls, indicating that they could discriminate between the calls and showed most interest in the calls most likely to be associated with successful fertilization. A more recent study has, however, failed to replicate these findings (Pfefferle *et al.*, 2008). Pfefferle and colleagues used hormonal analysis to identify the fertile period with accuracy, rather than relying on the heuristic of swelling size and subsequent acoustic analysis of calls and showed that 'copulation calls' do not reveal the timing of the fertile phase of the cycle. However, a playback experiment is essential to confirm the original finding that males seemed to respond differentially to calls given at different stages of the cycle, since it is possible that the males are responding to acoustic differences not captured in the acoustic analysis conducted by Pfefferle and colleagues (2008). A less controversial finding is that Barbary macaque calls differ depending on the mating outcome: ejaculatory copulations elicit calls with different structures to non-ejaculatory copulations (Pfefferle *et al.*, 2008). A playback study showed that males can extract information about the likely outcome of a copulation from the call alone: they responded more strongly to ejaculatory calls and subsequently spent more time around the females than after hearing non-ejaculatory calls (Pfefferle *et al.*, 2008).

Recent research into copulation call production in long-tailed macaques has revealed that the frequency of calling and the structure of calls vary systematically with the rank of the male partner, whether the male ejaculated, and whether the female was copulating with a partner who was mate guarding her (Engelhardt *et al.*, 2012). Call structure was not affected, however, by the fertile phase of the female. These calls therefore have the potential to provide complex information to listeners about the identity of the male partner and the circumstances and outcome of the copulation. Playback experiments systematically testing receiver sensitivity to acoustic variation determined by these factors still need to be completed, but it is suggested that these calls play a role in the manipulation of male mating behaviour.

Chimpanzees, like yellow baboons, produce copulation calls that are individually distinctive, but that do not differ according to the reproductive state of the female (Townsend, Deschner and Zuberbühler, 2011). Calls given during the fertile period, as defined by hormonal analysis, could not be distinguished from those given in non-fertile periods. Similarly, female bonobo copulation calls are individually distinctive, but the structure of these calls does not vary with the sex or rank of the partner or the size of the female's sexual swelling (Clay and Zuberbühler, 2011b). Playback experiments to test the sensitivity of the recipients to the identity cues available in both chimpanzee and bonobo calls are still required.

Many species of primate emit vocalizations during agonistic contests. The extent to which the acoustic structure of these vocalizations varies systematically with context and provides listeners with information about the ongoing fight has been examined in rhesus macaques and more recently in chimpanzees. Again, in this social context, the calls emitted seem to provide listeners with information about the nature of an interaction that the caller is actively engaged in, so the extent to which these calls are truly referring to something external to the signaller is highly debatable and thus some researchers reject the notion that these signals should be termed functionally referential.

Despite this, agonistic screams produced by rhesus macaques have been described as functionally referential. Observational work demonstrated that rhesus macaques produce five distinct call types in response to aggression from others (Gouzoules, Gouzoules and Marler, 1984). The type of call produced was dependent on the severity of the aggression received and the rank of the opponent. For instance, 'arched screams' were given in response to lower-ranking individuals who attacked without making physical contact, whereas 'noisy screams' were given more to physical contact from high-ranking individuals. Playback experiments tested listener understanding of these calls. Mothers were played 'screams' given by their offspring and their pattern of response indicated that they extracted information from the 'screams' about the severity of risk faced by their offspring and the rank of their opponent. Mothers showed the most interest, as measured by looking time at the speaker, in calls that indicated their offspring was receiving severe aggression or was being attacked by an individual of lower rank. To interpret these findings, it is important to emphasize that the social organization of rhesus macaques is based on the dominance hierarchy of matrilines: if a lower-ranking individual is threatening a member of a more dominant matriline, immediate action is needed by members of that matriline to defend their dominance status.

Chimpanzees frequently engage in agonistic interactions and they commonly produce 'screams' during these encounters. Systematic research into these calls and receiver responses to them have revealed that they provide listeners with a rich array of information about the ongoing interaction, but whether these signals are sufficiently external to the signaller to qualify as functionally referential signals is a matter of debate. Like rhesus macaques, chimpanzees commonly scream when the victim of aggression, but chimpanzees are also observed to scream in the role of aggressor, if attacking an individual of equal or higher rank. Observational work with wild chimpanzees of the Budongo Forest, Uganda, demonstrated that victim and aggressor screams were acoustically distinct call variants within their graded call repertoire (Slocombe and Zuberbühler, 2005b). Playback experiments were conducted in captivity to examine if listeners could understand the role of the caller in an agonistic encounter just from hearing their 'screams' (Slocombe et al., 2010a). Chimpanzees were presented with sequences of 'screams', simulating a fight occurring out of sight, which was either congruent or incongruent with the established dominance hierarchy. For instance, the incongruent sequence consisted of a low ranking 'aggressor scream' followed by a high-ranking 'victim scream', simulating a highly unlikely event. Chimpanzees looked significantly longer at the incongruent sequences than the congruent sequences, indicating that they were sensitive to the violation of the dominance hierarchy presented in this sequence. In order to process the incongruent nature of the sequence, listeners must have been able to identify callers from their 'screams' and critically to extract information about the role of each caller (victim or aggressor) in order to infer the direction of aggression. This ability to use 'screams' to make sense of fights that are out of sight is highly adaptive for this fission–fusion species in their natural habitat, where most fights occur out of sight.

When chimpanzee 'victim screams' were examined in more detail, it was found that these calls, like those of the rhesus macaques, were produced in a context-specific manner with regard to the severity of the aggression the caller was facing. Slocombe and Zuberbühler (2007) showed that wild chimpanzees facing mild and severe aggression produced acoustically distinct 'screams'. Listener understanding of these 'scream' variants was tested in the first intragroup playback experiment to be attempted with this species in the wild (Slocombe *et al.*, 2010a). Despite methodological difficulties in conducting realistic playbacks in a wild setting, which resulted in a small sample size, chimpanzees showed they could distinguish between severe and mild 'victim screams', looking longer at the speaker in response to severe 'screams'. Due to acoustic differences between the mild and severe 'victim screams', it was important to test if the listeners were simply attending to the most attention-grabbing, salient sound in the environment, or whether they were processing the likely eliciting event and reacting to that. In order to test this, each participant was also played a 'tantrum scream', given in frustration rather than in response to aggression. These screams matched the 'severe screams' very closely in acoustic structure, but the eliciting event was very different and of little interest to unrelated adult group members. Chimpanzees did not pay much attention to the 'tantrum screams', indicating that their prolonged orienting response to the 'severe screams' was not just a surface level response to an acoustically salient stimulus in the environment.

Facial expressions tend to be examined largely in social contexts, however, there is no evidence that facial expressions in this domain function referentially or provide more specific information about the nature of an interaction. As reviewed in Chapter 4, with our current methodologies this is partly because it is very difficult to test receiver understanding of visual signals (gestures and facial expressions).

9.5   What could a multimodal approach bring to this topic?

The vast majority of studies reviewed in this chapter focused on signals in a single modality. Our reliance on a unimodal framework means we have likely overlooked referential multimodal signals. First, whilst unimodal signals may be quite flexibly produced across contexts, when combined with another signal, the composite signal is often highly context specific. For example, Crockford and Boesch (2003) found generic chimpanzee 'barks' given in several contexts became highly context specific if their combination with 'drumming' signals were considered. Thus, a multimodal approach would likely help us to identify more context-specific signals that may be functioning in a referential manner. In terms of comprehension, using playback experiments to test the response of receivers to a vocal signal extracted from a multimodal signal may lead us to misunderstand the system. Unless living in a visually dense habitat, most primate receivers do not perceive vocal signals in isolation. They usually have access to visual signals, such as facial expressions, manual gestures and body postures that accompany the vocal signal to help them interpret the meaning or referent of the call. Thus, researchers may be surprised to find a highly unspecific alarm call system in a primate species, because they may have overlooked the possibility that receivers are using visual signals to disambiguate the call and a more specific vocal system is therefore unnecessary. A multimodal approach that provides receivers with composite vocal and visual signals (e.g. by using robotic models or video playbacks) could help overcome these issues. Finally, if we examine composite multimodal signals we may get a clearer understanding of the information or meaning of the signal extracted by receivers. Composite signals have a higher capacity for transmitting both referential and emotional information together and by focusing exclusively on the referential content we are likely missing out on the full meaning and even the real function of the signal. For instance, if a human points at an object and combines the gesture with a facial expression of disgust, the multimodal signal has not only a referential aspect, but also an important emotional component, which together provide highly relevant information to the receiver. Applying a multimodal approach to primate communication is thus likely to provide a richer understanding of both the referential and emotional content of signals.

Summary

Primates are capable of engaging in two types of referential behaviour: captive great apes point imperatively to objects primarily to communicate with human caregivers and many species of primates have vocalizations that function to refer to objects and events in their environment such as predators and food. As highlighted in the introduction, research into these two categories of behaviour have been strongly influenced by attempts to find precursors to two different human behaviours (pointing and linguistic reference). Although both these human behaviours direct the attention of recipients to external events and objects and are thus termed referential, the cognitive processes underlying them may be distinct. Both related behaviours in primates are very interesting, but perhaps it is not surprising that few successful attempts are made to examine a unified, cross-modal concept of 'reference' in primates that brings together the different definitions and criteria traditionally applied to pointing and vocal functional reference. It seems likely, given our current understanding of intentionality (see Chapter 8), that one key cross-modal difference is the occurrence of intentional reference in the case of ape pointing and functional reference in the vocal domain where the intentionality and cognitive mechanisms driving production are largely unknown (but see Schel, Machanda *et al.*, in press; Schel, Townsend *et al.*, in press).

An additional reason for the sharp modality-divide in terms of functionally referential signals are differences in methodology across modalities. First, with acoustic analysis techniques it is arguably easier to perform fine-grained analysis on vocal signals as opposed to manual gestures. This enables researchers to distinguish between two similar signals given in two different contexts. Thus, context-specific signals are more likely to be discovered in the vocal domain. Second, playback experiments then allow researchers to test the context independence of a signal with relative ease, thus testing the perception criterion of functional reference is possible. Analogous methods for visual signals are not widely available, although touch-screen methods in captivity have the potential to enable the testing of recipient understanding. Within vocalizations, functionally referential calls have been predominantly studied in evolutionarily important contexts, such as predator defence, food discovery, copulations and agonistic encounters. The study of functionally referential signals is very helpful in revealing how primates use their calls and the type of information receivers are capable of extracting from them.

Both ape pointing and functionally referential calls reveal some commonalities with human communication and some important differences (e.g. signaller's intent in production of functionally referential signals or the recipient's understanding of informative points). Research in this area is important for furthering our understanding of what is unique about human language and the evolutionary path of this remarkable ability. Adopting a multimodal approach may help researchers to identify more context-specific signals with the potential to function referentially, to examine the comprehension of signals in a more holistic and ecologically valid manner and to better understand both the emotional and referential content of signals.

Part IV

Approaches to the evolution of primate communication

10 A multimodal approach to the evolution of primate communication

Scientists who study primate communication, with the goal of understanding human communication, pursue two different, yet related, questions. Some scientists ask *phylogenetic* evolutionary questions (what was the historical pathway of a specific communicative ability?), and others ask *functional* evolutionary questions (what were the selection pressures that led to evolutionary changes in this domain?). These two types of question are both necessary to fully understand the evolution of communication, and one is not necessarily more important or useful than the other. Interestingly, integration between these two foci is rare, despite the potential benefits of integration. Integration between phylogenetic and functional questions could be highly informative when considering the evolution of communication, as understanding the reasons for change could help elucidate the specific process of change, and vice versa. Here, we argue that one way to bridge the gap between phylogenetic and functional questions could be to adopt a more multimodal approach to the study of primate communication, which is usually neglected in favour of a unimodal approach.

In this final chapter, we first summarize the general advantages of adopting a multimodal approach (see Chapter 5), regardless of whether the research questions are phylogenetic or functional. Second, we discuss the difference between phylogenetic and functional questions. Finally, we propose that integration between phylogenetic and functional questions would be helpful to move the field forward, and that a multimodal approach could be particularly useful in this endeavour (see also Waller *et al.*, 2013a).

10.1 Why adopt a multimodal approach?

Primate communication is usually broken down to its constituent parts and studied as distinct, unimodal systems, often as facial expression (e.g. Parr and Waller, 2006), gesture (e.g. Hobaiter and Byrne, 2011a; Liebal, Pika and Tomasello, 2006; Pika, Liebal and Tomasello, 2005), vocalization (e.g. Fitch and Hauser, 1995; Slocombe and Zuberbühler, 2005a) or olfaction (e.g. Heymann, 2006). Perhaps as a direct result, the study of these isolated modalities seems to have taken different research trajectories, where each modality attracts different methods and theoretical assumptions, as discussed in Chapters 4 and 5. To summarize previous chapters, a multimodal approach may offer several advantages over the dominant, unimodal approach, regardless of whether the specific research question is phylogenetic or functional. To fully understand how and why communication systems have changed during primate evolution (and to identify what human communication has built on) we need to be accurate in our assessment of primate signals. A multimodal approach could help scientists achieve this goal in three ways.

First, facial expression, gestures, vocalization and olfactory signals are studied in different ways in nonhuman primates, using radically different theoretical approaches and different methods (Slocombe, Waller and Liebal, 2011). Scientists *think* they know which modalities exhibit certain characteristics, and thus which were more likely to offer stepping-stones to the evolution of human language, but as they are studied in such different ways, it is possible that these conclusions are erroneous. One modality could exhibit characteristics that were co-opted into another at a later stage, of course, (e.g. gesture may have been a precursor to spoken language), but proposing such steps is largely irrelevant if we cannot make accurate comparisons between modalities. A multimodal approach is necessary to make better comparisons between modalities, and will give us a more complete picture of how these modalities operate differently (if at all).

Second, scientists often isolate a phenomenon from the holistic setting in order to understand the core properties. However, in the communication context, scientists could be making false conclusions by removing signals from the context in which they occur. Specifically, if a single signal is part of a composite signal, the meaning and characteristics of that signal could be entirely altered if it is removed from that composite signal and studied in isolation. For example, apes use a 'slap' gesture in playful and aggressive contexts, and it seems that the facial expression accompanying the gesture allows the receiver to respond appropriately (Rijksen, 1978). When the 'slap' is paired with a 'playface', it leads to play. Isolating the slap from the 'playface', will not, therefore, help us understand the signal better. In fact, it could lead to an incorrect characterization of this gesture. By examining the gesture alone, researchers may conclude that this signal is used flexibly across contexts. Yet if the composite signal is examined, researchers might conclude that the composite is context specific. Thus, studying signals as part of an integrated, multimodal and holistic system is essential to understand the characteristics of primate communication.

Third, combining and integrating signals has the potential to increase a signal repertoire exponentially (Partan and Marler, 1999). Therefore, communicative complexity might be less about how each single modality is used, and more about signal integration. In which case, the historical focus on single modalities might be missing an inherent feature of primate communication systems. Mapping socioecological variables onto facial, vocal, gestural and olfactory repertoires independently, may overlook important patterns. Instead, investigating how multimodality relates to variables such as group size, brain size and social structure, could be an important next step.

10.2   What is the difference between phylogenetic and functional questions?

The way in which scientists can answer questions at different levels was originally outlined by Niko Tinbergen (see Chapter 1). In his seminal paper, Tinbergen discussed four different ways in which a question about behaviour can be answered (Tinbergen,

1963), often referred to as Tinbergen's 'four questions' or 'four whys': *causation*, *survival value*, *ontogeny* and *evolution*. Explanations of behaviour at the level of causation and ontogeny concern the underlying physiological (or cognitive) and developmental causes, and are often termed proximate explanations, where proximate refers to the immediate, short-term mechanisms at work during the animal's lifetime (*how* it works). Explanations of behaviour in terms of survival value concern how the behaviour promotes an individual's ability to survive and reproduce (the function), and explanations in terms of evolution concern the phylogenetic history of the behaviour in terms of evolutionary change. The latter explanations are often termed ultimate, as they refer to past events in contrast to immediate ones (*why* it works that way). Tinbergen's framework is very helpful to understand why different explanations for a behaviour need not be in competition, indeed all are necessary to fully understand the manifestation of a trait in an individual. Confusion between the two types of explanation is common, however, which can result in questions being posed at one level, but being answered at another (see Scott-Phillips, Dickins and West, 2011, for a recent review of this issue).

The distinction between the two ultimate levels of explanation is less often discussed, but understanding of the distinction is nevertheless still relevant to avoid confusion. The difference between phylogenetic and functional questions bears some similarity to the theoretical distinction between comparative and evolutionary psychologists, where comparative psychologists are often interested in identifying similarities and differences between related species, and evolutionary psychologists are often interested in identifying the adaptive reasons for these differences. Scientists asking phylogenetic questions about communication study animal communication in an attempt to trace the evolutionary development of a specific communication system, often human, and often human language specifically (e.g. Arbib, Liebal and Pika, 2008; Fitch, 2000; Ouattara, Lemasson and Zuberbühler, 2009b). Scientists asking functional questions focus on what animal communication can reveal about the evolution of complex social systems in general, in an attempt to understand the fundamental principles of behavioural change, and the selection pressures underlying this change (e.g. Dobson, 2012; Gustison, le Roux and Bergman, 2012; McComb and Semple, 2005). Such an approach can also generate hypotheses about the evolution of human communication (Dunbar, 2003), but this is not necessarily the main focus.

The focus of phylogenetic questions is usually anthropocentric, where an attempt is made to identify which aspects of human communication are species unique, and which have been inherited from or built upon established primitive primate ground plans. The aim is to search for the building blocks that specific communication systems built on, in order to better understand the course of events over evolutionary time, since 'nonhuman primates are our closest living relatives, and their behaviour can be used to estimate the capacities of our extinct ancestors' (Fedurek and Slocombe, 2011). Study species tend to be very closely related to humans, and most often closely related nonhuman primates (e.g. chimpanzees).

A phylogenetic approach helps us understand what traits and skills were available for evolution to work with when new systems were developed. For example, it is clear that olfactory communication is less intensively used in humans compared to some other nonhuman primates, reflecting the relative reduction in size of the olfactory apparatus from prosimians to humans (Martin, 1990). In contrast, numerous studies have shown that chimpanzee and human facial expressions are produced using similar muscles and neural substrates (Burrows *et al.*, 2006; Sherwood *et al.*, 2004b, 2005) and are also processed in a similar manner to human facial expressions (Parr, Hopkins and de Waal, 1998). Human facial expression systems are thus very unlikely to be species unique. Instead, human facial expression must have built on an existing system of facial communication present in the shared ancestor. Indeed, similarity in facial musculature between humans and distantly related primates (Burrows and Smith, 2003) suggests that human facial expression is built on relatively archaic systems (Figure 10.1). Davila-Ross, Owren and Zimmerman (2009) conducted phylogenetic analysis on ape laughter vocalizations, and found that similarities mapped closely to genetic relationships between species. Thus, laughter is also unlikely to be a human-unique trait, and instead probably developed from homologous behaviours in the shared ancestor.

Some aspects of the human language system have also been identified in extant primate communication. For example, a form of referentiality, a key feature of human language, has been demonstrated in the vocalizations of monkeys (Seyfarth and Cheney, 1990) and apes (Slocombe and Zuberbühler, 2005a) (see Chapter 9 for a full discussion). Studies of apes have also suggested that intentionality, another key feature of human language, is a characteristic of ape gestural communication (Tomasello, 2008) (see Chapter 8). Whether this is equivalent to the referentiality and intentionality that abounds in human language and develops early and automatically in human ontogeny is a matter of debate. Importantly, of course, nonhuman primates do not develop language spontaneously in their natural environment, and attempts to teach nonhuman primates language in captivity have had far more success in comprehension than production (Savage-Rumbaugh and Lewin, 1994; see Table 10.1). Thus, any language-like skills identified in nonhuman primates are only potential precursors to the human language system, and not evidence that the species has been selected to produce language.

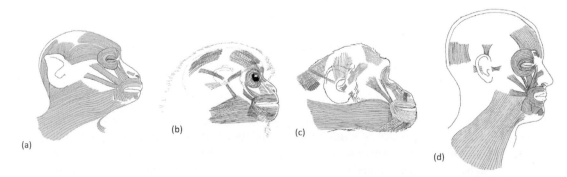

Figure 10.1 The basic ground plan of facial muscles in four primate species. (a) Rhesus macaques (from Waller *et al.*, 2008, adapted from Huber, 1931), (b) gibbons (from Burrows *et al.* 2011), chimpanzees (from Burrows *et al.* 2006) and humans (from Waller, Cray and Burrows, 2008). (Drawings: Tim Smith)

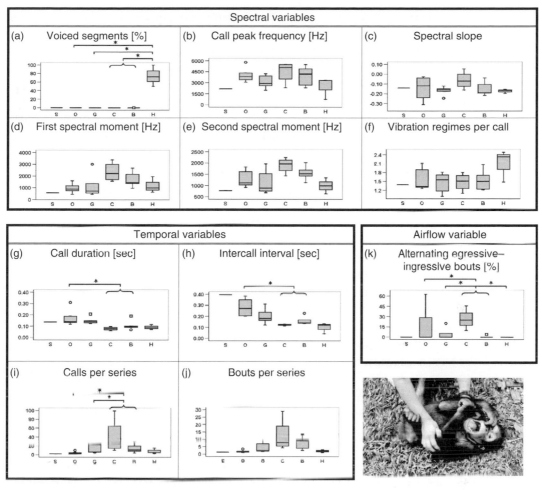

Figure 10.2 Species-specific comparisons for tickling-elicited sounds in apes and humans, including spectral (a–f), temporal (g–j), and airflow (k) of siamangs (S), great apes (O: orangutan; G: gorilla; C: chimpanzee; B: bonobo) and humans (H). Siamang values were closer to those of the orangutans than to any other species. Humans produced significantly more voiced sounds (with regular vocal-fold vibration) than any other species, whereas vocal-fold vibration patterns were irregular and noisy among the great apes (a). Statistical comparisons on other spectral variables showed no species-specific differences. Significant differences in temporal variables occurred only among the great apes. In comparison to orangutans, chimpanzees and bonobos exhibited shorter calls and intercall intervals (g, h) and chimpanzees and bonobos showed more calls per series than either gorillas or orangutans (i). All species could emit two or more bouts per series (j). Airflow results showed that all great apes produced both consecutive egressive calls (during exhalation phases) as well as alternating egressive–ingressive sounds (during exhalation-inhalation phases), with the latter being predominant in chimpanzees (k). In contrast, humans emitted exclusively egressive laughter. **O** = outlier; **□** = extreme value; **✱** = significant difference. (Adapted from: Davila Ross, Owren and Zimmermann, 2009)

Nevertheless, such findings suggest that human language could have built on existing systems rather than evolving from scratch.

Although phylogenetic questions are interested in an ultimate, evolutionary question (*what* was the path of evolutionary change), scientists are often interested in the proximate mechanisms of nonhuman primate communication, in order to find out which core processes were available for other communicative systems to develop from. When the focus is explicitly cognitive (as it often is in language evolution research), scientists face the question inherent in all animal cognition research: can the cognitive processes of other minds ever be fully exposed through scientific investigation? It is exceedingly difficult, particularly in studies of spontaneous communication, to exclude low-level interpretations of behaviour and design studies that inform us about mental processes. In many studies, even though the aim may be to increase our understanding of cognitive processes, it could be argued that we rarely succeed in addressing anything beyond surface behaviour (but see Cheney and Seyfarth 1988; Zuberbühler, Cheney and Seyfarth, 1999) where the goal was to distinguish between mental representation and stimulus response. One approach to tackling this issue is to elucidate whether nonhuman primates have the fundamental cognitive capacities to deal with different aspects of language, by trying to teach apes an artificial language system (as discussed above), and seeing which aspects can be learned and which cannot. As a whole, these studies have left us with the main message that primates do not have the capacity to grasp all aspects of human language, but that some of the key cognitive building blocks may be present (Savage-Rumbaugh and Lewin, 1996) (see Table 10.1).

Scientists interested in understanding *why* human communication systems have evolved focus on the evolutionary function of communication. Such an approach is, of course, still related to the pathway of phylogenetic change, but includes an additional question relating to the selection pressures that lead to these changes. Here, there is focus on a much wider range of species in order to elucidate general selective pressures on communication (e.g. the function of alarm calls in meerkats; Manser, Seyfarth and Cheney, 2002). Examples of convergent evolution are also helpful in order to identify selection pressures. Domestic dogs, for example, have emerged as an interesting model for the evolution of human-like communicative skills (e.g. Kaminski, Call and Fischer, 2004). Identifying the function of a behavioural trait, however, is not straightforward. Even if current fitness consequences are found (i.e. a behaviour leads to greater survival or reproduction), it does not always follow that this was why the behaviour evolved in the first place (Gould and Lewontin, 1979). Phylogenetic inertia may cause a trait and behaviour to be present in a species when it serves no current function (fitness neutral), or serves a different function (exaptation), but it has nevertheless been inherited from an ancestral species where the behaviour was originally selected due to bestowing a fitness advantage. Comparative modelling approaches often strive to factor out the effects of phylogenetic inertia in order to identify the real selective pressures (Shultz, Opie and Atkinson, 2011).

Analysing the social context and determining the social function of communicative signals is often crucial to identify any potential fitness consequences of performing the specific signal (e.g. what is the advantage of using this signal?). In some cases, the advantage to sender and receiver is clearer than others. Predator alarm vocalizations, for example, presumably help the listener avoid predators, and could benefit the sender in terms of kin selection or cooperative defence. The specific advantage of many social signals, however, can be less easy to predict and/or measure. One approach to solving this problem is to quantify immediate or short-term social effects of a signal (e.g. an increase in affiliative social contact) and extrapolate from this to infer social function (e.g. Waller and Dunbar, 2005). It can also be helpful to identify the socioecological variables associated with high levels of a communicative behaviour, and thus factors which could have acted as potential drivers for the evolution of the system. Comparative and modelling approaches are often employed to identify the relationships between different social and ecological factors across many species. Dobson (2009) found a positive relationship between social group size and facial mobility in nonhuman primates, suggesting that size of social group has driven the evolution of facial expression. Similarly, McComb and Semple (2005) found that increases in nonhuman primate vocal repertoire size were associated with increases in group size and time spent grooming. Both studies suggest that at least some communicative signals have coevolved with social bonding and have functioned to aid social cohesion. Such an interpretation fits well with the theory that language evolved as an efficient alternative to grooming when living in a large social group (Dunbar, 1996). In sum, as discussed in Chapter 1, evidence points towards a relationship between social complexity and communicative complexity, with communication evolving as a result of increased social complexity (Freeberg, Dunbar and Ord, 2012).

Group size is often used as a proxy for social complexity in comparative analyses, but whether this truly reflects the level of social complexity is a matter of debate. Arguably, it is the quality of relationships within a social group that is equally (if not more) important than the number. Emery et al. (2007) argue that the long-term monogamous and cooperative pair bonds exhibited by some species of birds are more cognitively demanding than other types of pair bond and may explain increased cognitive abilities in these species (which are often equivalent to primates' cognitive abilities). Similarly, it has been suggested that pair-bonded primates exhibit a social complexity, which is not captured by a group size proxy (Dunbar and Shultz, 2010; Shultz and Dunbar, 2007). It remains to be tested whether social bonds such as the close pair bonds seen in hylobatids also require complex communication (Waller et al., 2012).

10.3   How will a multimodal approach help us understand the evolution of communication?

Integration between phylogenetic and functional questions is not at all common. Consideration of findings at the two levels, could, however, be useful in better understanding the course of events that have led to the evolution of specific aspects of communication. The evolution of the human 'smile', for example, proposed as homologous to the primate 'bared-teeth' display (Bolwig, 1964; van Hooff, 1972), is difficult to understand unless phylogenetic and functional questions are considered

(a)   (b)

Figure 10.3 Human smile (a) and primate 'bared-teeth display' (b: crested macaque) are suggested to have evolved from the same ancestral display. (Photos: (a) from Olszanowski *et al*. (2008), www.emotional-face.org, (b) Jérôme Micheletta); reproduced with permission.

simultaneously (Figure 10.3). Based on FACS analysis (Ekman, Friesen and Hager, 2002; Vick *et al.*, 2007; see Chapter 4, section 4.3.2), the two displays are similar morphologically, but not identical. The human 'smile' is formed of AUs 6 + 12 + 25, whereas a typical primate 'bared-teeth' display is formed of AUs 10 + 12 + 16 + 25 (Parr *et al.*, 2007; Vick *et al.*, 2007). The phylogenetic question is whether these can be truly homologous displays if they have different muscular components. Consideration of functional questions, however, can be extremely helpful in better understanding the answer to this phylogenetic question.

Preuschoft and van Hooff (1995) proposed the *Power Asymmetry Hypothesis of Motivational Emancipation* (see Chapter 1, section 1.3.1 and Chapter 7, section 7.2.1) to explain the pattern of facial expression across primate species. Species with strict, linear dominance hierarchies use facial expressions (such as the 'bared-teeth' display) in narrow contexts and asymmetrically (from subordinates to dominants). In contrast, species with more relaxed dominance styles use the same facial expressions flexibly in broader contexts. Regardless of the immediate social context, the social outcome of the 'bared-teeth' display seems to be to reduce aggression and/or increase affinitive contact (Bout and Thierry, 2005; Flack and de Waal, 2007; Preuschoft, 1992; Waller and Dunbar, 2005). Likewise, another facial expression (the 'playface', proposed as a homologue of human 'laughter' face; van Hooff, 1972) is similarly emancipated from narrow usage in play when a species is less constrained by dominance hierarchies. Such species sometimes exhibit a facial expression, which seems a blended, or converged, display between the 'bared teeth' and 'playface' (Thierry *et al.*, 1989). As humans are characterized by relatively relaxed dominance, the human 'smile' may similarly represent convergence between the primate 'bared-teeth' display and the 'playface'. In which case, we might not expect the human smile to be physically identical to the 'bared-teeth' display in other primates, but to also bear some similarity to the primate 'playface' (AUs 12 + 25 + 26; Parr *et al.*, 2007). Note that the primate 'playface' does sometimes include upper teeth exposure, which could also result from convergence of two displays (Waller and Cherry, 2012).

Another reason to believe that the human 'smile' represents convergence between the two displays is that the adaptive functions of the 'bared-teeth display' and 'playface' (in nonhuman primates), as well as 'smiling' and 'laughter' (in humans), all seem to be broadly similar. Proximate mechanisms (cognitive, physiological, developmental bases) may differ, but all have been argued to have some sort of social bonding function (Dunbar, 2012; Mehu, Grammer and Dunbar, 2007; Waller and Dunbar, 2005). It is possible that occupying the same functional niche caused convergence to occur. In sum, it is only through consideration of function in humans and other species that we can truly understand how the human smile has become manifest in human social interaction, and how it is rooted in ancestral display. If primate gestures and vocalizations that are often proposed as precursors to human language (e.g. Arbib, Liebal and Pika, 2008; Cartmill and Byrne, 2007; Slocombe and Zuberbühler, 2007) were similarly considered in terms of function, it could become clearer which modality (if any) was a more likely precursor to human language.

Such integration of phylogenetic and functional questions when considering the evolution of communication is rare, however. A multimodal approach may help bridge this gap. First, an understanding of communication in its true, holistic form may reveal adaptive function when it is not clear from its components. Second, and perhaps more importantly, multimodality may itself have been an important precursor to more complex forms of communication, such as language, as it may support the simultaneous transmission of both emotional and cognitive information through different channels. Jablonka, Ginsburg and Dor (2012) argue that human language coevolved with human social emotions as part of a complex gene-culture coevolutionary framework. Crucial to this argument is that language built on the socio-communicative skills used in cooperative contexts (Tomasello, 2008), as collaborative social practice was necessary for the development of instructive communication, such as during tool making and alloparenting. The behaviours that primates use to facilitate cooperation and social bonding are not usually those that appear particularly cognitively based, such as referential vocal signals or intentional gestures, but instead are those considered to be more emotionally driven, such as laughter and facial expression (Dunbar, 2012). In which case, any consideration of what might have been the precursors to language in this scenario, would benefit from consideration of primate communication as a multimodal system. Multimodal communication may require incorporation of different

Table 10.1 What do ape 'language' projects tell us about human language? The table provides an overview of ape 'language' projects and their major findings in regard to the production and comprehension of different kinds of languages. Research on ape 'language' is a hotly debated area of research, as already indicated by usually referring to language in quotes (Parker and Gibson, 1990). The overarching main question of such projects is if primates other than humans are capable of learning and using linguistic symbols in ways similar to human language, both in terms of production and comprehension. Both similarities and differences between language abilities of human and nonhuman primates can shed light on the building blocks needed to acquire and use language to communicate with others. On the one hand, these projects indicate that great apes are capable of learning and using different kinds of linguistic symbols. The table demonstrates that individuals of different species learned to use different kinds of symbols, with some of them showing remarkable skills in regard to productivity, semanticity, combinatorial rules, and even cultural transmission and displacement. On the other hand, there are many more projects not listed in this table that failed to teach language to apes. The majority of individuals shown below did not spontaneously learn and use language (with the exception of the bonobo Kanzi and the chimpanzee Loulis) and it took long periods of extensive training to teach a comparably limited repertoire. Overall, their main motivation to produce signs was to communicate requests for certain objects or actions and thus their production of signs lacked the variety of intentions underlying language use in humans (Rivas, 2005). Most importantly, none of these projects could show that all criteria characterizing language use in humans (see Chapter 5, **Box 5.1**) were met by any of the language-trained apes. Therefore, some authors argue we should not to use the term language for any research in animal communication (Sebeok and Sebeok, 1991).

Medium	Individual	Species	Training	Repertoire (Production)	Production	Comprehension	Publications
Spoken English	Gua	Chimpanzee	No specific training; Raised like human child	-	-	Comprehension of approx. 58 spoken requests for certain actions, but not objects	Kellogg and Kellogg, 1933
	Vicky	Chimpanzee	To vocalize / imitate on command (rewarded with food); Molding of lips; Raised like human child	Mama, Papa, cup, up	Only those four words, supported by hands	Comprehension of a large number of spoken phrases, but only in the corresponding contexts/situations	Hayes, 1951; Hayes and Hayes, 1951
Plastic tokens	Sarah	Chimpanzee	Conditioning to learn association between token and referent; Food reward	100	Sequences; Displacement: Comments on past and future events	Comprehension of negation, name-of, if-then, same-different, greater-smaller	Premack, 1971; Premack & Premack, 1983
American Sign Language	Washoe	Chimpanzee	Not raised like human child, but in stimulating environment with close contact with humans; No spoken English allowed in her presence; Imitation (on command); Instrumental conditioning with tickling as reward	200	Novel meaningful combinations, considered semantic roles (e.g., agent-object); Use of attention-getting, noisy sounds to initiate interactions and to attract human's attention; Comments about internal states or external events; Modulation of signs (increase of size and speed, holding, reiteration and duplication of signs);	Not studied in very much detail; 'has been taken for granted' (Savage-Rumbaugh and Rumbaugh, 1998)	Bodamer and Gardner, 2002; Chalcraft and Gardner, 2005; Fouts, 1975; B. T. Gardner and Gardner, 1974; R. A. Gardner and Gardner, 1971; Leitten, Jensvold, Fouts and Wallin, (2012)

Name	Species	Raising conditions	Vocabulary	Achievements	Comprehension	References
Loulis	Chimpanzee	Humans did not sign in his presence (only seven signs); Learning by observing interactions of his adoptive mother Washoe with other chimpanzees or humans ('cultural transmission'	50 (at the age of 73 months)	Private signing; Contingent responses depending on the interlocutor's input Gestures developed gradually into signs; Production of two-sign combinations at 15 months of age	Some evidence that he responded to signs produced by Washoe	Gardner and Gardner, 1989
Nim Chimpsky	Chimpanzee	Not raised in stimulating environment; Several trainers	125 signs at age of 5 years	Mostly requests; Rarely initiated interactions; Mostly repetitions of the trainer's utterances; Mostly two-word combinations; No grammar	-	Terrace 1979, Terrace et al., 1979
Koko	Gorilla	First training took place in zoo-setting; later housed in trailer at university; Operant-conditioning and molding of hands; Received reward when she produced requested signs;	264 words at age of 5.6 years; today approx. 1000	Meaningful combinations, creation of compound names and metaphors; Invention of new signs; Differentiation of semantic relations of signs (e.g., agent-action, agent-object); Displacement: Comment on past and future events; Modulation of the meaning of signs (changing of articulation parameters such as motion, location, configuration, facial expression, or body posture); Communicates internal and emotional states and external environment; Private signing	Understood about 50% of 2-, 3-, or 4-element sentences in 'Assessment of Children's Language Comprehension' test; It is suggested that she comprehends approx. 2000 English words, but this is not systematically studied;	Patterson, 1980; Patterson, 1981; Patterson and Linden, 1981
Chantek	Orangutan	Raised in human environment, but not like human child;	Approx. 140	Combinations of signs; Invention of new signs; Displacement: Comments on past and future events;	Comprehension of English: understanding of action-object and	Miles, 1983, 1990

Table 10.1 (*cont.*)

Medium	Individual	Species	Training	Repertoire (Production)	Production	Comprehension	Publications
			No speech used in his presence; he only exposed to it after five years; Teaching of signs by imitation and molding of hands		Modulation of the meaning of signs	agent-action-combinations	Rumbaugh and Gill, 1977; Rumbaugh et al., 1977
Lexigrams	Lana	Chimpanzee	Lexical keyboard; Teaching of sequences of words ('stock phrases')	–	Reproduced taught sequences; Novel sentence constructions	Comprehension of meaning of lexigrams remained unclear	
Lexigrams, Spoken English	Kanzi	Bonobo	Language-rich environment shared with humans and bonobos; Learning by observing interactions of his mother with her trainers ('cultural transmission'); Unreinforced 'dialogue' with trainer	Several hundreds	Sequences based on consistent combinatorial rules (e.g., agent-action, action-object, lexigrams-gesture); Displacement: Comments on past and future events; Uses declaratives to name objects, to interact, and to comment	Understands meaning of all lexigrams he uses; Understands complex spoken sentences; Understands syntax based on categories like object, action, location; At age of 8, comprehension of spoken commands for actions similar to 2-year old child	Greenfield and Savage-Rumbaugh and 1990; Lyn et al., 2011; Savage-Rumbaugh, Murphy, Sevcik et al., 1993

emotional and cognitive systems (e.g. combining 'emotional' facial expressions with 'cognitive' gestures). In sum, such integration may have been a potentially important precursor to human language (which answers a phylogenetic question) precisely because integration had some advantage (which answers a functional question). Such speculations could fuel interesting avenues for future research.

Summary

Scientists studying the communication of non-human animals are often aiming to better understand the evolution of human communication, including human language. Some scientists take a phylogenetic perspective, where the goal is to trace the evolutionary history of communicative traits, while others take a functional perspective, where the goal is to understand the selection pressures underpinning specific traits. Both perspectives are necessary to fully understand the evolution of communication, but it is important to understand how the two perspectives differ and what they can and cannot tell us. Here, we suggest that integrating phylogenetic and functional questions can be fruitful in better understanding the evolution of communication. We also suggest that adopting a multimodal approach to communication might help to integrate phylogenetic and functional questions, and provide an interesting avenue for future research into language evolution.

Glossary

Accessory olfactory system (AOS): Part of the olfactory system that is involved in the perception of liquid-based odorants (smells), which often function as *pheromones*; associated with the vomeronasal organ.

Acoustic analysis: Measurement of the temporal and spectral structure of vocalizations.

Adaptation: (1) The state of being adapted to a specific environment or circumstance, or (2) the evolutionary process that results in a trait. A trait has evolved by means of natural selection and is an adaptation to the current environmental conditions, thus contributing to an individual's fitness and survival.

Affiliative: A non-aggressive, friendly social interaction or context, often characterized by close proximity between two individuals (e.g. grooming).

Affix: A *morpheme* that is attached to the word stem.

Allocare (alloparental care): Caretaking behaviour directed towards an infant by other conspecifics than the parents.

Analogy: Functional similarity of structure or behaviour, which is not based upon common evolutionary origins but upon similarity of use.

Anthropoids (Anthropoidea, 'higher primates'): Comprises *Old World monkeys*, *New World monkeys* and *apes*, including humans.

Anthropoidea: See *anthropoids*.

AOS: See *accessory olfactory system*.

Apes (Hominoidea): Taxonomic group of primates; comprises lesser apes (hylobatids: gibbons and siamangs) and great apes, including humans.

Apomorphic: A derived, or evolutionarily late, character state relative to its ancestral state.

Arboreal: Living in trees.

Axon: Long, thin process of a neuron that conducts electrochemical impulses (action potentials) away from the cell body of the neuron.

Axonal cytoskeleton: Consists of filament proteins that provide the scaffolding for axons.

Basilar membrane: Located in the cochlea of the inner ear; involved in the processing of auditory information; sound waves move parts of the basilar membrane resulting in neuronal activity of the attached receptor cells.

Behavioural ecology: Emerged from ethology; is the study of the evolutionary basis of animal behaviour as the result of ecological pressures.

Bicornuate: One form of mammalian uterus (compare to *simplex* for another type); the upper parts of the uterus remain separate, while the lower parts are fused, resulting in a 'heart-shaped' form.

Binocular visual field: Overlap of the visual fields of the two eyes; prerequisite for *stereoscopic vision*.

Blind spot: Area without photoreceptor cells in the vertebrate's eye, where the optic nerve passes through the retina and exits the eye.

Catarrhines (Catarrhini): Taxonomic group of primates, which comprises *Old World monkeys* and *apes*; together with *Platyrrhini/platyrrhines* (*New World monkeys*) they constitute the *Anthropoidea/Anthropoids*.

Catarrhini: See *Catarrhines*

Choroid: Layer between the *retina* and the *sclera* of the vertebrate eye; contains the majority of blood vessels of the eye to nourish the retina.

Cones: One type of photoreceptor cells; primarily found in the centre of the *retina*; responsible for colour vision; are more sensitive to finer details and respond faster than *rods*.

Conspecifics: Members of the same species.

Continuous behavioural sampling: Observational method that records a behaviour continuously over a certain period of time (as opposed to *time sampling*).

Cytoskeleton: Cellular scaffolding that plays an important role in the mechanical stability of a cell; organizes the shape and inner structure of a cell.

Declarative. Serves to declare or state; often refers to signals that are used to make a statement.

Despotic: A type of animal social structure; exhibits a rigid, asymmetric dominance hierarchy (in contrast to *egalitarian*)

Dichromatic: *Retina* contains two types of cones ('colour receptors'); each of them reacts to one wavelength of light; dichromatic species can see any mixture of these two pure spectral lights.

Discriminant function analysis: Statistical method used to classify entities into a priori known groups; to determine whether a set of variables ('predictor variables' = independent variables) is effective in predicting category membership ('grouping variable' = dependent variable).

Diurnal: Animals that are active during daylight and sleep/rest during the night.

Duetting: Vocalization performed by a pair-bonded male and female characterized by a relatively rigid pattern; probably serve various functions, including the strengthening of the pair bond or the defence of territory.

Egalitarian: A type of primate social structure; all individuals within one group have a similar social status and/or dominance hierarchy is relaxed (in contrast to *despotic*).

ELAN (EUDICO (European Distributed Corpora Project) Linguistic Annotator): Free software for the annotation/coding and management of video and audio data; developed at Max Planck Institute for Psycholinguistics, Nijmegen (http://tla.mpi.nl/tools/tla-tools/elan).

Encephalization: Tendency towards larger brains through evolutionary time; refers to the proportion of brain mass in relation to the animal's total body mass.

Enculturation: Process by which an individual acquires the values and behaviours of its surrounding social and cultural environment; here it specifically refers to nonhuman primates being raised in a human environment (e.g. as part of great ape 'language' projects).

Endocrine system: System of glands that secrete hormones into the bloodstream and regulate bodily functions (e.g. growth, reproduction and digestion).

Epiglottis: Flap of elastic cartilage tissue covered by mucous membrane attached to the entrance of the larynx; regulates the air passage through the larynx.

Ethogram: Catalogue or inventory of all discrete behaviours that characterize a species.

Extant: Living members of a taxonomic group.

Facial Action Coding System (FACS): A standardized, objective method to identify facial movements based on appearance changes of the face caused by specific muscle contractions.

FACS: See *Facial Action Coding System*.

Filament protein: See *neurofilament protein*.

Fission–fusion: Type of social structure; a group can temporarily divide into smaller subgroups with varying size and composition.

fMRI: See *functional magnetic resonance imaging*.

Focal animal sampling: Observational method with focus on one individual at a time for a defined period of time (bout).

Fovea (fovea centralis): Area in the centre of the *retina* with high density of *cones*; enables high resolution and thus maximum acuity.

Frugivorous: Species that primarily feed on fruits.

Functional magnetic resonance imaging (*fMRI*): Brain-imaging technique that measures brain activity based on the different magnetization of oxygen-rich versus oxygen-poor blood and thus changes in blood flow.

Functional reference: Discrete signals that are given reliably in response to a specific event and that evoke an adaptive response in listeners indicating that they understand the meaning of the call. The cognition underlying the production and perception of these signals is not considered in the definition.

Fundamental frequency: Acoustic parameter of a call or sound; lowest frequency of a periodic waveform.

GC-MS: See *gas chromatography–mass spectrometry*.

Gas chromatography–mass spectrometry (*GC-MS*): Two-component method to analyse the chemical composition of a substance (which is either gaseous or can be vaporized without decomposition) by separating its elements (gas chromatography), as well as identifying and measuring the physical quantities of the particles (mass spectrometry).

Gregarious: Tendency to associate with other conspecifics in some form of social system.

Gyrification: Process or extent of the folding of the cerebral cortex; results in the enlargement of the surface of the cerebral cortex.

Gyrus/gyri: Ridge or convex fold of the cerebral cortex, generally surrounded by sulci (see *sulcus*); result of *gyrification*.

Habituation–dishabituation: An experimental paradigm; consists of the repeated presentation of one stimulus (resulting in a decrease in response = habituation) followed by the presentation of a different stimulus (potentially resulting in the recovery of the response = dishabituation).

Haplorrhines (Haplorrhini): Suborder of primates that include tarsiers, Old World monkeys, New World monkeys and apes.

Homology: Similarity of structure, behaviour or development in different species based upon their descent from a common evolutionary ancestor.

Imitation: Form of social learning, which is based on the observation of another individual and the (more or less precise) replication of its behaviour.

Impact factor: Average number of citations of articles published in a journal; often used as a proxy for the relative importance of a journal within its field.

Imperative: Signal used as a request or command.

Intentional: In this book, defined as voluntarily produced, purposeful and goal-directed behaviours.

Interact: Commercial software by Mangold International for annotating, coding and analysing video and audio data (www.mangold-international.com).

Intrasexual: Interactions between individuals of the same sex.

Iodopsin: Photoreceptor proteins in the cone cells of the retina; important for colour vision.

Laterality: Dominance of one hemisphere over the other and as a consequence, the dominance of one side of the body over the other (e.g. right-hand preference).

Lexical syntax: Sequence of meaningful signals that are grouped together according to certain rules to form a meaningful overall structure.

Magnetic resonance imaging: Brain imaging technique; a magnetic field and pulses of radio waves are applied to influence the natural magnetization of the body's atoms; a scanner detects the resulting rotating magnetic fields of the atomic nuclei.

Main olfactory system (*MOS*): Part of the olfactory system that is involved in the perception of airborne volatile odorants.

Major histocompatibility complex (*MHC*): Family of genes present in all vertebrates; involved in immune responses and the production of olfactory signals that function as individual signatures.

Matriline (matrilineal): Derivation of lineage through the mother instead of the father; social group centres on female kinship.

Means-ends dissociation: The flexible use of one signal across different contexts or of several signals in the same context.

MHC: See *major histocompatibility complex.*

Microsmatic: Having a poorly developed sense of smell.

Mirror neurons: Neurons that fire both when an individual performs and observes a particular action, e.g. grasping.

Modality: In this book, it refers to different types or modes of primate communication, including olfactory signals, facial expressions, gestures and vocalizations; others have used this term to refer specifically to the sensory mode of the signal (olfactory, visual, auditory).

Monochromatic: *Retina* contains only one type of *cones* ('colour receptors'); monochromatic species can only differentiate shades of grey.

Monogamous: Type of mating system; individuals have only one mating partner; often in pair-bonded species.

Monophyletic: A taxonomic group that has a single common ancestor.

Morpheme: Smallest meaningful unit in a language, which carries semantic (grammatical or lexical) information.

Morphology: Anatomical appearance and form of taxa, organisms, organs or component parts thereof.

MOS: See *main olfactory system*.

MRI: See *magnetic resonance imaging*.

Multi-level societies: Type of primate social organization; several smaller social units (e.g. one-male units) assemble into larger groups (e.g. bands or troops).

Multimodal: One signal consists of different modalities (e.g. many vocalizations are inseparably accompanied by facial movements); a multimodal approach, however, refers to considering more than one modality at a time (e.g. measuring both vocalizations and gestures) and/or investigating them in an integrated way.

Myelination: Production of the *myelin sheaths* that cover the axons of neurons; is responsible for the electrical insulation of axons.

Myelin sheaths: Layers of myelin, which is an outgrowth of glia cells, that are wrapped around and along the axon of neurons.

Nascent: In the context of this book, developing sexual swellings of female primates in oestrus.

Neurofilament protein: A filament protein especially found in neurons; major component of the cytoskeleton that provides support for the radial growth of axons.

Neuropil: The spaces between neurons filled with axons, glia cells, and dendrites.

New World monkeys: See *Platyrrhines/Platyrrhini*.

Nocturnal: Animals that are active during night-time and sleep/rest during daylight.

Noyau: A social system where adults of differing sexes have separate home ranges and these ranges overlap with members of the opposite sex.

Object-choice: An experimental paradigm to investigate the subject's use of experimenter-given cues (pointing, tapping, gazing, etc.) to find a hidden object in a fixed number of possible locations.

Observer: Commercial software by Noldus Information Technology for annotating, coding and analysing of video and audio data (http://www.noldus.com).

Old World monkeys (Cercopithecoidae): Group of primates that live in Africa and Asia; members of the *Haplorrhini* suborder and the *Catarrhini* infraorder; includes cercopithecines and colobines.

Olfactory bulb: A part of the vertebrate forebrain and the olfactory system; involved in the perception of odours and pheromones.

Olfactory neuroepithelium: A tissue of several layers in the nasal cavity embedding *olfactory sensory neurons*.

Olfactory sensory neurons: Receptor cells located in the olfactory neuroepithelium, which are involved in the perception of olfactory signals.

Ontogenetic ritualization: Proposed mechanism of signal acquisition; a previously noncommunicative behaviour acquires a communicative function by being shaped through repeated interactions with another individual.

Ontogeny: Development of an individual throughout its lifespan.

Palmtop computers (PDA: personal digital assistant): Handheld devices used in observational methods to record animal behaviours.

PDA: See *palmtop computers*.

Peak frequency: Acoustic parameter of a call or sound; frequency with maximum energy.

Pectoral glands: Scent glands located at the chest.

PET: See *positron emission tomography*.

Pheromones: Relatively heavy, liquid-based and non-volatile hydrocarbon molecules that function as olfactory signals; primarily detected by the accessory olfactory system.

Philopatry: The practice of remaining in one's birthplace.

Phonology: The systematic organization of sounds in languages.

Photopic vision: Colour vision under well-lit conditions, mediated by *cones*.

Phylogenetic inertia: Persistence of (currently) fitness-neutral traits inherited from ancestral species.

Phylogeny: The ancestry of a taxonomic group.

Platyrrhines (Platyrrhini): include New World monkeys, which are restricted to Central and South America; together with *Catarrhines/Catarrhini*, they constitute the *Anthropoidea/Anthropoids*.

Plesiadapiforms: An extinct order of primate-like mammals, which first appeared 65 million years ago.

Polyandry: Type of mating system; one reproductively active female mates with several reproductively active males.

Polygamy: Type of mating system that includes more than two sexual partners (see *polygyny* and *polyandry*).

Polygyny: Type of mating system; one reproductively active male mates with several reproductively active females.

Pooling fallacy: See *pseudoreplication*.

Positron emission tomography: Nuclear imaging technique; based on a positron-emitting radionuclide (tracer), which is attached on a biologically active molecule and then introduced into the body; the emitted gamma rays are then detected.

Post-partum anoestrus: Period after giving birth in which females do not become sexually receptive and are not reproductively active.

Promiscuous: Type of mating system; in multi-male, multi-female groups where both sexes mate with several partners.

Prosimian (Prosimii, 'lower' or 'primitive' primates): include lemurs, lorises and tarsiers.

Prosimii: See *Prosimians*.

Proximate: Mechanistic and developmental explanations of behaviour with a focus on processes at work during an individual's lifetime (in contrast to *ultimate*).

Pseudoreplication: The pooling of multiple observations from each individual; common but fundamental error affecting the assumptions underlying random sampling; assumes that the purpose of data gathering is to obtain large samples of behaviour rather than samples of behaviour from a large number of individuals.

Radioactive tracer isotope: A chemical compound used in *positron emission tomography* (PET) to trace the path of biochemical reactions or the distribution of a substance within the body; one or more atoms of a biologically active molecule are replaced by a radioisotope and introduced to the body; the corresponding emission of gamma rays is then detected by the scanner.

Receiver psychology: Evolution of signals in response to the sensory and mental capabilities of the receiver to detect, discriminate and memorize signals; only those signallers that produced signals that were more easily received were selected.

Retina: Light-sensitive layer of the vertebrate eye; contains photoreceptor cells (*rods* and *cones*).

Rhodopsin: Pigment in photoreceptor cells of the retina; especially sensitive to light thus enabling vision in low-light conditions.

Rods: One type of photoreceptor cells; primarily in the periphery of the *retina*; responsible for *scotopic vision* and the discrimination of shades of grey.

Sclera: Outer, protective layer of the vertebrate eye.

Scotopic vision: Vision under low light conditions; involves mostly *rods*.

Semi-gregarious: Tendency to temporarily associate with other conspecifics in some form of social system.

Simplex: One form of the mammalian uterus; consists of one main cavity (compare to *bicornuate*).

Social referencing: The use of another's response to a (novel) stimulus to evaluate a situation and to guide one's own response.

Solitary: Type of primate social organization; individuals spend most of their time alone and only occasionally interact with other conspecifics.

Somatosensory: Includes the perception of touch and temperature.

SPSS (Statistical Package for Social Sciences): Commercial software used for statistical analyses.

Stereoscopic vision: An object can be perceived with two eyes at the same time enabling the perception of depth; is the result of the overlapping but slightly different visual fields of the two eyes (caused by their different position on the head) (see *binocular vision*).

Strepsirrhine (Strepsirrhini): Includes lemurs and lorises.

Strepsirrhini: See *Strepsirrhine*

Suffixation: Creation of a new word by placing a morpheme at the end of the unchanged word stem.

Sulcus: Fissure between *gyri* as result of the folding of the cerebral cortex.

Theory of Mind (*ToM*): The ability to attribute mental states like beliefs, intents, desires, pretending and knowledge to others and to know that those mental states may differ from one's own.

Time sampling methods: Observational method that records behaviour at defined points of time.

ToM: See *Theory of Mind*.

Trichromatic: *Retina* contains three types of *cones* ('colour receptors'); each of them reacts to one wavelength of light (blue, green, red); trichromatic species can see any mixture of these three pure spectral lights.

Ultimate: Phylogenetic and functional explanations of behaviour, focusing on processes at work over generations rather than during an individual's lifetime (in contrast to *proximate*).

Vestigial vomeronasal organ: Rudimentary form of the vomeronasal organ; involved in perception of olfactory signals.

Visual acuity: The ability to continue to see the details of an object separately and unblurred as those details are made smaller and closer together.

VNO: See *vomeronasal organ*.

Vomeronasal organ (Jacobson's organ): Part of the accessory olfactory system; used to detect pheromones particularly in the context of reproductive behaviour.

Vocal learning: The ability to modify aspects of the acoustic structure of existing vocalizations in the repertoire.

Vocal plasticity: The ability to generate novel vocalizations.

References

Abbott, D. H. (1984). Behavioral and physiological suppression of fertility in subordinate marmoset monkeys. *American Journal of Primatology*, **6**(3), 169–186.

Abbott, D. H., Barrett, J. and George, L. M. (1993). Comparative aspects of the social suppression of reproduction in female marmosets and tamarins. In A. B. Rylands (ed.), *Marmosets and Tamarins: Systematics, Behaviour, and Ecology.* Oxford: Oxford University Press, pp. 152–163.

Abbott, D. H., Saltzman, W., Schultz-Darken, N. J. and Smith, T. E. (1997). Specific neuroendo-crine mechanisms not involving generalized stress mediate social regulation of female reproduction in cooperatively breeding marmoset monkeys. *Annals of the New York Academy of Sciences*, **807**(1), 219–238.

Aboitiz, F., García, R. R., Bosman, C. and Brunetti, E. (2006). Cortical memory mechanisms and language origins. *Brain and Language*, **98**(1), 40–56.

Aitken, P. G. (1981). Cortical control of conditioned and spontaneous vocal behavior in rhesus monkeys. *Brain and Language*, **13**(1), 171–184.

Alcock, J. (1984). *Animal Behavior.* Sunderland, MA: Sinauer.

Allison, T., Puce, A. and McCarthy, G. (2000). Social perception from visual cues: role of the STS region. *Trends in Cognitive Sciences*, **4**(7), 267–278.

Altmann, J. (1974). Observational study of behavior: sampling methods. *Behaviour*, **49**, 227–267.

Altmann, S. A. (1962). Social behavior of anthropoid primates: analysis of recent concepts. In E. L. Bliss (ed.), *Roots of Behavior.* New York: Harper, pp. 277–285.

Anderson, J. R., Kuroshima, H., Hattori, Y. and Fujita, K. (2010). Flexibility in the use of requesting gestures in squirrel monkeys (*Saimiri sciureus*). *American Journal of Primatology*, **72**(8), 707–714.

Anderson, J.R., Kuroshima, H., Kuwahata, H., Fujita K., Vick, S.-J. (2001). Training squirrel monkeys (*Saimiri sciureus*) to deceive: acquisition and analysis of behavior toward cooperative and competitive trainers. *Journal of Comparative Psychology*, **115**(3), 282–293.

Anderson, J. R., Kuwahata, H. and Fujita, K. (2007). Gaze alternation during 'pointing' by squirrel monkeys (*Saimiri sciureus*)? *Animal Cognition*, **10**(2), 267–271.

Andrew, R. J. (1963). Evolution of facial expression. *Science*, **142**(3595), 1034–1041.

Ankel-Simons, F. (2000). *Primate Anatomy* (2nd edn). San Diego, CA: Academia Press.

Arbib, M. A. (2005). From monkey-like action recognition to human language: An evolutionary framework for neurolinguistics. *Behavioral and Brain Sciences*, **28**(2), 105–124.

Arbib, M. A., Liebal, K. and Pika, S. (2008). Primate vocalization, gesture, and the evolution of human language. *Current Anthropology*, **49**(6), 1053–1076.

Arnold, K. and Zuberbühler, K. (2006). Language evolution: Semantic combinations in primate calls. *Nature*, **441**(7091), 303.

Arnold, K. and Zuberbühler, K. (2008). Meaningful call combinations in a non-human primate. *Current Biology*, **18**(5), R202–R203.

Arnold, K. and Zuberbühler, K. (2012). Call combinations in monkeys: Compositional or idiomatic expressions? *Brain and Language*, **120**(3), 303–309.

Arnold, K., Pohlner, Y. and Zuberbühler, K. (2008). A forest monkey's alarm call series to predator models. *Behavioral Ecology and Sociobiology*, **62**(4), 549–559.

Arnold, K., Pohlner, Y. and Zuberbühler, K. (2011). Not words but meanings? Alarm calling behaviour in a forest guenon. In V. Sommer and C. Ross (eds), *Primates of Gashaka*, Vol. **35**. New York: Springer, pp. 437–468.

Atsalis S. (2008). *A Natural History of the Brown Mouse Lemur*. Upper Saddle River, NJ: Prentice Hall.

Aujard, F. (1997). Effect of vomeronasal organ removal on male socio-sexual responses to female in a prosimian primate (*Microcebus murinus*). *Physiology and Behavior*, **62**(5), 1003–1008.

Baldwin, L. A. and Teleki, G. (1976). Patterns of gibbon behavior on Hall's Island, Bermuda: a preliminary ethogram for *Hylobates lar*. In D. M. Rumbaugh (ed.) *Gibbon and Siamang*, vol. **4**, pp. 21–105.

Bamshad, M. and Albers, E. (1996). Neural circuitry controlling vasopressin-stimulated scent marking in Syrian hamsters (*Mesocricetus auratus*). *The Journal of Comparative Neurology*, **369**(2), 252–263.

Bard, K. A. (1992). Intentional behavior and intentional communication in young free-ranging orangutans. *Child Development*, **63**(5), 1186–1197.

Bard, K. A. (2003). Development of emotional expressions in chimpanzees (*Pan troglodytes*). *Annals of the New York Academy of Sciences*, **1000**(1), 88–90.

Bard, K. A. (2007). Neonatal imitation in chimpanzees (*Pan troglodytes*) tested with two paradigms. *Animal Cognition*, **10**(2), 233–242.

Baron-Cohen, S. (1992). How monkeys do things with 'words'. *Behavioral and Brain Sciences*, **15**(1), 148–149.

Barraclough, N. E. and Perrett, D. I. (2011). From single cells to social perception. *Philosophical Transactions of the Royal Society B: Biological Sciences*, **366**(1571), 1739–1752.

Barraclough, N. E., Xiao, D., Baker, C. I., Oram, M. W. and Perrett, D. I. (2005). Integration of visual and auditory information by superior temporal sulcus neurons responsive to the sight of actions. *Journal of Cognitive Neuroscience*, **17**(3), 377–391.

Barrett, J., Abbott, D. H. and George, L. M. (1993). Sensory cues and the suppression of reproduction in subordinate female marmoset monkeys, *Callithrix jacchus*. *Journal of Reproduction and Fertility*, **97**(1), 301–310.

Bartlett, T. Q. (2008). *The Gibbons of Khao Yai: Seasonal Variation in Behavior and Ecology*. Upper Saddle River, NJ: Prentice Hall.

Barton, R. A. (2006). Olfactory evolution and behavioral ecology in primates. *American Journal of Primatology*, **68**(6), 545–558.

Barton, R. A. and Harvey, P. H. (2000). Mosaic evolution of brain structure in mammals. *Nature*, **405**(6790), 1055–1057.

Bates, E., Benigni, L., Bretherton, I., Camaioni, L. and Volterra, V. (1979). *The Emergence of Symbols: Cognition and Communication in Infancy*. New York: Academic Press.

Bates, E., Camaioni, L. and Volterra, V. (1975). The acquisition of performatives prior to speech. *Merrill-Palmer Quarterly of Behavior and Development*, **21**(3), 205–226.

Bauer, H. R. and Philip, M. M. (1983). Facial and vocal individual recognition in the common chimpanzee. *The Psychological Record*, **33**(2), 161–170.

Bearder, S. K. (1987). Lorises, bush babies, and tarsiers: Diverse societies in solitary foragers. In B. B. Smuts, D. L. Cheney, R. M. Seyfarth, R. W. Wrangham and T. T. Struhsaker (eds), *Primate Societies*. Chicago, IL: University of Chicago Press, pp. 12–24.

Beauchamp, G. K. and Yamazaki, K. (2005). Individual differences and the chemical senses. *Chemical Senses*, **30**(suppl. 1), i6–i9.

Bekoff, M. (1974). Social play and play-soliciting by infant canids. *American Zoologist*, **14**(1), 323–340.

Belcher, A., Epple, G., Küderling, I. and Smith, A. B. (1988). Volatile components of scent material from cotton-top tamarin (*Saguinus o. oedipus*). *Journal of Chemical Ecology*, **14**(5), 1367–1384.

Beletsky, L. D. (1983). Aggressive and pair bond maintenance songs of female red-winged blackbirds (*Agelaius phoeniceus*). *Zeitschrift für Tierpsychologie*, **62**(1), 47–54.

Benga, O. (2005). Intentional communication and the anterior cingulate cortex. *Interaction Studies*, **6**(2), 201–221.

Benz, J. J. (1993). Food-elicited vocalizations in golden lion tamarins: design-features for representational communication. *Animal Behaviour*, **45**(3), 443–455.

Benz, J. J., Leger, D. W. and French, J. A. (1992). Relation between food preference and food-elicited vocalizations in golden lion tamarins (*Leontopithecus rosalia*). *Journal of Comparative Psychology*, **106**(2), 142–149.

Bergman, T. (2013). Speech-like vocalized lip-smacking in geladas. *Current Biology*, **23**(7), R268–269.

Bering, J. M. (2004). A critical review of the 'enculturation hypothesis': the effects of human rearing on great ape social cognition. *Animal Cognition*, **7**(4), 201–212.

Berkson, G. and Becker, J. D. (1975). Facial expressions and social responsiveness of blind monkeys. *Journal of Abnormal Psychology*, **84**(5), 519.

Bickerton, D. (1992). *Language and Species*. Chicago, IL: University of Chicago Press.

Blaschke, M. and Ettlinger, G. (1987). Pointing as an act of social communication by monkeys. *Animal Behaviour*, **35**(5), 1520–1523.

Bloch, J. I. and Silcox, M. T. (2006). Cranial anatomy of the Paleocene plesiadapiform *Carpolestes simpsoni* (Mammalia, Primates) using ultra high-resolution X-ray computed tomography, and the relationships of plesiadapiforms to Euprimates. *Journal of Human Evolution*, **50**(1), 1–35.

Bloch, J. I., Silcox, M. T., Boyer, D. M. and Sargis, E. J. (2007). New Paleocene skeletons and the relationship of plesiadapiforms to crown-clade primates. *Proceedings of the National Academy of Sciences*, **104**(4), 1159–1164.

Bodamer, M. D. and Gardner, R. A. (2002). How cross-fostered chimpanzees (*Pan troglodytes*) initiate and maintain conversations. *Journal of Comparative Psychology*, **116**(1), 12–26.

Bolwig, N. (1964). Facial expression in primates with remarks on a parallel development in certain carnivores (a preliminary report on work in progress). *Behaviour*, **22**(3/4), 167–192.

Bonnie, K. E. and de Waal, F. B. M. (2006). Affiliation promotes the transmission of a social custom: handclasp grooming among captive chimpanzees. *Primates*, **47**(1), 27–34.

Bouchet, H., Blois-Heulin, C. and Lemasson, A. (2012). Age- and sex-specific patterns of vocal behavior in De Brazza's monkeys (*Cercopithecus neglectus*). *American Journal of Primatology*, **74**(1), 12–28.

Bouchet, H., Blois-Heulin, C., Pellier, A. S., Zuberbühler, K. and Lemasson, A. (2012). Acoustic variability and individual distinctiveness in the vocal repertoire of red-capped mangabeys (*Cercocebus torquatus*). *Journal of Comparative Psychology*, **126**(1), 45–56.

Bout, N. and Thierry, B. (2005). Peaceful meaning for the silent bared-teeth displays of mandrills. *International Journal of Primatology*, **26**(6), 1215–1228.

Bradbury, J. W. and Vehrencamp, S. L. (1998). *Principles of Animal Communication*. Sunderland, MA: Sinauer.

Brentano, F. (1874 [1973]). *Psychologie vom empirischen Standpunkt*. Hamburg: Felix Meiner Verlag.

Bressem, J. (2008). Notating gestures: proposal for a form based notation system of coverbal gestures. Unpublished manuscript.

Brosch, M. and Scheich, H. (2003). Neural representation of sound patterns in the auditory cortex of monkeys. In A. A. Ghazanfar (ed.), *Primate Audition: Ethology and Neurobiology*. Boca Raton: CRC Press, pp. 151–175.

Brown, C. H., Gomez, R. and Waser, P. M. (1995). Old World monkey vocalizations: adaptation to the local habitat? *Animal Behaviour*, **50**(4), 945–961.

Bruner, J. S. (1981). Intention in the structure of action and interaction. In L. Lipsitt (ed.), *Advances in Infancy Research*, Vol. **1**. Norwood, NJ: Ablex, pp. 41–56.

Bullinger, A. F., Zimmermann, F., Kaminski, J. and Tomasello, M. (2011). Different social motives in the gestural communication of chimpanzees and human children. *Developmental Science*, **14**(1), 58–68.

Burling, R. (2005). *The Talking Ape: How Language Evolved*. Oxford, UK: Oxford University Press.

Burrows, A. M. (2008). The facial expression musculature in primates and its evolutionary significance. *Bioessays*, **30**(3), 212–225.

Burrows, A. M. and Smith, T. D. (2003). Muscles of facial expression in Otolemur, with a comparison to Lemuroidea. *Anatomical Record Part A: Discoveries in Molecular Cellular and Evolutionary Biology*, **274**(1), 827–836.

Burrows, A. M., Diogo, R., Waller, B. M., Bonar, C. J. and Liebal, K. (2011). Evolution of the muscles of facial expression in a monogamous ape: Evaluating the relative influences of ecological and phylogenetic factors in hylobatids. *The Anatomical Record: Advances in Integrative Anatomy and Evolutionary Biology*, **294**(4), 645–663.

Burrows, A. M., Waller, B. M. and Parr, L. A. (2009). Facial musculature in the rhesus macaque (*Macaca mulatta*): evolutionary and functional contexts with comparisons to chimpanzees and humans. *Journal of Anatomy*, **215**(3), 320–334.

Burrows, A. M., Waller, B. M., Parr, L. A. and Bonar, C. J. (2006). Muscles of facial expression in the chimpanzee (*Pan troglodytes*): descriptive, comparative and phylogenetic contexts. *Journal of Anatomy*, **208**(2), 153–167.

Butterworth, G. (1998). What is special about pointing in babies? In F. Simion and G. Butterworth (eds), *The Development of Sensory, Motor and Cognitive Capacities in Early Infancy: From Perception to Cognition*. East Sussex, UK: Psychology Press, pp. 171–187.

Byrne, R. W. and Bates, L. A. (2006). Why are animals cognitive? *Current Biology*, **16**(12), R445–448.

Caeiro, C., Waller, B. M., Zimmerman, E., Burrows, A. M. and Davila-Ross, M. (2013). OrangFACS: a muscle-based facial movement coding system for orangutans (*Pongo* spp.). *International Journal of Primatology*, **34**(1), 115–129.

Caine, N. G., Addington, R. L. and Windfelder, T. L. (1995). Factors affecting the rates of food calls given by red-bellied tamarins. *Animal Behaviour*, **50**(1), 53–60.

Call, J. and Tomasello, M. (1994). Production and comprehension of referential pointing by orangutans (*Pongo pygmaeus*). *Journal of Comparative Psychology*, **108**(4), 307–317.

Call, J. and Tomasello, M. (eds). (2007). *The Gestural Communication of Apes and Monkeys*. Mahwah, NJ: Lawrence Erlbaum Associates.

Call, J., Agnetta, B. and Tomasello, M. (2000). Cues that chimpanzees do and do not use to find hidden objects. *Animal Cognition*, **3**(1), 23–34.

Calvert, G. A. (2001). Crossmodal processing in the human brain: Insights from functional neuroimaging studies. *Cerebral Cortex*, **11**(12), 1110–1123.

Campbell, M. W. and de Waal, F. B. M. (2011). Ingroup-outgroup bias in contagious yawning by chimpanzees supports link to empathy. *PLoS One*, **6**(4), e18283.

Candiotti, A., Zuberbühler, K. and Lemasson, A. (2012a). Convergence and divergence in Diana monkey vocalizations. *Biology Letters*, **8**(3), 382–385.

Candiotti, A., Zuberbühler, K. and Lemasson, A. (2012b). Context-related call combinations in female Diana monkeys. *Animal Cognition*, **15**(3), 327–339.

Candland, D. K., Blumer, E. S. and Mumford, M. D. (1980). Urine as a communicator in a New World primate, *Saimiri sciureus*. *Animal Learning and Behavior*, **8**(3), 468–480.

Cantalupo, C. and Hopkins, W. D. (2001). Asymmetric Broca's area in great apes: a region of the ape brain is uncannily similar to one linked with speech in humans. *Nature*, **414**(6863), 505–505.

Carpenter, C. R. (1940) *A Field Study in Siam of the Behavior and Social Relations of the Gibbon (Hylobates lar)*. New York: The Johns Hopkins Press pp. 1–212.

Carpenter, M., Nagell, K. and Tomasello, M. (1998). Social cognition, joint attention, and communicative competence from 9 to 15 months of age. *Monographs of the Society for Research in Child Development*, **63**(4), 1–174.

Cartmill, E. A. and Byrne, R. W. (2007). Orangutans modify their gestural signalling according to their audience's comprehension. *Current Biology*, **17**(15), 1345–1348.

Cartmill, E. A. and Byrne, R. W. (2010). Semantics of primate gestures: intentional meanings of orangutan gestures. *Animal Cognition*, **13**(6), 793–804.

Cartmill, M. (1972). Arboreal adaptations and the origin of the order Primates. In R. H. Tuttle (ed.), *The Functional and Evolutionary Biology of Primates*. Chicago, IL: Aldine-Atherton, pp. 97–122.

Cartmill, M. (1974). Rethinking primate origins. *Science*, **184**(4135), 436–443.

Cartmill, M. (1992). New views on primate origins. *Evolutionary Anthropology: Issues, News, and Reviews*, **1**(3), 105–111.

Cäsar, C., Byrne, R., Young, R. J. and Zuberbühler, K. (2012a). The alarm call system of wild black-fronted titi monkeys, *Callicebus nigrifrons*. *Behavioral Ecology and Sociobiology*, **66**(5), 653–667.

Cäsar, C., Byrne, R. W., Hoppitt, W., Young, R. J. and Zuberbühler, K. (2012b). Evidence for semantic communication in titi monkey alarm calls. *Animal Behaviour*, **84**(2), 405–411.

Catani, M. and Jones, D. K. (2005). Perisylvian language networks of the human brain. *Annals of Neurology*, **57**(1), 8–16.

Chalcraft, V. J. and Gardner, R. A. (2005). Cross-fostered chimpanzees modulate signs of American Sign Language. *Gesture*, **5**(1–2), 107–132.

Chandrasekaran, C., Lemus, L., Trubanova, A., Gondan, M. and Ghazanfar, A. A. (2011). Monkeys and humans share a common computation for face/voice integration. *PLoS Computational Biology*, **7**(9), e1002165.

Chapman, C. A. (1990). Association patterns of spider monkeys: the influence of ecology and sex on social organization. *Behavioral Ecology and Sociobiology*, **26**(6), 409–414.

Charles-Dominique, P. (1977). *Ecology and Behaviour of Nocturnal Primates*. New York: Columbia University Press.

Charles-Dominique, P. (1978). Solitary and gregarious prosimians: evolution of social structure in primates. In D. J. Chivers and K. A. Joysey (eds), *Recent Advances in Primatology*, Vol. **3**. London: Academic Press, pp. 139–149.

Cheney, D. L. (1981). Inter-group encounters among free-ranging vervet monkeys. *Folia Primatologica*, **35**(2–3), 124–146.

Cheney, D. L. and Seyfarth, R. M. (1985). Vervet monkey alarm calls: manipulation through shared information? *Behaviour*, **94**(1–2), 150–166.

Cheney, D. L. and Seyfarth, R. M. (1988). Assessment of meaning and the detection of unreliable signals by vervet monkeys. *Animal Behaviour*, **36**(2), 477–486.

Cheney, D. L. and Seyfarth, R. M. (1990). *How Monkeys See the World: Inside the Mind of Another Species*. Chicago, IL: University of Chicago Press.

Cheney, D. L. and Seyfarth, R. M. (2005). Constraints and preadaptations in the earliest stages of language evolution. *Linguistic Review*, **22**(2–4), 135–159.

Cheney, D. L. and Wrangham, R. W. (1987). Predation. In B. B. Smuts, D. L. Cheney, R. M. Seyfarth, R. W. Wrangham and T. T. Struhsaker (eds), *Primate Societies*. Chicago, IL, US: University of Chicago Press, pp. 227–239.

Cheney, D. L., Seyfarth, R. M. and Palombit, R. (1996). The function and mechanisms underlying baboon 'contact' barks. *Animal Behaviour*, **52**(3), 507–518.

Chevalier-Skolnikoff, S. (1973). Facial expression of emotion in nonhuman primates. In P. Ekman (ed.), *Darwin and Facial Expression: A Century of Research in Review*. Oxford: Academic Press, pp. 11–89.

Chevalier-Skolnikoff, S. (1974). Ontogenetic development of communication patterns in stump-tail monkeys (*Macaca speciosa*). *American Journal of Physical Anthropology*, **40**(1), 133–133.

Chivers, D. J. (1976). Communication within and between family groups of siamang (*Symphalangus syndactylus*). *Behaviour*, **57**(1–2), 116–135.

Chomsky, N. (1966). *Cartesian Linguistics: A Chapter in the History of Rationalist Thought*. Cambridge: Cambridge University Press.

Christiansen, M. H. and Kirby, S. (2003). Language evolution: consensus and controversies. *Trends in Cognitive Sciences*, **7**(7), 300–307.

Clark, A. B. (1982). Scent marks as social signals in *Galago crassicaudatus* I. Sex and reproductive status as factors in signals and responses. *Journal of Chemical Ecology*, **8**(8), 1133–1151.

Clarke, E., Reichard, U. H. and Zuberbühler, K. (2006). The syntax and meaning of wild gibbon songs. *PLoS One*, **1**(1), e73.

Clarke, P. M. R., Barrett, L. and Henzi, S. P. (2009). What role do olfactory cues play in chacma baboon mating? *American Journal of Primatology*, **71**(6), 493–502.

Clay, Z. and Zuberbühler, K. (2009). Food-associated calling sequences in bonobos. *Animal Behaviour*, **77**(6), 1387–1396.

Clay, Z. and Zuberbühler, K. (2011a). Bonobos extract meaning from call sequences. *PLoS One*, **6**(4), e18786.

Clay, Z. and Zuberbühler, K. (2011b). The structure of bonobo copulation calls during reproductive and non-reproductive sex. *Ethology*, **117**(12), 1158–1169.

Cleveland, J. and Snowdon, C. T. (1982). The complex vocal repertoire of the adult cotton-top tamarin (*Saguinus oedipus oedipus*). *Zeitschrift für Tierpsychologie*, **58**(3), 231–270.

Clutton-Brock, T. and Janson, C. (2012). Primate socioecology at the crossroads: past, present, and future. *Evolutionary Anthropology*, **21**(4), 136–150.

Clutton-Brock, T. H. and Harvey, P. H. (1977). Primate ecology and social organization. *Journal of Zoology*, **183**(1), 1–39.

Clutton-Brock, T. H., Brotherton, P. N. M., O'Riain, M. J. *et al.* (2000). Individual contributions to babysitting in a cooperative mongoose, *Suricata suricatta*. *Proceedings of the Royal Society of London. Series B: Biological Sciences*, **267**(1440), 301–305.

Cohen, C. and Haun, D. B. M. (2013). The development of tag-based cooperation via a socially acquired trait. *Evolution and Human Behavior*, **34**(3), 230–235.

Cohen, E. (2012). The evolution of tag-based cooperation in humans. *Current Anthropology*, **53**(5), 588–616.

Coleman, S. W., Patricelli, G. L. and Borgia, G. (2004). Variable female preferences drive complex male displays. *Nature*, **428**(6984), 742–745.

Colishaw, G. (1992). Song function in gibbons. *Behaviour*, **121**(1–2), 131–153.

Colonnesi, C., Stams, G. J. J. M., Koster, I. and Noom, M. J. (2010). The relation between pointing and language development: a meta-analysis. *Developmental Review*, **30**(4), 352–366.

Colquhoun, I. C. (2011). A review and interspecific comparison of nocturnal and cathemeral Strepsirrhine primate olfactory behavioural ecology. *International Journal of Zoology*, **2011**, 1–11.

Conroy, G. C. (1990). *Primate Evolution*. New York: W.W. Norton and Company.

Cooper, B. G. and Goller, F. (2004). Multimodal signals: enhancement and constraint of song motor patterns by visual display. *Science*, **303**(5657), 544–546.

Corballis, M. C. (1992). *The Lopsided Brain: Evolution of the Generative Mind*. New York: Oxford University Press.

Corballis, M. C. (2002). *From Hand to Mouth, the Origins of Language*. Princeton, NJ: Princeton University Press.

Corballis, M. C. (2003). From mouth to hand: gesture, speech, and the evolution of right-handedness. *Behavioral and Brain Sciences*, **26**(2), 199–208.

Cords, M. (1987). Forest guenos and patas monkeys: male-male competition in one-male groups. In B. B. Smuts, D. L. Cheney, R. M. Seyfarth, R. W. Wrangham and T. T. Struhsaker (eds), *Primate Societies*. Chicago: University of Chicago Press, pp. 98–111.

Cords, M. (2002). When are there influxes in blue monkey groups? In M. E. Glenn and M. Cords (eds), *The Guenons: Diversity and Adaptation in African Monkeys*. New York: Kluwer Academic/Plenum Publishers, pp. 189–201.

Crockford, C. and Boesch, C. (2003). Context-specific calls in wild chimpanzees, *Pan troglodytes verus*: analysis of barks. *Animal Behaviour*, **66**(1), 115–125.

Crockford, C. and Boesch, C. (2005). Call combinations in wild chimpanzees. *Behaviour*, **142**(4), 397–421.

Crockford, C., Herbinger, I., Vigilant, L. and Boesch, C. (2004). Wild chimpanzees produce group-specific calls: a case for vocal learning? *Ethology*, **110**(3), 221–243.

Crockford, C., Wittig, R. M., Mundry, R. and Zuberbühler, K. (2012). Wild chimpanzees inform ignorant group members of danger. *Current Biology*, **22**(2), 142–146.

Cussins, A. (1992). Content, embodiment and objectivity: the theory of cognitive trails. *Mind*, **101** (404), 651–688.

Darwin, C. (1872). *The Expression of the Emotions in Man and Animals*. New York: Oxford University Press.

Davila-Ross, M., Allcock, B., Thomas, C. and Bard, K. A. (2011). Aping expressions? Chimpanzees produce distinct laugh types when responding to laughter of others. *Emotion*, **11**(5), 1013–1020.

Davila-Ross, M., J Owren, M. and Zimmermann, E. (2009). Reconstructing the evolution of laughter in great apes and humans. *Current Biology*, **19**(13), 1106–1111.

Dawkins, R. and Krebs, J. R. (1978). Animal signals: information or manipulation. In J. R. Krebs and N. B. Davies (eds), *Behavioural Ecology: An Evolutionary Approach*. Oxford: Blackwell Scientific Publications, pp. 282–309.

De Boer, B., Sandler, W. and Kirby, S. (2012). New perspectives on duality of patterning: Introduction to the special issue. *Language and Cognition*, **4**(4), 251–259.

De Marco, A. and Visalberghi, E. (2007). Facial displays in young tufted capuchin monkeys (*Cebus apella*): appearance, meaning, context and target. *Folia Primatologica*, **78**(2), 118–137.

de Saussure, F. (1983). *Course in General Linguistics*. La Salle, IL: Open Court.

de Waal, F. B. M. (1988). The communicative repertoire of captive bonobos (*Pan paniscus*), compared to that of chimpanzees. *Behaviour*, **106**(3/4), 183–251.

de Waal, F. B. M. (1989). Dominance 'style' and primate social organization. In V. Standen and R. A. Foley (eds), *Comparative Socioecology: The Behavioural Ecology of Humans and Other Mammals*. Oxford: Blackwell, pp. 243–263.

de Waal, F. B. M. (1995). Sex as an alternative to aggression in the bonobo. In P. R. Abramson and S. D. Pinkerton (eds), *Sexual Nature, Sexual Culture*. Chicago, IL: University Press of Chicago, pp. 37–56.

Decety, J., Grezes, J., Costes, N. *et al.*, (1997). Brain activity during observation of actions. Influence of action content and subject's strategy. *Brain*, **120**(10), 1763–1777.

delBarco-Trillo, J., Sacha, C. R., Dubay, G. R. and Drea, C. M. (2012). Eulemur, me lemur: the evolution of scent-signal complexity in a primate clade. *Philosophical Transactions of the Royal Society B: Biological Sciences*, **367**(1597), 1909–1922.

Dennett, D. C. (1983). Intentional systems in cognitive ethology: the 'Panglossian paradigm' defended. *Behavioral and Brain Sciences*, **6**(3), 343–390.

Dennett, D. C. (1987). *The Intentional Stance*. Cambridge, MA: The MIT Press.

Deschner, T., Heistermann, M., Hodges, K. and Boesch, C. (2004). Female sexual swelling size, timing of ovulation, and male behavior in wild West African chimpanzees. *Hormones and Behavior*, **46**(2), 204–215.

DeWitt, I. and Rauschecker, J. P. (2012). Phoneme and word recognition in the auditory ventral stream. *Proceedings of the National Academy of Sciences*, **109**(8), E505–E514.

Di Bitetti, M. S. (2003). Food-associated calls of tufted capuchin monkeys (*Cebus apella nigritus*) are functionally referential signals. *Behaviour*, **140**(5), 565–592.

Di Fiore, A. and Campbell, C. J. (2007). The atelines: variation in ecology, behavior, and social organization. In C. J. Campbell, A. Fuentes, K. C. MacKinnon, M. Panger and S. K. Bearder (eds), *Primates in Perspective*. New York: Oxford University Press, pp. 155–185.

di Pellegrino, G., Fadiga, L., Fogassi, L., Gallese, V. and Rizzolatti, G. (1992). Understanding motor events: a neurophysiological study. *Experimental Brain Research*, **91**(1), 176–180.

Digby, L. J. (1995). Social organization in a wild population of *Callithrix jacchus*: II. Intragroup social behavior. *Primates*, **36**(3), 361–375.

Digby, L. J., Ferrari, S. F. and Saltzman, W. (2007). Callitrichines: The role of competition in cooperatively breeding species. In C. J. Campbell, A. Fuentes, K. C. MacKinnon, M. Panger and S. K. Bearder (eds), *Primates in Perspective*. New York: Oxford University Press, pp. 85–106.

Digweed, S. M., Fedigan, L. M. and Rendall, D. (2005). Variable specificity in the anti-predator vocalizations and behaviour of the white-faced capuchin, *Cebus capucinus*. *Behaviour*, **142**(8), 997–1021.

Diogo, R., Wood, B. A., Aziz, M. A. and Burrows, A. (2009). On the origin, homologies and evolution of primate facial muscles, with a particular focus on hominoids and a suggested unifying nomenclature for the facial muscles of the Mammalia. *Journal of Anatomy*, **215**(3), 300–319.

Dittus, W. P. J. (1984). Toque macaque food calls: semantic communication concerning food distribution in the environment. *Animal Behaviour*, **32**(2), 470–477.

Dixson, A. F. (1983). Observations on the evolution and behavioral significance of 'sexual skin' in female primates. In J. S. Rosenblatt, R. A. Hinde, C. Beer and M.-C. Busnel (eds), *Advances in the Study of Behavior* Vol. **13**. New York: Academic Press, pp. 63–106.

Dixson, A. F. (1998). *Primate Sexuality: Comparative Studies of the Prosimians, Monkeys, Apes, and Human Beings*. Oxford: Oxford University Press.

Dobson, S. D. (2009). Socioecological correlates of facial mobility in nonhuman anthropoids. *American Journal of Physical Anthropology*, **139**(3), 413–420.

Dobson, S. D. (2012). Coevolution of facial expression and social tolerance in macaques. *American Journal of Primatology*, **74**(3), 229–235.

Dobson, S. D. and Sherwood, C. C. (2011). Correlated evolution of brain regions involved in producing and processing facial expressions in anthropoid primates. *Biology Letters*, **7**(1), 86–88.

Dominy, N. J. and Lucas, P. W. (2001). Ecological importance of trichromatic vision to primates. *Nature*, **410**(6826), 363–366.

Dominy, N. J., Ross, C. F. and Smith, T. D. (2004). Evolution of the special senses in primates: past, present, and future. *The Anatomical Record Part A: Discoveries in Molecular, Cellular, and Evolutionary Biology*, **281**(1), 1078–1082.

Dressnandt, J. and Jürgens, U. (1992). Brain stimulation-induced changes of phonation in the squirrel monkey. *Experimental Brain Research*, **89**(3), 549–559.

Duchenne de Boulogne, G. B. (1990 [first published in French in 1862]). *The Mechanisms of Human Facial Expression*. R. A. Cuthbertson (ed. and trans.). Cambridge: Cambridge University Press.

Dulac, C. and Torello, A. T. (2003). Molecular detection of pheromone signals in mammals: from genes to behaviour. *Nature Reviews Neuroscience*, **4**(7), 551–562.

Dunbar, R. I. M. (1987). Habitat quality, population dynamics, and group composition in colobus monkeys (*Colobus guereza*). *International Journal of Primatology*, **8**(4), 299–329.

Dunbar, R. I. M. (1988). *Primate Social Systems*. New York: Cornell University Press.

Dunbar, R. I. M. (1992). Neocortex size as a constraint on group size in primates. *Journal of Human Evolution*, **22**(6), 469–493.

Dunbar, R. I. M. (1993). Coevolution of neocortical size, group-size and language in humans. *Behavioral and Brain Sciences*, **16**(4), 681–694.

Dunbar, R. I. (1995). Neocortex size and group size in primates: a test of the hypothesis. *Journal of Human Evolution*, **28**(3), 287–296.

Dunbar, R. I. M. (1996). *Grooming, Gossip, and the Evolution of Language*. London: Faber and Faber.

Dunbar, R. I. M. (1998). The social brain hypothesis. *Evolutionary Anthropology*, **6**(5), 178–190.

Dunbar, R. I. M. (2003). The social brain: mind, language, and society in evolutionary perspective. *Annual Review of Anthropology*, **32**, 163–181.

Dunbar R. I. M. (2011). Evolutionary basis of the social brain. In J. Decety and J. J. Cacioppo (eds), *Oxford Handbook of Social Neuroscience*. Oxford, UK: Oxford University Press, pp. 28–38.

Dunbar, R. I. M. (2012). Bridging the bonding gap: the transition from primates to humans. *Philosophical Transactions of the Royal Society B: Biological Sciences*, **367**(1597), 1837–1846.

Dunbar, R. I. M. and Shultz, S. (2010). Bondedness and sociality. *Behaviour*, **147**(7), 775–803.

Egbert, A. L. and Stokes, A. W. (1976). The social behavior of brown bears on an Alaskan stream. In M. R. Pelton, J. W. Lentfer and G. E. Folk (eds), *Bears: Their Biology and Management: Proceedings of the 3rd International Conference on Bear Research and Management*. Morges, Switzerland: International Union for the Conservation of nature and Natural Resources, pp. 41–56.

Eggert, F., Ferstl, R. and Müller-Ruchholtz, W. (1999). MHC and olfactory communication in humans. In R. E. Johnston, D. Müller-Schwarze and P. W. Sorensen (eds), *Advances in Chemical Signals in Vertebrates*. New York: Kluwer Academic/Plenum Publishers, pp. 181–188.

Ekman, P. (1979). About brows: emotional and conversational signals. In M. von Cranach, F. Koppa, W. Lepenies and D. Ploog (eds), *Human Ethology: Claims and Limits of a New Discipline*. New York: Cambridge University Press, pp. 169–202.

Ekman, P. (1984). Expression and the nature of emotion. In K. Scherer and P. Ekman (eds), *Approaches to Emotion*. Hillsdale, NJ: Erlbaum, pp. 19–344.

Ekman, P. (1992). An argument for basic emotions. *Cognition and Emotion*, **6**(3–4), 169–200.

Ekman P. (1994). Moods, emotions, and traits. In P. Ekman and R. J. Davidson (eds), *The Nature of Emotion: Fundamental Questions*. New York: Oxford University Press, pp. 56–58.

Ekman, P. and Friesen, W. V. (1978). *Facial Action Coding System: A Technique for the Measurement of Facial Movement*. Palo Alto, CA: Consulting Psychologists Press.

Ekman, P., Davidson, R. J. and Friesen, W. V. (1990). The Duchenne smile: emotional expression and brain physiology: II. *Journal of Personality and Social Psychology*, **58**(2), 342–353.

Ekman, P., Friesen, W. V. and Hager, J. C. (2002). *The Facial Action Coding System* (2nd edn). Salt Lake City, UT: Research Nexus.

Elliot Smith, G. (1927). *The Evolution of Man*. Oxford: Oxford University Press.

Elowson, A. M. and Snowdon, C. T. (1994). Pygmy marmosets, *Cebuella pygmaea*, modify vocal structure in response to changed social environment. *Animal Behaviour*, **47**(6), 1267–1277

Elowson, A. M., Tannenbaum, P. L. and Snowdon, C. T. (1991). Food-associated calls correlate with food preferences in cotton-top tamarins. *Animal Behaviour*, **42**(6), 931–937.

Embick, D., Marantz, A., Miyashita, Y., O'Neil, W. and Sakai, K. L. (2000). A syntactic specialization for Broca's area. *Proceedings of the National Academy of Sciences*, **97**(11), 6150–6154.

Emery, N. J. (2000). The eyes have it: The neuroethology, function and evolution of social gaze. *Neuroscience and Biobehavioral Reviews*, **24**(6), 581–604.

Emery, N. J., Seed, A. M., von Bayern, A. M. P. and Clayton, N. S. (2007). Cognitive adaptations of social bonding in birds. *Philosophical Transactions of the Royal Society B: Biological Sciences*, **362**(1480), 489–505.

Engelhardt, A., Fischer, J., Neumann, C., Pfeifer, J. B. and Heistermann, M. (2012). Information content of female copulation calls in wild long-tailed macaques (*Macaca fascicularis*). *Behavioral Ecology and Sociobiology*, **66**(1), 121–134.

Engh, A. L., Hoffmeier, R. R., Cheney, D. L. and Seyfarth, R. M. (2006). Who, me? Can baboons infer the target of vocalizations? *Animal Behaviour*, **71**(2), 381–387.

Enstam, K. L. and Isbell, L. A. (2007). The guenons (genus *Cercopithecus*) and their allies. *Primates in Perspective*. New York: Oxford University Press, pp. 252–274.

Epple, G. (1986). Communication by chemical signals. In G. Mitchell and J. Erwin (eds), *Comparative Primate Biology: Behavior, Conservation, and Ecology*, Vol. **2**. New York: Alan R. Liss, pp. 531–580.

Epple, G. and Katz, Y. (1984). Social influences on estrogen excretion and ovarian cyclicity in saddle back tamarins (*Saguinus fuscicollis*). *American Journal of Primatology*, **6**(3), 215–227.

Epple, G., Belcher, A. M., Kuderling, I. *et al.* (1993). Making sense out of scents: species differences in scent glands, scent-marking behaviour, and scent-mark composition in the Callitrichidae. In A. B. Rylands (ed.), *Marmosets and Tamarins: Systematics, Behaviour, and Ecology*. Oxford: Oxford University Press, pp. 123–151.

Evans, C. S. (1997). Referential signals. In D. H. Owings, M. D. Beacher and N. S. Thompson (eds), *Perspectives in Ethology*. New York: Plenum Press, pp. 99–143.

Evans, C. S. and Evans, L. (1999). Chicken food calls are functionally referential. *Animal Behaviour*, **58**(2), 307–319.

Ey, E., Rahn, C., Hammerschmidt, K. and Fischer, J. (2009). Wild female olive baboons adapt their grunt vocalizations to environmental conditions. *Ethology*, **115**(5), 493–503.

Fedurek, P. and Slocombe, K. E. (2011). Primate vocal communication: a useful tool for understanding human speech and language evolution? *Human Biology*, **83**(2), 153–173.

Fedurek, P. and Slocombe, K. E. (2013). The social function of food-associated calls in male chimpanzees. *American Journal of Primatology*.

Fedurek, P., Schel, A. M. and Slocombe, K. E. (2013). The acoustic structure of pant-hooting facilitates chorusing. *Behavioral Ecology and Sociobiology*, doi: 10.1007/s00265-013-1585-7.

Fernández-Carriba, S., Loeches, Á., Morcillo, A. and Hopkins, W. D. (2002). Asymmetry in facial expression of emotions by chimpanzees. *Neuropsychologia*, **40**(9), 1523–1533.

Ferrari, P. F., Gallese, V., Rizzolatti, G. and Fogassi, L. (2003). Mirror neurons responding to the observation of ingestive and communicative mouth actions in the monkey ventral premotor cortex. *European Journal of Neuroscience*, **17**(8), 1703–1714.

Ferrari, P. F., Paukner, A., Ionica, C. and Suomi, S. J. (2009). Reciprocal face-to-face communication between rhesus macaque mothers and their newborn infants. *Current Biology*, **19**(20), 1768–1772.

Ferrari, P. F., Vanderwert, R. E., Paukner, A., Bower, S., Suomi, S. J. and Fox, N. A. (2012). Distinct EEG amplitude suppression to facial gestures as evidence for a mirror mechanism in newborn monkeys. *Journal of Cognitive Neuroscience*, **24**(5), 1165–1172.

Ferrari, P. F., Visalberghi, E., Paukner, A., Fogassi, L., Ruggiero, A. and Suomi, S. J. (2006). Neonatal imitation in rhesus macaques. *PLoS Biology*, **4**(9), e302.

Fichtel, C. and Kappeler, P. M. (2002). Anti-predator behavior of group-living Malagasy primates: mixed evidence for a referential alarm call system. *Behavioral Ecology and Sociobiology*, **51**(3), 262–275.

Finlay, B. L. and Darlington, R. B. (1995). Linked regularities in the development and evolution of mammalian brains. *Science*, **268**(5217), 1578–1584.

Firestein, S. (2001). How the olfactory system makes sense of scents. *Nature* **413**(6852), 211–218.

Fischer, J. (1998). Barbary macaques categorize shrill barks into two call types. *Animal Behaviour*, **55**(4), 799–807.

Fischer, J. (2003). Developmental modifications in the vocal behavior of nonhuman primates. In A. A. Ghazanfar (ed.), *Primate Audition: Ethology and Neurobiology*. Boca Raton: CRC Press, pp. 109–125.

Fischer, J., Cheney, D. L. and Seyfarth, R. M. (2000). Development of infant baboons' responses to graded bark variants. *Proceedings of the Royal Society of London. Series B: Biological Sciences*, **267**(1459), 2317–2321.

Fitch, W. T. (2000). The evolution of speech: a comparative review. *Trends in Cognitive Sciences*, **4**(7), 258–267.

Fitch, W. T. and Hauser, M. D. (1995). Vocal production in nonhuman primates: acoustics, physiology, and functional constraints on 'honest' advertisement. *American Journal of Primatology*, **37**(3), 191–219.

Fitch, W. T. and Reby, D. (2001). The descended larynx is not uniquely human. *Proceedings of the Royal Society of London. Series B: Biological Sciences*, **268**(1477), 1669–1675.

Flack, J. C. and de Waal, F. B. M. (2007). Context modulates signal meaning in primate communication. *Proceedings of the National Academy of Sciences of the United States of America*, **104**(5), 1581–1586.

Fleagle, J. G. (1999). *Primate Adaptation and Evolution* (2nd edn). San Diego, CA: Academic Press.

Fogassi, L. and Ferrari, P. F. (2007). Mirror neurons and the evolution of embodied language. *Current Directions in Psychological Science*, **16**(3), 136–141.

Font, E. and Carazo, P. (2010). Animals in translation: Why there is meaning (but probably no message) in animal communication. *Animal Behaviour*, **80**(2), e1–e6.

Fontaine, R. (1984). Imitative skills between birth and six months. *Infant Behavior and Development*, **7**(3), 323–333.

Forrester, G. S. (2008). A multidimensional approach to investigations of behaviour: Revealing structure in animal communication signals. *Animal Behaviour*, **76**(5), 1749–1760.

Fouts, R. (1975). Capacities for language in great apes. In R. H. Tuttle (ed.), *Socioecology and Psychology of Primates*. The Hague: Mouton, pp. 371 390.

Fox, G. J. (1977). *Social Dynamics in Siamangs. Unpublished Doctoral dissertation*, University of Wisconsin, Milwaukee, WI.

Fox, G. J. (1982). Potentials for pheromones in chimpanzee vaginal fatty acids. *Folia Primatologica*, **37**(3–4), 255–266.

Freeberg, T. M., Dunbar, R. I. M. and Ord, T. J. (2012). Social complexity as a proximate and ultimate factor in communicative complexity. *Philosophical Transactions of the Royal Society B: Biological Sciences*, **367**(1597), 1785–1801.

Freiwald, W. and Tsao, D. (2011). Taking apart the neural machinery of face processing. In A. J. Calder, G. Rhodes, M. H. Johnston and J. V. Haxby (eds), *The Oxford Handbook of Face Perception*. New York: Oxford University Press, pp. 707–718.

French, J. A. (1997). Proximate regulation of singular breeding in callitrichid primates. In N. G. Solomon and J. A. French (eds), *Cooperative Breeding in Mammals*. Cambridge: Cambridge University Press, pp. 34–75.

Fridlund, A. J. (1991). Sociality of solitary smiling: potentiation by an implicit audience. *Journal of Personality and Social Psychology*, **60**(2), 229–240.

Fridlund, A. J. (1992). Darwin's anti-Darwinism in 'The expression of the emotions in man and animals'. In K. T. Strongman (ed.), *International Review of Studies on Emotion*, Vol. **2**, New York: Wiley, pp. 117–137.

Friederici, A. D. and Alter, K. (2004). Lateralization of auditory language functions: a dynamic dual pathway model. *Brain and Language*, **89**(2), 267–276.

Furuichi, T. and Ihobe, H. (1994). Variation in male relationships in bonobos and chimpanzees. *Behaviour*, **130**(3–4), 211–228.

Galdikas, B. F. and Vasey, P. (1992). Why are orangutans so smart? Ecological and social hypotheses. In F. D. Burton (ed.), *Social Processes and Mental Abilities in Nonhuman Primates*. Lewiston, New York: Edwin Mellen Press, pp. 183–224.

Galdikas, B. M. F. (1985). Orangutan sociality at Tanjung Puting. *American Journal of Primatology*, **9**(2), 101–119.

Gallese, V., Fadiga, L., Fogassi, L. and Rizzolatti, G. (1996). Action recognition in the premotor cortex. *Brain*, **119**(2), 593–609.

Garber, P. A. (1988). Diet, foraging patterns, and resource defense in a mixed species troop of *Saguinus mystax* and *Saguinus fuscicollis* in Amazonian Peru. *Behaviour*, **105**(112), 18–34.

Garber, P. A. (1993). Feeding ecology and behavior of the genus Saguinus. In A. B. Rylands (ed.), *Marmosets and Tamarins: Systematics, Behaviour, and Ecology*. Oxford: Oxford University Press, pp. 273–295.

Gardner, B. T. and Gardner, R. A. (1974). Comparing the early utterances of child and chimpanzee. In A. Pick (ed.), *Minnesota Symposia of Child Psychology*, Vol. **8**. Minneapolis: University of Minnesota Press, pp. 3–23.

Gardner, R. A., Gardner, B. T. and Van Cantford (eds) (1989). *Teaching Sign Language to Chimpanzees*. New York: State University of New York Press.

Gardner, R. A. and Gardner, B. T. (1969). Teaching sign language to a chimpanzee. *Science*, **165** (3894), 664–672.

Gardner, R. A. and Gardner, B. T. (1971). Two-way communication with an infant chimpanzee. In A. Schrier and F. Stollnitz (eds), *Behavior of Nonhuman Primates*, Vol. **4**. New York: Academic Press, pp. 117–184.

Geissmann, T. (1984). Inheritance of song parameters in the gibbon song, analyzed in 2 hybrid gibbons (*Hylobates pileatus x H. lar*). *Folia Primatologica*, **42**(3–4), 216–235.

Geissmann, T. (1999). Duet songs of the siamang, *Hylobates syndactylus*: II. Testing the pair-bonding hypothesis during a partner exchange. *Behaviour*, **136**, 1005–1039.

Gemba, H., Miki, N. and Sasaki, K. (1995). Cortical field potentials preceding vocalization and influences of cerebellar hemispherectomy upon them in monkeys. *Brain Research*, **697**(1–2), 143–151.

Gentilucci, M. and Corballis, M. C. (2006). From manual gesture to speech: A gradual transition. *Neuroscience and Biobehavioral Reviews*, **30**(7), 949–960.

Genty, E. and Byrne, R. W. (2010). Why do gorillas make sequences of gestures? *Animal Cognition*, **13**(2), 287–301.

Genty, E., Breuer, T., Hobaiter, C. and Byrne, R. W. (2009). Gestural communication of the gorilla (*Gorilla gorilla*): repertoire, intentionality and possible origins. *Animal Cognition*, **12**(3), 527–546.

Ghazanfar, A. A. and Hauser, M. D. (1999). The neuroethology of primate vocal communication: Substrates for the evolution of speech. *Trends in Cognitive Sciences*, **3**(10), 377–384.

Ghazanfar, A. A. and Logothetis, N. K. (2003). Neuroperception: facial expressions linked to monkey calls. *Nature*, **423**(6943), 937–938.

Ghazanfar, A. A. and Rendall, D. (2008). Evolution of human vocal production. *Current Biology*, **18**(11), R457–R460.

Ghazanfar, A. A. and Schroeder, C. E. (2006). Is neocortex essentially multisensory? *Trends in Cognitive Sciences*, **10**(6), 278–285.

Ghazanfar, A. A., Chandrasekaran, C. and Morrill, R. J. (2010). Dynamic, rhythmic facial expressions and the superior temporal sulcus of macaque monkeys: implications for the evolution of audiovisual speech. *European Journal of Neuroscience*, **31**(10), 1807–1817.

Ghazanfar, A. A., Maier, J. X., Hoffman, K. L. and Logothetis, N. K. (2005). Multisensory integration of dynamic faces and voices in rhesus monkey auditory cortex. *The Journal of Neuroscience*, **25**(20), 5004–5012.

Ghazanfar, A. A., Takahashi, D. Y., Mathur, N. and Fitch, W. (2012). Cineradiography of monkey lip-smacking reveals putative precursors of speech dynamics. *Current Biology*, **22**(13), 1176–1182.

Gil-da-Costa, R., Martin, A., Lopes, M. A., Muñoz, M., Fritz, J. B. and Braun, A. R. (2006). Species-specific calls activate homologs of Broca's and Wernicke's areas in the macaque. *Nature Neuroscience*, **9**(8), 1064–1070.

Gillette, R. G., Brown, R., Herman, P., Vernon, S. and Vernon, J. (1973). The auditory sensitivity of the lemur. *American Journal of Physical Anthropology*, **38**(2), 365–370.

Glowa, J. R. and Newman, J. D. (1986). Benactyzine increases alarm call rates in the squirrel monkey. *Psychopharmacology*, **90**(4), 457–460.

Goldfoot, D. A., Essock-Vitale, S. M., Asa, C. S., Thornton, J. E. and Leshner, A. I. (1978). Anosmia in male rhesus monkeys does not alter copulatory activity with cycling females. *Science*, **199**(4333), 1095–1096.

Goldin-Meadow, S. (2003). Beyond words: the importance of gesture to researchers and learners. *Child Development*, **71**(1), 231–239.

Gómez, J. C. (1990). The emergence of intentional communication as a problem-solving strategy in the gorilla. In S. T. Parker and K. R. Gibson (eds), *'Language' and Intelligence in Monkeys and Apes*. Cambridge: Cambridge University Press, pp. 333–355.

Gómez, J. C. (1996). Ostensive behavior in great apes: The role of eye contact. In A. E. Russon, S. T. Parker and K. A. Bard (eds.), *Reaching into Thought: The Minds of the Great Apes*. Cambridge: Cambridge University Press, pp. 131–151.

Gómez, J. C. (2007). Pointing behaviors in apes and human infants: a balanced interpretation. *Child Development*, **78**(3), 729–734.

Goodall, J. (1986). *The Chimpanzees of Gombe: Patterns of Behavior*. Cambridge, MA: The Belknap Press of Harvard University.

Goodall, J., Bandora, A., Bergmann, F. *et al.* (1979). Intercommunity interactions in the chimpanzee population of the Gombe National Park. In D. Hamburg and E. R. McCown (eds), *The Great Apes*. Menlo Park, California: Benjamin/Cummings, pp. 13–53.

Goossens, B., Dekleva, M., Reader, S. M., Sterck, E. H. M. and Bolhuis, J. J. (2008). Gaze following in monkeys is modulated by observed facial expressions. *Animal Behaviour*, **75**(5), 1673–1681.

Gordon, S. D. and Uetz, G. W. (2011). Multimodal communication of wolf spiders on different substrates: evidence for behavioural plasticity. *Animal Behaviour*, **81**(2), 367–375.

Gothard, K. M., Battaglia, F. P., Erickson, C. A., Spitler, K. M. and Amaral, D. G. (2007). Neural responses to facial expression and face identity in the monkey amygdala. *Journal of Neurophysiology*, **97**(2), 1671–1683.

Gottlieb, G. (1971). Ontogenesis of sensory function in birds and mammals. In E. Tobach, L. Aronson and E. Shaw (eds), *The Biopsychology of Development*. New York. Academic Press, pp. 67–128.

Gould, J. L. and Gould, C. G. (1988). *The Honey Bee*. New York: W. H. Freeman.

Gould, S. J. and Lewontin, R. C. (1979). The spandrels of San Marco and the Panglossian paradigm: a critique of the adaptationist programme. *Proceedings of the Royal Society of London. Series B: Biological Sciences*, **205**(1161), 581–598.

Gouzoules, S., Gouzoules, H. and Marler, P. (1984). Rhesus monkey (*Macaca mulatta*) screams: representational signalling in the recruitment of agonistic aid. *Animal Behaviour*, **32**(1), 182–193.

Gouzoules, H., Gouzoules, S. and Marler, P. (1985). External reference in mammalian vocal communication. In G. Zivin (ed.), *The Development of Expressive Behavior: Biology–Environment Interactions*. New York: Academic Press, pp. 77–101.

Grafton, S. T., Arbib, M. A., Fadiga, L. and Rizzolatti, G. (1996). Localization of grasp representations in humans by positron emission tomography. 2. Observation compared with imagination. *Experimental Brain Research*, **112**(1), 103–111.

Greenfield, P. M. and Savage-Rumbaugh, E. S. (1990). Grammatical combination in *Pan paniscus*: processes of learning and invention in the evolution and development of language. In S. T. Parker and K. R. Gibson (eds), *'Language' and Intelligence in Monkeys and Apes. Comparative Developmental Perspectives*. New York: Cambridge University Press, pp. 540–578.

Grice, H. P. (1957). Meaning. *The Philosophical Review*, **66**(3), 377–388.

Gros-Louis, J. (2004a). Responses of white-faced capuchins (*Cebus capucinus*) to naturalistic and experimentally presented food-associated calls. *Journal of Comparative Psychology*, **118**(4), 396–402.

Gros-Louis, J. (2004b). The function of food-associated calls in white-faced capuchin monkeys, *Cebus capucinus*, from the perspective of the signaller. *Animal Behaviour*, **67**(3), 431–440.

Gross, L. (2006). Evolution of neonatal imitation. *PLoS Biology*, **4**(9), e311.

Groswasser, Z., Korn, C., Groswasser-Reider, I. and Solzi, P. (1988). Mutism associated with buccofacial apraxia and bihemispheric lesions. *Brain and Language*, **34**(1), 157–168.

Groves, C. (2001). *Primate Taxonomy*. Washington DC: Smithsonian Institution Press.

Guilford, T. and Dawkins, M. S. (1991). Receiver psychology and the evolution of animal signals. *Animal Behaviour*, **42**(1), 1–14.

Gursky, S. (2003). Territoriality in the spectral tarsier, *Tarsius spectrum*. In P. C. Wright, E. L. Simons and S. L. Gursky (eds), *Tarsiers: Past, Present and Future*. New Brunswick, NJ: Rutgers University Press, pp. 221–236.

Gustison, M. L., le Roux, A. and Bergman, T. J. (2012). Derived vocalizations of geladas (*Theropithecus gelada*) and the evolution of vocal complexity in primates. *Philosophical Transactions of the Royal Society B: Biological Sciences*, **367**(1597), 1847–1859.

Hadj-Bouziane, F., Bell, A. H., Knusten, T. A., Ungerleider, L. G. and Tootell, R. B. H. (2008). Perception of emotional expressions is independent of face selectivity in monkey inferior temporal cortex. *Proceedings of the National Academy of Sciences*, **105**(14), 5591–5596.

Hagey, L. R., Fry, B. G. and Fitch-Snyder, H. (2007). Talking defensively, a dual use for the brachial gland exudate of slow and pygmy lorises. In S. L. Gursky and K. A. I. Nekaris (eds), *Primate Anti-predator Strategies*. New York: Springer, pp. 253–272.

Halpern, M. (1987). The organization and function of the vomeronasal system. *Annual Review of Neuroscience*, **10**, 325–362.

Halpin, Z. T. (1980). Individual odors and individual recognition: review and commentary. *Biology of Behaviour*, **5**(3), 233–248.

Hammerschmidt, K., Newman, J. D., Champoux, M. and Suomi, S. J. (2000). Changes in rhesus macaque 'coo' vocalizations during early development. *Ethology*, **106**(10), 873–886.

Hardus, M. E., Lameira, A. R., Singleton, I. *et al.* (2009a). A description of the orangutans, vocal and sound repertoire, with a focus on geographic variation. In S. A. Wich, S. S. U. Atmoki, T. M. Setia and C. P. van Schaik (eds), *Orangutans: Geographical Variation in Behavioral Ecology and Conservation*, Vol. **1**. Oxford: Oxford University Press, pp. 49–65.

Hardus, M. E., Lameira, A. R., van Schaik, C. P. and Wich, S. A. (2009b). Tool use in wild orang-utans modifies sound production: a functionally deceptive innovation? *Proceedings of the Royal Society B: Biological Sciences*, **276**(1673), 3689–3694.

Hare, B. and Tomasello, M. (2004). Chimpanzees are more skilful in competitive than in cooperative cognitive tasks. *Animal Behaviour*, **68**(3), 571–581.

Hare, B., Brown, M., Williamson, C. and Tomasello, M. (2002). The domestication of social cognition in dogs. *Science*, **298**(5598), 1634–1636.

Harlow, H. F. and Harlow, M. (1966). Learning to love. *American Scientist*, **54**(3), 244–272.

Harris, T. R. and Monfort, S. L. (2003). Behavioral and endocrine dynamics associated with infanticide in a black and white colobus monkey (*Colobus guereza*). *American Journal of Primatology*, **61**(3), 135–142.

Hasselmo, M. E., Rolls, E. T. and Baylis, G. C. (1989). The role of expression and identity in the face-selective responses of neurons in the temporal visual cortex of the monkey. *Behavioural Brain Research*, **32**(3), 203–218.

Hattori, Y., Kano, F. and Tomonaga, M. (2010). Differential sensitivity to conspecific and allospecific cues in chimpanzees and humans: A comparative eye-tracking study. *Biology Letters*, **6**(5), 610–613.

Hattori, Y., Kuroshima, H. and Fujita, K. (2010). Tufted capuchin monkeys (*Cebus apella*) show understanding of human attentional states when requesting food held by a human. *Animal Cognition*, **13**(1), 87–92.

Hauser, M. D. (1989). Ontogenetic changes in the comprehension and production of vervet monkey (*Cercopithecus aethiops*) vocalizations. *Journal of Comparative Psychology*, **103**(2), 149–158.

Hauser, M. D. (1993). Right hemisphere dominance for the production of facial expression in monkeys. *Science*, **261**(5120), 475–477.

Hauser, M. D. (1998). Functional referents and acoustic similarity: field playback experiments with rhesus monkeys. *Animal Behaviour*, **55**(6), 1647–1658.

Hauser, M. D. and Akre, K. (2001). Asymmetries in the timing of facial and vocal expressions by rhesus monkeys: implications for hemispheric specialization. *Animal Behaviour*, **61**(2), 391–400.

Hauser, M. D. and Marler, P. (1993). Food-associated calls in rhesus macaques (*Macaca mulatta*): II. Costs and benefits of call production and suppression. *Behavioral Ecology*, **4**(3), 206–212.

Hauser, M. D., Chomsky, N. and Fitch, W. T. (2002). The faculty of language: what is it, who has it, and how did it evolve? *Science*, **298**(5598), 1569–1579.

Hauser, M. D., Teixidor, P., Field, L. and Flaherty, R. (1993). Food-elicited calls in chimpanzees: Effects of food quantity and divisibility. *Animal Behaviour*, **45**(4), 817–819.

Haxby, J. V., Hoffman, E. A. and Gobbini, M. I. (2000). The distributed human neural system for face perception. *Trends in Cognitive Sciences*, **4**(6), 223–233.

Haxby, J. V., Hoffman, E. A. and Gobbini, M. I. (2002). Human neural systems for face recognition and social communication. *Biological Psychiatry*, **51**(1), 59–67.

Hayes, C. (1951). *The Ape in our House*. New York: Harper and Brothers.

Hayes, K. J. and Hayes, C. (1951). The intellectual development of a home-raised chimpanzee. *Proceedings of the American Philosophical Society*, **95**(2), 105–109.

Hayes, R. A., Morelli, T. L. and Wright, P. C. (2004). Anogenital gland secretions of *Lemur catta* and *Propithecus verreauxi coquereli*. A preliminary chemical examination. *American Journal of Primatology*, **63**(2), 49–62.

Hebets, E. A. and Papaj, D. R. (2005). Complex signal function: developing a framework of testable hypotheses. *Behavioral Ecology and Sociobiology*, **57**(3), 197–214.

Heesy, C. P. (2003). *The evolution of orbit orientation in mammals and the function of the primate postorbital bar*. Unpublished PhD dissertation, State University of New York at Stony Brook, Stony Brook, NY.

Heesy, C. P. and Ross, C. F. (2001). Evolution of activity patterns and chromatic vision in primates: morphometrics, genetics and cladistics. *Journal of Human Evolution*, **40**(2), 111–149.

Heffner, H. E., Ravizza, R. J. and Masterton, B. (1969). Hearing in primitive mammals, IV: bushbaby (*Galago senegalensis*). *Journal of Auditory Research*, **9**, 19–23.

Heffner, R. S. (2004). Primate hearing from a mammalian perspective. *The Anatomical Record Part A: Discoveries in Molecular, Cellular, and Evolutionary Biology*, **281A**(1), 1111–1122.

Henglmüller, S. M. and Ladichm, F. (1999). Development of agonistic behaviour and vocalization in croaking gouramis. *Journal of Fish Biology*, **54**(2), 380–395.

Hennessy, D. F., Owings, D. H., Rowe, M. P., Coss, R. G. and Leger, D. W. (1981). The information afforded by a variable signal: constraints on snake-elicited tail flagging by California ground squirrels. *Behaviour*, **78**(3/4), 188–226.

Herbinger, I., Papworth, S., Boesch, C. and Zuberbühler, K. (2009). Vocal, gestural and locomotor responses of wild chimpanzees to familiar and unfamiliar intruders: a playback study. *Animal Behaviour*, **78**(6), 1389–1396.

Hesler, N. and Fischer, J. (2007). Gestural communication in Barbary macaques (*Macaca sylvanus*): an overview. In M. Tomasello and J. Call (eds), *The Gestural Communication of Apes and Monkeys*. Mahwah, NJ: Lawrence Erlbaum Associates, pp. 159–195.

Hewes, G. W. (1973). Primate communication and gestural origin of language. *Current Anthropology*, **14**(1–2), 5–24.

Heymann, E. W. (2006). The neglected sense: olfaction in primate behavior, ecology and evolution. *American Journal of Primatology*, **68**(6), 519–524.

Hickok, G. and Poeppel, D. (2007). The cortical organization of speech processing. *Nature Reviews Neuroscience*, **8**(5), 393–402.

Hill, C. M. (1994). The role of female Diana monkeys, *Cercopithecus diana*, in territorial defence. *Animal Behaviour*, **47**(2), 425–431.

Hinde, R. A. (1976). Interactions, relationships and social structure. *Man*, **11**, 1–17.

Hobaiter, C. and Byrne, R. W. (2011a). The gestural repertoire of the wild chimpanzee. *Animal Cognition*, **14**(5), 745–767.

Hobaiter, C. and Byrne, R. W. (2011b). Serial gesturing by wild chimpanzees: its nature and function for communication. *Animal Cognition*, **14**(6), 827–838.

Hockett, C. F. (1960). The origin of speech. *Scientific American*, **203**, 89–97.

Hook-Costigan, M. A. and Rogers, L. J. (1998). Lateralized use of the mouth in production of vocalizations by marmosets. *Neuropsychologia*, **36**(12), 1265–1273.

Hoover, K. C. (2010). Smell with inspiration: The evolutionary significance of olfaction. *American Journal of Physical Anthropology*, **143**(S51), 63–74.

Hopkins, W. D. and Cantalupo, C. (2005). Individual and setting differences in the hand preferences of chimpanzees (*Pan troglodytes*): a critical analysis and some alternative explanations. *Laterality*, **10**(1), 65–80.

Hopkins, W. D. and Leavens, D. A. (1998). Hand use and gestural communication in chimpanzees (*Pan troglodytes*). *Journal of Comparative Psychology*, **112**(1), 95–99.

Hopkins, W. D. and Nir, T. M. (2010). Planum temporale surface area and grey matter asymmetries in chimpanzees (*Pan troglodytes*): the effect of handedness and comparison with findings in humans. *Behavioural Brain Research*, **208**(2), 436–443.

Hopkins, W. D., Cantalupo, C. and Taglialatela, J. (2007). Handedness is associated with asymmetries in gyrification of the cerebral cortex of chimpanzees. *Cerebral Cortex*, **17**(8), 1750–1756.

Hopkins, W. D., Pika, S., Liebal, K. *et al.* (2012). Handedness for manual gestures in great apes. A meta-analysis. In S. Pika and K. Liebal (eds), *Developments in Primate Gesture Research*. Amsterdam: John Benjamins Publishing Company, pp. 93–112.

Hopkins, W. D., Taglialatela, J. P. and Leavens, D. A. (2011). Do chimpanzees have voluntary control of their facial expressions and vocalizations? In A. Vilain, J.-L. Schwartz, C. Abry and J. Vauclair (eds.), *Primate Communication and Human Language: Vocalisation, Gestures, Imitation and Deixis in Humans and Non-humans*. Amsterdam: John Benjamins Publishing Company, pp. 71–88.

Hopkins, W. D., Taglialatela, J. P. and Leavens, D. A. (2007). Chimpanzees differentially produce novel vocalizations to capture the attention of a human. *Animal Behaviour*, **73**(2), 281–286.

Hostetter, A. B., Cantero, M. and Hopkins, W. D. (2001). Differential use of vocal and gestural communication by chimpanzees (*Pan troglodytes*) in response to the attentional status of a human (*Homo sapiens*). *Journal of Comparative Psychology*, **115**(4), 337–343.

Huber, E. (1931). *Evolution of Facial Musculature and Facial Expression*. Oxford: Oxford University Press.

Hurford, J. R. (2011). *The Origins of Grammar: Language in the Light of Evolution II*. Oxford: Oxford University Press.

Hurlbert, S. H. (1984). Pseudoreplication and the design of ecological field experiments. *Ecological Monographs*, **54**(2), 187–211.

Huxley, J. S. (1914). The courtship-habits of the great crested grebe (*Podiceps cristatus*); with an addition to the theory of sexual selection. *Proceedings of the Zoological Society of London*, **84**(3), 491–562.

Isbell, L. A. (1991). Contest and scramble competition: patterns of female aggression and ranging behavior among primates. *Behavioral Ecology*, **2**(2), 143–155.

Isbell, L. A. (1994). The vervets' year of doom. *Natural History*, **103**(8), 48–55.

Iverson, J. M. and Goldin-Meadow, S. (1998). Why people gesture when they speak. *Nature*, **396**(6708), 228–230.

Jablonka, E., Ginsburg, S. and Dor, D. (2012). The co-evolution of language and emotions. *Philosophical Transactions of the Royal Society B: Biological Sciences*, **367**(1599), 2152–2159.

Jacob, S., McClintock, M. K., Zelano, B. and Ober, C. (2002). Paternally inherited HLA alleles are associated with women's choice of male odor. *Nature Genetics*, **30**(2), 175–179.

Jaeggi, A. V., Burkart, J. M. and van Schaik, C. P. (2010). On the psychology of cooperation in humans and other primates: combining the natural history and experimental evidence of prosociality *Philosophical Transactions of the Royal Society B: Biological Sciences*, **365** (1553), 2723–2735.

Janson, C. H. (2000). Primate socio-ecology: the end of a golden age. *Evolutionary Anthropology: Issues, News, and Reviews*, **9**(2), 73–86.

Janson, C. H. and Chapman, C. A. (1999). Resources as determinants of primate community structure. In J. G. Fleagle, C. H. Janson and K. E. Reed (eds), *Primate Communities*. Cambridge: Cambridge University Press, pp. 237–267.

Jia, C., Chen, W. R. and Shepherd, G. M. (1999). Synaptic organization and neurotransmitters in the rat accessory olfactory bulb. *Journal of Neurophysiology*, **81**(1), 345–355.

Johnstone, R. A. (1996). Multiple displays in animal communication: 'Backup signals' and 'multiple messages'. *Philosophical Transactions of the Royal Society of London. Series B: Biological Sciences*, **351**(1337), 329–338.

Jolly, C. J. (2007). Baboons, mandrills, and mangabeys. Afro papionin socioecology in a phylogentic perspective. In C. J. Campbell, A. Fuentes, K. C. MacKinnon, M. Panger and S. Bearder (Eds.), *Primates in perspective*. New York: Oxford University Press, pp. 240–251.

Joly, O., Ramus, F., Pressnitzer, D., Vanduffel, W. and Orban, G. A. (2012). Interhemispheric differences in auditory processing revealed by fMRI in awake rhesus monkeys. *Cerebral Cortex*, **22**(4), 838–853.

Jones, C. B. and Van Cantford, T. E. (2007a). A schema for multimodal communication applied to male mantled howler monkeys (*Alouatta palliata*). *Laboratory Primate Newsletter*, **46**(2), 6–9.

Jones, C. B. and Van Cantford, T. E. (2007b). Multimodal communication by male mantled howler monkeys (*Alouatta palliata*) in sexual contexts: a descriptive analysis. *Folia Primatologica*, **78**(3), 166–185.

Jung-Beeman, M. (2005). Bilateral brain processes for comprehending natural language. *Trends in Cognitive Sciences*, **9**(11), 512–518.

Jürgens, U. (1976). Projections from the cortical larynx area in the squirrel monkey. *Experimental Brain Research*, **25**(4), 401–411.

Jürgens, U. (2002). Neural pathways underlying vocal control. *Neuroscience and Biobehavioral Reviews*, **26**(2), 235–258.

Jürgens, U., Kirzinger, A. and von Cramon, D. (1982). The effects of deep-reaching lesions in the cortical face area on phonation. A combined case report and experimental monkey study. *Cortex; A Journal Devoted to the Study of the Nervous System and Behavior*, **18**(1), 125–139.

Kaas, J. H., Hackett, T. A. and Tramo, M. J. (1999). Auditory processing in primate cerebral cortex. *Current Opinion in Neurobiology*, **9**(2), 164–170.

Kaminski, J., Call, J. and Fischer, J. (2004). Word learning in a domestic dog: Evidence for 'fast mapping'. *Science*, **304**(5677), 1682–1683.

Kaminski, J., Call, J. and Tomasello, M. (2004). Body orientation and face orientation: two factors controlling apes' begging behavior from humans. *Animal Cognition*, **7**, 216–223.

Kano, F., Call, J. and Tomonaga, M. (2012). Face and eye scanning in gorillas (*Gorilla gorilla*), orangutans (*Pongo pygmaeus*), and humans (*Homo sapiens*): Unique eye-viewing patterns in humans among hominids. *Journal of Comparative Psychology*, **126**(4), 388–398.

Kano, T. (1992). *The Last Ape: Pygmy Chimpanzee Behavior and Ecology*. Stanford, CA: Stanford University Press.

Kaplan, G. and Rogers, L. J. (2002). Patterns of gazing in orangutans (*Pongo pygmaeus*). *International Journal of Primatology*, **23**(3), 501–526.

Kappeler, P. M. (1997). Determinants of primate social organization: comparative evidence and new insights from Malagasy lemurs. *Biological Reviews*, **72**(1), 111–151.

Kappeler, P. M. (1998). To whom it may concern: the transmission and function of chemical signals in *Lemur catta*. *Behavioral Ecology and Sociobiology*, **42**(6), 411–421.

Kappeler, P. M. and van Schaik, C. P. (2002). Evolution of primate social systems. *International Journal of Primatology*, **23**(4), 707–740.

Kawai, M., Ohsawa, H., Mori, U. and Dunbar, R. (1983). Social organization of gelada baboons: social units and definitions. *Primates*, **24**(1), 13–24.

Kellogg, W. N. and Kellogg, L. A. (1933). *The Ape and the Child*. New York: McGraw-Hill.

Kendon, A. (2004). *Gesture: Visible Action as Utterance*. Cambridge: Cambridge University Press.

Keverne, E. B. (1999). The vomeronasal organ. *Science*, **286**(5440), 716–720.

Keysers, C. and Gazzola, V. (2009). Expanding the mirror: vicarious activity for actions, emotions and sensations. *Current Opinion in Neurobiology*, **19**(6), 666–671.

Kimura, D. (1993). *Neuromotor Mechanisms in Human Communication*. New York: Oxford University Press.

King, A. J. and Palmer, A. R. (1985). Integration of visual and auditory information in bimodal neurons in the guinea-pig superior colliculus. *Experimental Brain Research*, **60**(3), 492–500.

King, B. J. (1999). *The Origin of Language: What Non-human Primates Can Tell Us*. Santa Fe, NM: School of American Research Press.

King, B. J. (2004). *The Dynamic Dance: Nonvocal Communication in African Great Apes*. Cambridge, MA: Harvard University Press.

Kinzey, W. G. (1981). The Titi Monkeys, Genus *Callicebus*. In A. F. Coimbra-Filho and R. A. Mittermeier (eds), *Ecology and Behavior of Neotropical Primates*, Vol. **1**. Rio de Janeiro: Academia Brasileira de Ciências, pp. 241–276.

Kirchhof, J. and Hammerschmidt, K. (2006). Functionally referential alarm calls in tamarins (*Saguinus fuscicollis* and *Saguinus mystax*): evidence from playback experiments. *Ethology*, **112**(4), 346–354.

Kirk, E. C. and Kay, R. F. (2004). The evolution of high visual acuity in the Anthropoidea. In C.F. Ross and R.F. Kay (eds), *Anthropoid Origins*. New York: Kluwer Academic/Plenum Publishers, pp. 539–602.

Kirk, E. C., Cartmill, M., Kay, R. F. and Lemelin, P. (2003). Comment on 'Grasping primate origins'. *Science*, **300**(5620), 741.

Kirzinger, A. and Jürgens, U. (1985). The effects of brainstem lesions on vocalization in the squirrel monkey. *Brain Research*, **358**(1–2), 150–162.

Kita, S. (2003). *Pointing: Where Language, Culture and Cognition Meet*. Mahwah, NJ: Erlbaum.

Kitzmann, C. D. and Caine, N. G. (2009). Marmoset (*Callithrix geoffroyi*) food-associated calls are functionally referential. *Ethology*, **115**(5), 439–448.

Kleiman, D. G. (1977). Monogamy in mammals. *Quarterly Review of Biology*, **52**(1), 39–69.

Klein, J. (1986). *Natural History of the Major Histocompatibility Complex*. New York: John Wiley and Sons.

Knapp, L. A., Robson, J. and Waterhouse, J. S. (2006). Olfactory signals and the MHC: a review and a case study in *Lemur catta*. *American Journal of Primatology*, **68**(6), 568–584.

Kobayashi, H. and Kohshima, S. (1997). Unique morphology of the human eye. *Nature*, **387**(6635), 767–768.

Kobayashi, H. and Kohshima, S. (2001). Unique morphology of the human eye and its adaptive meaning: Comparative studies on external morphology of the primate eye. *Journal of Human Evolution*, **40**(5), 419–435.

Koenig, A. (2002). Competition for resources and its behavioral consequences among female primates. *International Journal of Primatology*, **23**(4), 759–783.

Kohler, E., Keysers, C., Umilta, M. A. *et al.* (2002). Hearing sounds, understanding actions: action representation in mirror neurons. *Science*, **297**(5582), 846–848.

Kojima, S., Izumi, A. and Ceugniet, M. (2003). Identification of vocalizers by pant hoots, pant grunts and screams in a chimpanzee. *Primates*, **44**(3), 225–230.

Krebs, J. R. and Davies, N. B. (1993). *An Introduction to Behavioural Ecology*. Oxford: Blackwell Publishing.

Kroodsma, D. E., Byers, B. E., Goodale, E., Johnson, S. and Liu, W. C. (2001). Pseudoreplication in playback experiments, revisited a decade later. *Animal Behaviour*, **61**, 1029–1033.

Kruuk, H. (1972). *The Spotted Hyena: A Study of Predation and Social Behavior*. Chicago, IL: University of Chicago Press.

Kumashiro, M., Ishibashi, H., Itakura, S., Iriki, A. (2002). Bidirectional communication between a Japanese monkey and a human through eye gaze and pointing. *Current Psychology of Cognition*, **21**(1), 3–32.

Kummer, H. (1968). *Social Organization of Hamadryas Baboons: A Field Study*. Basel, Switzerland: Karger.

Kummer, H. (1997). *In Quest of the Sacred Baboon: A Scientist's Journey*. Princeton, NJ: Princeton University Press.

Kummer, H. and Kurt, F. (1965). A comparison of social behavior in captive and wild hamadryas baboons. In H. Vagtborg (ed.) *The Baboon in Medical Research*, Vol. **1**. Austin, TX: University of Texas Press, pp. 65–80.

Kuroda, S. (1980). Social behavior of the pygmy chimpanzees. *Primates*, **21**(2), 181–197.

Kuypers, H. G. J. M. (1958). Corticobulbar connexions to the pons and lower brainstem in man. An anatomical study. *Brain*, **81**, 364–388.

Laidre, M. E. (2008). Do captive mandrills invent new gestures? *Animal Cognition*, **11**(2), 179–187.

Laidre, M. E. (2009). Informative breath: olfactory cues sought during social foraging among Old World monkeys (*Mandrillus sphinx*, *M. leucophaeus* and *Papio anubis*). *Journal of Comparative Psychology*, **123**(1), 34–44.

Laidre, M. E. (2011). Meaningful gesture in monkeys? Investigating whether mandrills create social culture. *PLoS One*, **6**(2), e14610.

Laporte, M. N. C. and Zuberbühler, K. (2010). Vocal greeting behaviour in wild chimpanzee females. *Animal Behaviour*, **80**(3), 467–473.

Laporte, M. N. C. and Zuberbühler, K. (2011). The development of a greeting signal in wild chimpanzees. *Developmental Science*, **14**(5), 1220–1234.

Laska, M., Seibt, A. and Weber, A. (2000). 'Microsmatic' primates revisited: Olfactory sensitivity in the squirrel monkey. *Chemical Senses*, **25**(1), 47–53.

Le Gros Clark, W. E. (1959). *The Antecedents of Man*. New York: Harper and Row.

Leavens, D. A. (2004). Manual deixis in apes and humans. *Interaction Studies*, **5**(3), 387–408.

Leavens, D. A. and Hopkins, W. D. (1998). Intentional communication by chimpanzees (*Pan troglodytes*): a cross-sectional study of the use of referential gestures. *Developmental Psychology*, **34**(5), 813–822.

Leavens, D. A. and Hopkins, W. D. (1999). The whole-hand point: the structure and function of pointing from a comparative perspective. *Journal of Comparative Psychology*, **113**(4), 417–425.

Leavens, D. A., Hopkins, W. D. and Bard, K. A. (1996). Indexical and referential pointing in chimpanzees (*Pan troglodytes*). *Journal of Comparative Psychology*, **110**(4), 346–353.

Leavens, D. A., Hopkins, W. D. and Bard, K. A. (2005). Understanding the point of chimpanzee pointing epigenesis and ecological validity. *Current Directions in Psychological Science*, **14**(4), 185–189.

Leavens, D. A., Hopkins, W. D. and Thomas, R. K. (2004). Referential communication by chimpanzees (*Pan troglodytes*). *Journal of Comparative Psychology*, **118**(1), 48–57.

Leavens, D. A., Hostetter, A. B., Wesley, M. J. and Hopkins, W. D. (2004). Tactical use of unimodal and bimodal communication by chimpanzees, *Pan troglodytes*. *Animal Behaviour*, **67** (3), 467–476.

Leavens, D. A., Russell, J. L. and Hopkins, W. D. (2005). Intentionality as measured in the persistence and elaboration of communication by chimpanzees (*Pan troglodytes*). *Child Development*, **76**(1), 291–306.

Leavens, D. A., Russell, J. L. and Hopkins, W. D. (2010). Multimodal communication by captive chimpanzees (*Pan troglodytes*). *Animal Cognition*, **13**(1), 33–40.

Leitten, L., Jensvold, M. L. A., Fouts, R. S. and Wallin, J. M. (2012). Contingency in requests of signing chimpanzees (*Pan troglodytes*). *Interaction Studies*, **13**(2), 147–164.

Leliveld, L. M. C., Scheumann, M. and Zimmermann, E. (2010). Effects of caller characteristics on auditory laterality in an early primate (*Microcebus murinus*). *PLoS One*, **5**(2), e9031.

Lemasson, A., Gandon, E. and Hausberger, M. (2010a). Attention to elders' voice in non-human primates. *Biology Letters*, **6**, 325–328.

Lemasson, A., Koda, H., Kato, A. *et al*. (2010). Influence of sound specificity and familiarity on Japanese macaques' (*Macaca fuscata*) auditory laterality. *Behavioural Brain Research*, **208**(1), 286–289.

Liberman, A. M., Harris, K. S., Hoffman, H. S. and Griffith, B. C. (1957). The discrimination of speech sounds within and across phoneme boundaries. *Journal of Experimental Psychology*, **54**(5), 358–368.

Liebal, K. and Call, J. (2012). The origins of non-human primates' manual gestures. *Philosophical Transactions of the Royal Society B: Biological Sciences*, **367**(1585), 118–128.

Liebal, K., Behne, T., Carpenter, M. and Tomasello, M. (2009). Infants use shared experience to interpret pointing gestures. *Developmental Science*, **12**(2), 264–271.

Liebal, K., Bressem, J. and Müller, C. (2010). Simultaneous structures of gestures in non-human primates. Paper presented at the 3rd Conference of the International Society of Gesture Studies.

Liebal, K., Call, J. and Tomasello, M. (2004a). Use of gesture sequences in chimpanzees. *American Journal of Primatology*, **64**(4), 377–396.

Liebal, K., Pika, S., Call, J. and Tomasello, M. (2004b). To move or not to move: how apes adjust to the attentional state of others. *Interaction Studies*, **5**(2), 199–219.

Liebal, K., Pika, S. and Tomasello, M. (2004c). Social communication in siamangs (*Symphalangus syndactylus*): use of gestures and facial expressions. *Primates*, **45**(1), 41–57.

Liebal, K., Pika, S. and Tomasello, M. (2006). Gestural communication of orangutans (*Pongo pygmaeus*). *Gesture*, **6**(1), 1–38.

Lieberman, P. (1984). *The Biology and Evolution of Language*. Cambridge, MA: Harvard University Press.

Lieberman, P. (2012). Vocal tract anatomy and the neural bases of talking. *Journal of Phonetics*, **40**(4), 608–622.

Lieberman, P., Klatt, D. H. and Wilson, W. H. (1969). Vocal tract limitations on the vowel repertoires of rhesus monkey and other nonhuman primates. *Science*, **164**(3884), 1185–1187.

Liotti, M., Gay, C. T. and Fox, P. T. (1994). Functional imaging and language: evidence from positron emission tomography. *Journal of Clinical Neurophysiology: Official publication of the American Electroencephalographic Society*, **11**(2), 175–190.

Liszkowski, U., Brown, P., Callaghan, T., Takada, A. and de Vos, C. (2012). A prelinguistic gestural universal of human communication. *Cognitive Science*, **36**(4), 698–713.

Liszkowski, U., Carpenter, M., Henning, A., Striano, T. and Tomasello, M. (2004). Twelve-month-olds point to share attention and interest. *Developmental Science*, **7**(3), 297–307.

Liszkowski, U., Carpenter, M., Striano, T. and Tomasello, M. (2006). 12- and 18-month-olds point to provide information for others. *Journal of Cognition and Development*, **7**(2), 173–187.

Liszkowski, U., Carpenter, M. and Tomasello, M. (2007). Reference and attitude in infant pointing. *Journal of Child Language*, **34**(1), 1–20.

Liszkowski, U., Schaefer, M., Carpenter, M. and Tomasello, M. (2009). Prelinguistic infants, but not chimpanzees, communicate about absent entities. *Psychological Science*, **20**(5), 654–660.

Losin, E. A. R., Russell, J. L., Freeman, H., Meguerditchian, A. and Hopkins, W. D. (2008). Left hemisphere specialization for oro-facial movements of learned vocal signals by captive chimpanzees. *PLoS One*, **3**(6), e2529.

Luckett, W. P. (1976). Cladistic relationships among primate higher categories: Evidence of the fetal membranes and placenta. *Folia Primatologica*, **25**(4), 245–276.

Luef, E. M. and Liebal, K. (2012). Infant-directed communication in lowland gorillas (*Gorilla gorilla*): Do older animals scaffold communicative competence in infants? *American Journal of Primatology*, **74**(9), 841–852.

Lüthe, L., Häusler, U. and Jürgens, U. (2000). Neuronal activity in the medulla oblongata during vocalization. A single-unit recording study in the squirrel monkey. *Behavioural Brain Research*, **116**(2), 197–210.

Lyn, H., Greenfield, P. M., Savage-Rumbaugh, S., Gillespie-Lynch, K. and Hopkins, W. D. (2011). Nonhuman primates do declare! A comparison of declarative symbol and gesture use in two children, two bonobos, and a chimpanzee. *Language and Communication*, **31**(1), 63–74.

Ma, W. and Klemm, W. R. (1997). Variations of equine urinary volatile compounds during the oestrous cycle. *Veterinary Research Communications*, **21**(6), 437–446.

Macedonia, J. M. (1990). What is communicated in the antipredator calls of lemurs: evidence from playback experiments with ringtailed and ruffed lemurs. *Ethology*, **86**(3), 177–190.

Macedonia, J. M. and Evans, C. S. (1993). Essay on contemporary issues in ethology: variation among mammalian alarm call systems and the problem of meaning in animal signals. *Ethology*, **93**(3), 177–197.

Machlis, L., Dodd, P. W. D. and Fentress, J. C. (1985). The pooling fallacy: problems arising when individuals contribute more than one observation to the data set. *Zeitschrift für Tierpsychologie*, **68**(3), 201–214.

Maciej, P., Fischer, J. and Hammerschmidt, K. (2011). Transmission characteristics of primate vocalizations: implications for acoustic analyses. *PLoS One*, **6**(8), e23015.

MacKay, D. M. (1972). Formal analysis of communicative processes. In R. A. Hinde (ed.), *Non-verbal Communication*. Cambridge: Cambridge University Press, pp. 3–26.

MacKinnon, J. (1974). The behaviour and ecology of wild orang-utans (*Pongo pygmaeus*). *Animal Behaviour*, **22**(1), 3–74.

MacNeilage, P. F., Studdert-Kennedy, M. G. and Lindblom, B. (1987). Primate handedness reconsidered. *Behavioral and Brain Sciences*, **10**(2), 247–303.

Maestripieri, D. (1996a). Gestural communication and its cognitive implications in pigtail macaques (*Macaca nemestrina*). *Behaviour*, **133**(13–14), 997–1022.

Maestripieri, D. (1996b). Social communication among captive stump-tailed macaques (*Macaca arctoides*). *International Journal of Primatology*, **17**(5), 785–802.

Maestripieri, D. (1997). Gestural communication in macaques: Usage and meaning of nonvocal signals. *Evolution of Communication*, **1**(2), 193–222.

Maestripieri, D. (1999). Primate social organization, gestural repertoire size, and communication dynamics: a comparative study of macaques. In B. J. King (ed.), *The Evolution of Language: Assessing the Evidence from Nonhuman Primates*. Santa Fe, NM: School of American Research, pp. 55–77.

Maille, A., Engelhart, L., Bourjade, M. and Blois-Heulin, C. (2012). To beg, or not to beg? That is the question: mangabeys modify their production of requesting gestures in response to human's attentional states. *PLoS One*, **7**(7), e41197.

Mann, J. (2000). Unraveling the dynamics of social life: long term studies and observational methods. In J. Mann, R. C. Connor, P. L. Tyack and H. Whitehead (eds), *Cetacean Societies: Field Studies of Dolphins and Whales*. Chicago, IL: University of Chicago Press, pp. 45–64.

Manser, M. B., Seyfarth, R. M. and Cheney, D. L. (2002). Suricate alarm calls signal predator class and urgency. *Trends in Cognitive Sciences*, **6**(2), 55–57.

Mariani, C., Spinnler, H., Sterzi, R. and Vallar, G. (1980). Bilateral perisylvian softenings: bilateral anterior opercular syndrome (Foix–Chavany–Marie syndrome). *Journal of Neurology*, **223**(4), 269–284.

Marler, P. (1965). Communication in monkeys and apes. In I. DeVore (ed.), *Primate Behavior: Field Studies of Monkeys and Apes*. New York: Holt, Rinehart, and Winston, pp. 544–584.

Marler, P. (1976). Social organization, communication and graded signals: the chimpanzee and the gorilla. In P. P. G. Bateson and R. A. Hinde (eds), *Growing Points*. Cambridge: Cambridge University Press, pp. 239–280.

Marler, P. (1977). The structure of animal communication sounds. In T. Bullock (ed.), *Recognition of complex acoustic signals*. Berlin: Dahlem Konferenzen, pp. 17–35.

Marler, P., Evans, C. S. and Hauser, M. D. (1992). Animal signals: Motivational, referential, or both? In H. Papoušek and U. Jürgens (Eds.), *Nonverbal Vocal Communication: Comparative and Developmental Approaches*. Cambridge: Cambridge University Press, pp. 66–86.

Marshall, A. J., Wrangham, R. W. and Arcadi, A. C. (1999). Does learning affect the structure of vocalizations in chimpanzees? *Animal Behaviour*, **58**(4), 825–830.

Martin, P. and Bateson, P. (2007). *Measuring Behaviour: An Introductory Guide* (3rd edn). Cambridge: Cambridge University Press.

Martin, R. D. (1973). A review of the behaviour and ecology of the lesser mouse lemur (*Microcebus murinus* J.F. Miller, 1777). In R. P. Michael and J. H. Crook (eds), *Comparative Ecology and Behaviour of Primates*. London: Academic Press, pp. 1–68.

Martin, R. D. (1990). *Primate Origins and Evolution: A Phylogenetic Reconstruction*. Princeton, NJ: Princeton University Press.

Martin, R. D. and Bearder, S. K. (1979). Radio bush baby. *Natural History*, **88**, 76–81.

Mason, W. A. (1960). The effects of social restriction on the behavior of rhesus monkeys: I. Free social behavior. *Journal of Comparative and Physiological Psychology*, **53**(6), 582–589.

Masterton, B., Heffner, H. and Ravizza, R. (1969). The evolution of human hearing. *The Journal of the Acoustical Society of America*, **45**, 966–985.

Matsumoto-Oda, A., Oda, R., Hayashi, Y. *et al.* (2003). Vaginal fatty acids produced by chimpanzees during menstrual cycles. *Folia Primatologica*, **74**(2), 75–79.

Matsumura, S. (1999). The evolution of 'egalitarian' and 'despotic' social systems among macaques. *Primates*, **40**(1), 23–31.

Mayr, E. (1961). Cause and effect in biology. *Science*, **134**(3489), 1501–1506.

Mayr, E. (1963). *Animal Species and Evolution*. Cambridge, MA: The Belknap Press of Harvard University Press.

Mayr, E. (1969). *Principles of Systematic Zoology*. New York: McGraw-Hill.

McComb, K. and Semple, S. (2005). Coevolution of vocal communication and sociality in primates. *Biology Letters*, **1**(4), 381–385.

McDowell, J. (1994). *Mind and World*. Cambridge, MA: Harvard University Press.

McGrew, W. C. and Tutin, C. E. G. (1978). Evidence for a social custom in wild chimpanzees? *Man*, **13**, 234–251.

McGurk, H. and MacDonald, J. (1976). Hearing lips and seeing voices. *Nature*, **264**(5588), 746–748.

McNeill, D. (2002). Gesture and language dialectic. *Acta Linguistica Hafniensia*, **34**(1), 7–37.

Mech, L. D., Adams, L. G., Meier, T. J., Burch, J. W. and Dale, B. W. (1998). *The Wolves of Denali*. Minneapolis, MN: University of Minnesota Press.

Meguerditchian, A. and Vauclair, J. (2006). Baboons communicate with their right hand. *Behavioural Brain Research*, **171**(1), 170–174.

Meguerditchian, A., Gardner, M. J., Schapiro, S. J. and Hopkins, W. D. (2012). The sound of one-hand clapping: handedness and perisylvian neural correlates of a communicative gesture in chimpanzees. *Proceedings of the Royal Society B: Biological Sciences*, **279**(1735), 1959–1966.

Meguerditchian, A., Molesti, S. and Vauclair, J. (2011). Right-handedness predominance in 162 baboons (*Papio anubis*) for gestural communication: Consistency across time and groups. *Behavioral Neuroscience*, **125**(4), 653–660.

Meguerditchian, A., Vauclair, J. and Hopkins, W. D. (2010). Captive chimpanzees use their right hand to communicate with each other: implications for the origin of the cerebral substrate for language. *Cortex*, **46**(1), 40–48.

Mehu, M., Grammer, K. and Dunbar, R. I. M. (2007). Smiles when sharing. *Evolution and Human Behavior*, **28**(6), 415–422.

Meredith, M. (1991). Sensory processing in the main and accessory olfactory systems: comparisons and contrasts. *The Journal of Steroid Biochemistry and Molecular Biology*, **39**(4), 601–614.

Meredith, M. A. and Stein, B. E. (1983). Interactions among converging sensory inputs in the superior colliculus. *Science*, **221**(4608), 389–391.

Meredith, M. A. and Stein, B. E. (1986). Spatial factors determine the activity of multisensory neurons in cat superior colliculus. *Brain Research*, **365**(2), 350–354.

Meunier, H., Prieur, J. and Vauclair, J. (2013). Olive baboons communicate intentionally by pointing. *Animal Cognition*, **16**(2), 155–163.

Michael, R. P. and Zumpe, D. (1982). Influence of olfactory signals on the reproductive behaviour of social groups of rhesus monkeys (*Macaca mulatta*). *Journal of Endocrinology*, **95**(2), 189–205.

Micheletta, J., Engelhardt, A., Matthews, L. E. E., Agil, M. and Waller, B. M. (2013). Multi-component and multimodal lipsmacking in crested macaques (*Macaca nigra*). *American Journal of Primatology*, **75**(7), 763–773.

Michelsen, A., Andersen, B. B., Storm, J., Kirchner, W. H. and Lindauer, M. (1992). How honeybees perceive communication dances, studied by means of a mechanical model. *Behavioral Ecology and Sociobiology*, **30**(3), 143–150.

Miklósi, Á. and Soproni, K. (2006). A comparative analysis of animals' understanding of the human pointing gesture. *Animal Cognition*, **9**(2), 81–93.

Miles, H. L. (1983). Apes and language: the search for communicative competence. In J. de Luce and H. T. Wilder (eds), *Language in Primates. Perspectives and Implications*. New York: Springer, pp. 43–61.

Miles, H. L. (1990). The cognitive foundations for reference in a signing orangutan. In S. T. Parker and K. R. Gibson (eds), *Language and Intelligence in Monkeys and Apes*. Cambridge: Cambridge University Press, pp. 511–539.

Miller, L. E. and Treves, A. (2007). Predation on primates: Past studies, current challenges, and directions for the future. In C. J. Campbell, A. Fuentes, J. MacKinnon, M. Panger and S. K. Bearder (eds), *Primates in Perspective*. New York: Oxford University Press, pp. 525–542.

Miller, R. E., Caul, W. F. and Mirsky, I. A. (1967). Communication of affects between feral and socially isolated monkeys. *Journal of Personality and Social Psychology*, **7**(3), 231–239.

Milton, K. (1975). Urine-rubbing behaviour in the mantled howler monkey *Alouatta palliata*. *Folia Primatologica*, **23**(1–2), 105–112.

Mitani, J. C. (1988). Male gibbon (*Hylobates agilis*) singing behavior: natural history, song variations and function. *Ethology*, **79**(3), 177–194.

Mitani, J. C. and Brandt, K. L. (1994). Social factors influence the acoustic variability in the long-distance calls of male chimpanzees. *Ethology*, **96**(3), 233–252.

Mitani, J. C. and Gros-Louis, J. (1998). Chorusing and call convergence in chimpanzees: tests of three hypotheses. *Behaviour*, **135**, 1041–1064.

Mitani, J. C. and Marler, P. (1989). A phonological analysis of male gibbon singing behavior. *Behaviour*, **109**, 20–45.

Mitani, J. C. and Nishida, T. (1993). Contexts and social correlates of long-distance calling by male chimpanzees. *Animal Behaviour*, **45**, 735–746.

Mitani, J. C., Hasegawa, T., Gros-Louis, J., Marler, P. and Byrne, R. (1992). Dialects in wild chimpanzees? *American Journal of Primatology*, **27**(4), 233–243.

Mitani, J. C., Hunley, K. L. and Murdoch, M. E. (1999). Geographic variation in the calls of wild chimpanzees: a reassessment. *American Journal of Primatology*, **47**(2), 133–151.

Mitchell, R. W. and Anderson, J. R. (1997). Pointing, withholding information, and deception in capuchin monkeys (*Cebus apella*). *Journal of Comparative Psychology*, **111**(4), 351–361.

Møller, G. W., Harlow, H. F., and Mitchell, G. D. (1968). Factors affecting agonistic communication in rhesus monkeys (*Macaca mulatta*). *Behaviour*, **31**(3/4), 339–357.

Mollon, J. D. (1989). 'Tho' she kneel'd in that place where they grew. . .' The uses and origins of primate colour vision. *Journal of Experimental Biology*, **146**(1), 21–38.

Moore, C. and Corkum, V. (1994). Social understanding at the end of the first year of life. *Developmental Review*, **14**(4), 349–372.

Moore, R. (in press). Ontogenetic constraints on Grice's Theory of Communication. In D. Matthews (ed.), *Pragmatic Development in First Language Acquisition*. Amsterdam: John Benjamins.

Morecraft, R. J., Louie, J. L., Herrick, J. L. and Stilwell-Morecraft, K. S. (2001). Cortical innervation of the facial nucleus in the non-human primate. *Brain*, **124**(1), 176–208.

Morimoto, Y. and Fujita, K. (2011). Capuchin monkeys (*Cebus apella*) modify their own behaviors according to a conspecific's emotional expressions. *Primates*, **52**(3), 279–286.

Morrill, R. J., Paukner, A., Ferrari, P. F. and Ghazanfar, A. A. (2012). Monkey lipsmacking develops like the human speech rhythm. *Developmental Science*, **15**(4), 557–568.

Morton, E. S. (1975). Ecological sources of selection on avian sounds. *The American Naturalist*, **109**(965), 17–34.

Mulcahy, N. J. and Call, J. (2009). The performance of bonobos (*Pan paniscus*), chimpanzees (*Pan troglodytes*), and orangutans (*Pongo pygmaeus*) in two versions of an object-choice task. *Journal of Comparative Psychology*, **123**(3), 304–309.

Mulcahy, N. J. and Hedge, V. (2012). Are great apes tested with an abject object-choice task? *Animal Behaviour*, **83**(2), 313–321.

Munger, S. D., Leinders-Zufall, T. and Zufall, F. (2009). Subsystem organization of the mammalian sense of smell. *Annual Review of Physiology*, **71**, 115–140.

Munoz, N. E. and Blumstein, D. T. (2012). Multisensory perception in uncertain environments. *Behavioral Ecology*, **23**(3), 457–462.

Myowa, M. (1996). Imitation of facial gestures by an infant chimpanzee. *Primates*, **37**(2), 207–213.

Myowa-Yamakoshi, M., Tomonaga, M., Tanaka, M. and Matsuzawa, T. (2004). Imitation in neonatal chimpanzees (*Pan troglodytes*). *Developmental Science*, 7(4), 437–442.

Narins, P. M. and Capranica, R. R. (1976). Sexual differences in the auditory system of the tree frog *Eleutherodactylus coqui*. *Behavioral and Brain Science*, **192**(4237), 378–380.

Narins, P. M., Grabul, D. S., Soma, K. K., Gaucher, P. and Hödl, W. (2005). Cross-modal integration in a dart-poison frog. *Proceedings of the National Academy of Sciences of the United States of America*, **102**(7), 2425–2429.

Nieuwenhuys, R., Donkelaar, H. J. and Nicholson, C. (eds). (1998). *The Central Nervous System of Vertebrates*, Vol. **3**. Heidelberg, Germany: Springer Verlag.

Nishida, T. (1968). The social group of wild chimpanzees in the Mahali Mountains. *Primates*, **9**(3), 167–224.

Nishida, T. (1979). *The Social Structure of the Chimpanzees of the Mahali Mountains*. Menlo Park, CA: Benjamin-Cummings Publishing Company.

Nishida, T. (1980). The leaf-clipping display: a newly discovered expressive gesture in wild chimpanzees. *Journal of Human Evolution*, **9**(2), 117–128.

Nishida, T. and Hiraiwa-Hasegawa, M. (1987). Chimpanzees and bonobos: Cooperative relationships among males. In B. B. Smuts, D. L. Cheney, R. M. Seyfarth, R. W. Wrangham and T. T. Struhsaker (eds), *Primate Societies*. Chicago, IL: University of Chicago Press, pp. 165–177.

Nishida, T., Hiraiwa-Hasegawa, M., Hasegawa, T. and Takahata, Y. (1985). Group extinction and female transfer in wild chimpanzees in the Mahale National Park, Tanzania. *Zeitschrift für Tierpsychologie*, **67**(1–4), 284–301.

Notman, H. and Rendall, D. (2005). Contextual variation in chimpanzee pant hoots and its implications for referential communication. *Animal Behaviour*, **70**(1), 177–190.

O'Connell, S. (1994). Joint attention in chimpanzees, bonobos and squirrel monkeys. Paper presented at the Hang Seng Center for Cognitive Studies conference on Theories of Mind.

Olzanowslei, M., Pochwatko, G., Kukliński, K., Ścibor-Rylski, M. and Ohme, R. (2008). Warsaw set of emotional facial expression pictures: validation study. Opatija, Croatia: EAESP General Meeting.

Onishi, K. H., Baillargeon, R. and Leslie, A. M. (2007). 15-month-old infants detect violations in pretend scenarios. *Acta Psychologica*, **124**(1), 106–128.

Ouattara, K., Lemasson, A. and Zuberbühler, K. (2009a). Campbell's monkeys use affixation to alter call meaning. *PLoS One*, **4**(11), e7808.

Ouattara, K., Lemasson, A. and Zuberbühler, K. (2009b). Campbell's monkeys concatenate vocalizations into context-specific call sequences. *Proceedings of the National Academy of Sciences*, **106**(51), 22026–22031.

Owings, D. H. and Virginia, R. A. (1978). Alarm calls of California ground squirrels (*Spermophilus beecheyi*). *Zeitschrift für Tierpsychologie*, **46**(1), 58–70.

Owren, M. J. and Rendall, D. (2001). Sound on the rebound: Bringing form and function back to the forefront in understanding nonhuman primate vocal signaling. *Evolutionary Anthropology: Issues, News, and Reviews*, **10**(2), 58–71.

Owren, M. J., Dieter, J. A., Seyfarth, R. M. and Cheney, D. L. (1992). 'Food' calls produced by adult female rhesus (*Macaca mulatta*) and Japanese (*M. fuscata*) macaques, their normally-raised offspring, and offspring cross-fostered between species. *Behaviour*, **120**(3/4), 218–231.

Owren, M. J., Dieter, J. A., Seyfarth, R. M. and Cheney, D. L. (1993). Vocalizations of rhesus (*Macaca mulatta*) and Japanese (*M. fuscata*) macaques cross-fostered between species show evidence of only limited modification. *Developmental Psychobiology*, **26**(7), 389–406.

Palagi, E. and Mancini, G. (2011). Playing with the face: playful facial 'chattering' and signal modulation in a monkey species (*Theropithecus gelada*). *Journal of Comparative Psychology*, **125**(1), 11–21.

Palagi, E. and Norscia, I. (2009). Multimodal signaling in wild *Lemur catta*: Economic design and territorial function of urine marking. *American Journal of Physical Anthropology*, **139**(2), 182–192.

Palagi, E., Dapporto, L., and Borgognini Tarli, S. B. (2005). The neglected scent: on the marking function of urine in *Lemur catta*. *Behavioral Ecology and Sociobiology*, **58**(5), 437–445.

Palombit, R. A. (1996). Pair bonds in monogamous apes: a comparison of the siamang *Hylobates syndactylus* and the white-handed gibbon *Hylobates lar*. *Behaviour*, **133**, 321–356.

Panksepp, J. (1998). The periconscious substrates of consciousness: affective states and the evolutionary origins of the SELF. *Journal of Consciousness Studies*, **5**(5–6), 566–582.

Panksepp, J. and Burgdorf, J. (2003). 'Laughing' rats and the evolutionary antecedents of human joy? *Physiology and Behavior*, **79**(3), 533–547.

Parker, S. T. and Gibson, K. R.(eds) (1990). *Language and Intelligence in Monkeys and Apes: Comparative Developmental Perspectives*. Cambridge: Cambridge University Press.

Parr, L. A. (2001). Cognitive and physiological markers of emotional awareness in chimpanzees (*Pan troglodytes*). *Animal Cognition*, **4**(3–4), 223–229.

Parr, L. A. (2004). Perceptual biases for multimodal cues in chimpanzee (*Pan troglodytes*) affect recognition. *Animal Cognition*, **7**(3), 171–178.

Parr, L. A. and Hecht, E. E. (2011). Facial perception in non-human primates. In A. J. Calder, G. Rhodes, M. H. Johnston and J. V. Haxby (eds), *The Oxford Handbook of Face Perception*. New York: Oxford University Press, pp. 691–706.

Parr, L. A. and Waller, B. M. (2006). Understanding chimpanzee facial expression: insights into the evolution of communication. *Social Cognitive and Affective Neuroscience*, **1**(3), 221–228.

Parr, L. A., Cohen, M. and de Waal, F. B. M. (2005a). Influence of social context on the use of blended and graded facial displays in chimpanzees. *International Journal of Primatology*, **26** (1), 73–103.

Parr, L. A., Hecht, E., Barks, S. K., Preuss, T. M. and Votaw, J. R. (2009). Face processing in the chimpanzee brain. *Current Biology*, **19**(1), 50–53.

Parr, L. A., Hopkins, W. D. and de Waal, F. B. M. (1998). The perception of facial expressions by chimpanzees, *Pan troglodytes*. *Evolution of Communication*, **2**(1), 1–23.

Parr, L. A., Waller, B. M., Burrows, A. M., Gothard, K. M. and Vick, S. J. (2010). Brief communication. MaqFACS: a muscle-based facial movement coding system for the rhesus macaque. *American Journal of Physical Anthropology*, **143**(4), 625–630.

Parr, L. A., Waller, B. M. and Fugate, J. (2005b). Emotional communication in primates: implications for neurobiology. *Current Opinion in Neurobiology*, **15**(6), 716–720.

Parr, L. A., Waller, B. M. and Heintz, M. (2008). Facial expression categorization by chimpanzees using standardized stimuli. *Emotion*, **8**(2), 216–231.

Parr, L. A., Waller, B. M., Vick, S. J. and Bard, K. A. (2007). Classifying chimpanzee facial expressions using muscle action. *Emotion*, **7**(1), 172–181.

Partan, S. and Marler, P. (1999). Communication goes multimodal. *Science*, **283**(5406), 1272–1273.

Partan, S., Yelda, S., Price, V. and Shimizu, T. (2005). Female pigeons, *Columba livia*, respond to multisensory audio/video playbacks of male courtship behaviour. *Animal Behaviour*, **70**(4), 957–966.

Partan, S. R. (1998). *Multimodal communication: The integration of visual and acoustic signals by macaques*. Unpublished PhD thesis, University of California, Davis.

Partan, S. R. (2002). Single and multichannel signal composition: Facial expressions and vocalizations of rhesus macaques (*Macaca mulatta*). *Behaviour*, **139**, 993–1028.

Partan, S. R. and Marler, P. (2005). Issues in the classification of multimodal communication signals. *The American Naturalist*, **166**(?), 231–245.

Partan, S. R., Fulmer, A. G., Gounard, M. A. M. and Redmond, J. E. (2010). Multimodal alarm behavior in urban and rural gray squirrels studied by means of observation and a mechanical robot. *Current Zoology*, **56**(3), 313–326.

Partan, S. R., Larco, C. P. and Owens, M. J. (2009). Wild tree squirrels respond with multisensory enhancement to conspecific robot alarm behaviour. *Animal Behaviour*, **77**(5), 1127–1135.

Patel, E. R. and Girard-Buttoz, C. (2008). Non-nutritive tree gouging in wild Milne-Edwards' sifakas (*Propithecus edwardsi*): description and potential communicative functions. *Primate Eye*, **96**, 283.

Patterson, F. G. (1978). The gestures of a gorilla: language acquisition in another pongid. *Brain and Language*, **5**(1), 72–97.

Patterson, F. G. (1980). Innovative uses of language by a gorilla: A case study. In K. E. Nelson (ed.), *Children's Language*, Vol. **2**. New York: Gardner Press, pp. 497–561.

Patterson, F. (1981). Ape language. *Science*, **211**(4477), 86–87.

Patterson, F. G. and Linden, F. (1981). *The Education of Koko*. New York: Holt, Rinehardt and Winston.

Paukner, A. and Anderson, J. R. (2006). Video-induced yawning in stumptail macaques (*Macaca arctoides*). *Biology Letters*, **2**(1), 36–38.

Paukner, A., Ferrari, P. F. and Suomi, S. J. (2011). Delayed imitation of lipsmacking gestures by infant rhesus macaques (*Macaca mulatta*). *PLoS One*, **6**(12), e28848.

Pelé, M., Dufour, V., Thierry, B. and Call, J. (2009). Token transfers among great apes (*Gorilla gorilla, Pongo pygmaeus, Pan paniscus*, and *Pan troglodytes*): species differences, gestural requests, and reciprocal exchange. *Journal of Comparative Psychology*, **123**(4), 375–384.

Penn, D. J. and Potts, W. K. (1999). The evolution of mating preferences and major histocompatibility complex genes. *The American Naturalist*, **153**(2), 145–164.

Pereira, M. E. and Macedonia, J. M. (1991). Ringtailed lemur anti-predator calls denote predator class, not response urgency. *Animal Behaviour*, **41**(3), 543–544.

Peres, C. A. (1989). Costs and benefits of territorial defense in wild golden lion tamarins, *Leontopithecus rosalia*. *Behavioral Ecology and Sociobiology*, **25**(3), 227–233.

Perret, M. (1995). Chemocommunication in the reproductive function of mouse lemurs. In L. Alterman, G. A. Doyle and M. K. Izard (eds), *Creatures of the Dark: The Nocturnal Prosimians*. New York Plenum Press, pp. 377–392.

Perrett, D. I. and Mistlin, A. J. (1990). Perception of facial characteristics by monkeys. In W. C. Stebbins and M. A. Berkley (eds), *Comparative Perception, Complex Signals*, Vol **2**. New York: Wiley, pp. 187–215.

Perry, S. (2011). Social traditions and social learning in capuchin monkeys (*Cebus*). *Philosophical Transactions of the Royal Society B: Biological Sciences*, **366**(1567), 988–996.

Perry, S. and Manson, J. H. (2003). Traditions in monkeys. *Evolutionary Anthropology: Issues, News, and Reviews*, **12**(2), 71–81.

Petersen, M. R., Beecher, M. D., Zoloth, S. R., Green, S., Marler, P. R., Moody, D. B. and Stebbins, W. C. (1984). Neural lateralization of vocalizations by Japanese macaques: Communicative significance is more important than acoustic structure. *Behavioral Neuroscience*, **98**(5), 779–790.

Petersen, M. R., Beecher, M. D., Zoloth, S. R., Moody, D. B. and Stebbins, W. C. (1978). Neural lateralization of species-specific vocalizations by Japanese macaques (*Macaca fuscata*). *Science*, **202**(4365), 324–327.

Petkov, C. I., Kayser, C., Steudel, T., Whittingstall, K., Augath, M. and Logothetis, N. K. (2008). A voice region in the monkey brain. *Nature Neuroscience*, **11**(3), 367–374.

Petrulis, A., Peng, M. and Johnston, R. E. (1999). Effects of vomeronasal organ removal on individual odor discrimination, sex-odor preference, and scent marking by female hamsters. *Physiology & Behavior*, **66**(1), 73–83.

Pfefferle, D., Brauch, K., Heistermann, M., Hodges, J. K. and Fischer, J. (2008). Female Barbary macaque (*Macaca sylvanus*) copulation calls do not reveal the fertile phase but influence mating outcome. *Proceedings of the Royal Society B: Biological Sciences*, **275**(1634), 571–578.

Pfefferle, D., Heistermann, M., Hodges, J. K. and Fischer, J. (2008). Male Barbary macaques eavesdrop on mating outcome: a playback study. *Animal Behaviour*, **75**(6), 1885–1891.

Phillips, K. A., Buzzell, C. A., Holder, N. and Sherwood, C. C. (2011). Why do capuchin monkeys urine wash? An experimental test of the sexual communication hypothesis using fMRI. *American Journal of Primatology*, **73**(6), 578–584.

Pierce, Jr, J. D. (1985). A review of attempts to condition operantly alloprimate vocalizations. *Primates*, **26**(2), 202–213.

Pika, S. and Mitani, J. (2006). Referential gestural communication in wild chimpanzees (*Pan troglodytes*). *Current Biology*, **16**(6), R191–R192.

Pika, S., Liebal, K. and Tomasello, M. (2003). Gestural communication in young gorillas (*Gorilla gorilla*): gestural repertoire, learning, and use. *American Journal of Primatology*, **60**(3), 95–111.

Pika, S., Liebal, K. and Tomasello, M. (2005). Gestural communication in subadult bonobos (*Pan paniscus*): repertoire and use. *American Journal of Primatology*, **65**(1), 39–61.

Pilcher, D. L., Hammock, E. A. D. and Hopkins, W. D. (2001). Cerebral volumetric asymmetries in non-human primates: a magnetic resonance imaging study. *Laterality: Asymmetries of Body, Brain and Cognition*, **6**(2), 165–179.

Plooij, F. X. (1978). Some basic traits of language in wild chimpanzees? In A. Lock (ed.), *Action, Gesture and Symbol: The Emergence of Language*. London: Academic Press, pp. 111–131.

Plooij, F. X. (1979). How wild chimpanzee babies trigger the onset of mother-infant play – and what the mother makes of it. In M. Bullowa (ed.), *Before Speech: The Beginning of Interpersonal Communication*. Cambridge: Cambridge University Press, pp. 223–244.

Plooij, F. X. (1984). The behavioral development of free-living chimpanzee babies and infants. Norwood, NJ: Ablex.

Pollard, K. A. and Blumstein, D. T. (2012). Evolving communicative complexity: Insights from rodents and beyond. *Philosophical Transactions of the Royal Society B: Biological Sciences*, **367**(1597), 1869–1878.

Pollick, A. S. and de Waal, F. B. M. (2007). Ape gestures and language evolution. *Proceedings of the National Academy of Sciences*, **104**(19), 8184–8189.

Poremba, A., Malloy, M., Saunders, R. C., Carson, R. E., Herscovitch, P. and Mishkin, M. (2004). Species-specific calls evoke asymmetric activity in the monkey's temporal poles. *Nature*, **427**(6973), 448–451.

Porter, R. H. (1998). Olfaction and human kin recognition. *Genetica*, **104**(3), 259–263.

Poss, S., Kuhar, C., Stoinski, T. S. and Hopkins, W. D. (2006). Differential use of attentional and visual communicative signaling by orangutans (*Pongo pygmaeus*) and gorillas (*Gorilla gorilla*) in response to the attentional status of a human. *American Journal of Primatology*, **68**(10), 978–992.

Povinelli, D. J., Bering, J. M. and Giambrone, S. (2000). Towards a science of other minds: escaping the argument of analogy. *Cognitive Science*, **24**(3), 501–541.

Povinelli, D. J. and Eddy, T. J. (1996). Factors influencing young chimpanzees' (*Pan troglodytes*) recognition of attention. *Journal of Comparative Psychology*, **110**(4), 336–345.

Povinelli, D. J., Bierschwale, D. T. and Cech, C. G. (1999). Comprehension of seeing as a referential act in young children, but not juvenile chimpanzees. *British Journal of Developmental Psychology*, **17**(1), 37–60.

Powzyk, J. A. and Mowry, C. B. (2003). Dietary and feeding differences between sympatric *Propithecus diadema diadema* and *Indri indri*. *International Journal of Primatology*, **24**(6), 1143–1162.

Premack, D. (1971). Language in chimpanzees. *Science*, **172**, 808–822.

Premack, D. and Premack, A. J. (1983). *The Mind of an Ape*: Norton New York.

Preuschoft, S. (1992). Laughter and smile in Barbary macaques (*Macaca sylvanus*). *Ethology*, **91**(3), 220–236.

Preuschoft, S. (2004). Power and communication. In B. Thierry, M. Singh and W. Kaumanns (eds) *Macaque Societies: A Model for the Study of Social Organization*. Cambridge: Cambridge University Press, pp. 56–60.

Preuschoft, S. and van Hooff, J. A. R. A. M. (1995). Homologizing primate facial displays: a critical review of methods. *Folia Primatologica*, **65**(3), 121–137.

Pullen, S. L., Bearder, S. K. and Dixson, A. F. (2000). Preliminary observations on sexual behavior and the mating system in free-ranging lesser galagos (*Galago moholi*). *American Journal of Primatology*, **51**(1), 79–88.

Rahlfs, M. and Fichtel, C. (2010). Anti-predator behaviour in a nocturnal primate, the grey mouse lemur (*Microcebus murinus*). *Ethology*, **116**(5), 429–439.

Rajanarayanan, S. and Archunan, G. (2011). Identification of urinary sex pheromones in female buffaloes and their influence on bull reproductive behaviour. *Research in Veterinary Science*, **91**(2), 301–305.

Rasoloharijaona, S., Randrianambinina, B. and Joly-Radko, M. (2010). Does nonnutritive tree gouging in a rainforest-dwelling lemur convey resource ownership as does loud calling in a dry forest-dwelling lemur? *American Journal of Primatology*, **72**(12), 1062–1072.

Rauschecker, J. P. and Scott, S. K. (2009). Maps and streams in the auditory cortex: nonhuman primates illuminate human speech processing. *Nature Neuroscience*, **12**(6), 718–724.

Rauschecker, J. P. and Tian, B. (2000). Mechanisms and streams for processing of 'what' and 'where' in auditory cortex. *Proceedings of the National Academy of Sciences*, **97**(22), 11800–11806.

Redshaw, M. and Locke, K. (1976). The development of play and social behaviour in two lowland gorilla infants. *Journal of the Jersey Wildlife Preservation Trust, 13th Annual Report*, 71–86.

Reichard, U. H. and Boesch, C. (eds) (2003). *Monogamy Mating Strategies and Partnerships in Birds, Humans and other Mammals*. Cambridge: Cambridge University Press.

Reichert, K. E., Heistermann, M., Hodges, J. K., Boesch, C. and Hohmann, G. (2002). What females tell males about their reproductive status: Are morphological and behavioural cues reliable signals of ovulation in bonobos (*Pan paniscus*)? *Ethology*, **108**(7), 583–600.

Rekwot, P. I., Ogwu, D., Oyedipe, E. O. and Sekoni, V. O. (2001). The role of pheromones and biostimulation in animal reproduction. *Animal Reproduction Science*, **65**(3–4), 157–170.

Rendall, D., Owren, M. J. and Ryan, M. J. (2009). What do animal signals mean? *Animal Behaviour*, **78**(2), 233–240.

Restrepo, D., Arellano, J., Oliva, A. M., Schaefer, M. L. and Lin, W. (2004). Emerging views on the distinct but related roles of the main and accessory olfactory systems in responsiveness to chemosensory signals in mice. *Hormones and Behavior*, **46**(3), 247–256.

Riede, T., Bronson, E., Hatzikirou, H. and Zuberbühler, K. (2005). Vocal production mechanisms in a non-human primate: morphological data and a model. *Journal of Human Evolution*, **48**(1), 85–96.

Rijksen, H. D. (1978). *A Fieldstudy on Sumatran Orangutans (Pongo pygmaeus abelii, Lesson 1827): Ecology, Behaviour And Conservation*. Wageningen, the Netherlands: Veenman, H. and Zonen B. V.

Rilling, J. K. (2006). Human and nonhuman primate brains: are they allometrically scaled versions of the same design? *Evolutionary Anthropology: Issues, News, and Reviews*, **15**(2), 65–77.

Rinberg, D. and Gelperin, A. (2006). Olfactory neuronal dynamics in behaving animals. *Seminars in Cell and Developmental Biology*, **17**(4), 454–461.

Rinberg, D., Koulakov, A. and Gelperin, A. (2006). Sparse odor coding in awake behaving mice. *Journal of Neuroscience*, **26**(34), 8857–8865.

Rinn, W. E. (1984). The neuropsychology of facial expression: a review of the neurological and psychological mechanisms for producing facial expressions. *Psychological Bulletin*, **95**(1), 52–77.

Rivas, E. (2005). Recent use of signs by chimpanzees (*Pan troglodytes*) in interactions with humans. *Journal of Comparative Psychology*, **119**(4), 404–417.

Rizzolatti, G. and Arbib, M. A. (1998). Language within our grasp. *Trends in Neuroscience*, **21**(5), 188–194.

Rizzolatti, G. and Craighero, L. (2004). The mirror-neuron system. *Annual Review of Neuroscience*, **27**, 169–192.

Rizzolatti, G., Fadiga, L., Gallese, V. and Fogassi, L. (1996). Premotor cortex and the recognition of motor actions. *Cognitive Brain Research*, **3**(2), 131–141.

Rizzolatti, G., Fogassi, L. and Gallese, V. (2001). Neurophysiological mechanisms underlying the understanding and imitation of action. *Nature Reviews Neuroscience*, **2**(9), 661–670.

Roberts, A. I., Vick, S.-J. and Buchanan-Smith, H. M. (2012a). Usage and comprehension of manual gestures in wild chimpanzees. *Animal Behaviour*, **84**(2), 459–470.

Roberts, A. I., Vick, S.-J., Roberts, S. G. B., Buchanan-Smith, H. M. and Zuberbühler, K. (2012b). A structure-based repertoire of manual gestures in wild chimpanzees: statistical analyses of a graded communication system. *Evolution and Human Behavior*, **33**(5), 578–589.

Roberts, J. A., Taylor, P. W. and Uetz, G. W. (2007). Consequences of complex signaling: predator detection of multimodal cues. *Behavioral Ecology*, **18**(1), 236–240.

Robinson, J. G. (1979). An analysis of the organization of vocal communication in the titi monkey *Callicebus moloch*. *Zeitschrift für Tierpsychologie*, **49**(4), 381–405.

Robinson, J. G. (1984). Syntactic structures in the vocalizations of wedge-capped capuchin monkeys, *Cebus olivaceus*. *Behaviour*, **90**(1/3), 46–79.

Robinson, J. G., Wright, P. C. and Kinzey, W. G. (1987). Monogamous cebids and their relatives: intergroup calls and spacing. In B. B. Smuts, D. L. Cheney, R. M. Seyfarth, R. W. Wrangham and T. T. Struhsaker (eds), *Primate Societies*. Chicago, IL: University of Chicago Press, pp. 44–53.

Rogers, L. J. and Kaplan, G. (2000). *Songs, Roars, and Rituals: Communication in Birds, Mammals, and Other Animals*. Cambridge, MA: Harvard University Press.

Romanski, L. M. (2012). Integration of faces and vocalizations in ventral prefrontal cortex: Implications for the evolution of audiovisual speech. *Proceedings of the National Academy of Sciences*, **109**(Supplement 1), 10717–10724.

Rosenthal, G. G., Rand, A. S. and Ryan, M. J. (2004). The vocal sac as a visual cue in anuran communication: an experimental analysis using video playback. *Animal Behaviour*, **68**(1), 55–58.

Ross, C. F. (1995). Allometric and functional influences on primate orbit orientation and the origins of the Anthropoidea. *Journal of Human Evolution*, **29**(3), 201–227.

Rosvold, H. E., Mirsky, A. F. and Pribram, K. H. (1954). Influence of amygdalectomy on social behavior in monkeys. *Journal of Comparative and Physiological Psychology*, **47**(3), 173–178.

Roush, R. S. and Snowdon, C. T. (2000). Quality, quantity, distribution and audience effects on food calling in cotton-top tamarins. *Ethology*, **106**(8), 673–690.

Rowe, C. (1999). Receiver psychology and the evolution of multicomponent signals. *Animal Behaviour*, **58**(5), 921–931.

Rowe, C. and Guilford, T. (1996). Hidden colour aversions in domestic chicks triggered by pyrazine odours of insect warning displays. *Nature*, **383**(6600), 520–522.

Rumbaugh, D. M. and Gill, T. V. (1977). Lana's acquisition of language skills. In D. M. Rumbaugh (ed.), *Language Learning by a Chimpanzee: The Lana Project*. New York: Academic Press, pp. 165–192.

Rumbaugh, D. M., Warner, H. and von Glasersfeld, E. (1977). The LANA project: Origins and tactics. In D. M. Rumbaugh (ed.), *Language Learning by a Chimpanzee: The Lana Project*. New York: Academic Press, pp. 87–90.

Russell, C. L., Bard, K. A. and Adamson, L. B. (1997). Social referencing by young chimpanzees (*Pan troglodytes*). *Journal of Comparative Psychology*, **111**(2), 185–191.

Russell, J. L., Lyn, H., Schaeffer, J. A. and Hopkins, W. D. (2011). The role of socio-communicative rearing environments in the development of social and physical cognition in apes. *Developmental Science*, **14**(6), 1459–1470.

Rylands, A. B. (1986). Ranging behaviour and habitat preference of a wild marmoset group, *Callithrix humeralifer* (Callitrichidae, Primates). *Journal of Zoology*, **210**(4), 489–514.

Rylands, A. B. (1993). *Marmosets and Tamarins: Systematics, Behaviour, and Ecology*. Oxford: Oxford University Press.

Savage, A., Ziegler, T. E. and Snowdon, C. T. (1988). Sociosexual development, pair bond formation, and mechanisms of fertility suppression in female cotton-top tamarins (*Saguinus oedipus oedipus*). *American Journal of Primatology*, **14**(4), 345–359.

Savage-Rumbaugh, E. S. and Lewin, R. (1994). *Kanzi: The Ape at the Brink of the Human Mind*. New York: Wiley.

Savage-Rumbaugh, E. S., Murphy, J., Sevcik, R. A., Brakke, K. E., Williams, S. L., Rumbaugh, D. M. and Bates, E. (1993). Language comprehension in ape and child. *Monographs of the Society for Research in Child Development*, **58**(314), 1–252.

Savage-Rumbaugh, E. S. and Rumbaugh, D. M. (1998). Perspectives on consciousness, language, and other emergent processes in apes and humans. In S. J. Hameroff, A. W. Kaszniak and A. C. Scott (eds), *Towards a Science of Consciousness II: The Second Tuscon Discussions and Debates*. Cambridge, MA: MIT Press, pp. 533–549.

Scarantino, A. (2010). Animal communication between information and influence. *Animal Behaviour*, **79**(6), E1–E5.

Schaller, G. B. (1963). *The Mountain Gorilla: Ecology and Behavior*. Chicago, IL: University of Chicago Press.

Schaller, G. B. (1972). *The Serengeti Lion: A Study of Predator – Prey Relations*. Chicago, IL: University of Chicago Press.

Schel, A. M., Candiotti, A. and Zuberbühler, K. (2010). Predator-deterring alarm call sequences in Guereza colobus monkeys are meaningful to conspecifics. *Animal Behaviour*, **80**(5), 799–808.

Schel, A., Machanda, Z., Townsend, S., Zuberbühler, K. and Slocombe, K. (in press). Chimpanzee food calls are directed at specific individuals. *Animal Behaviour*.

Schel, A., Townsend, S., Machanda, Z., Zuberbühler, K. and Slocombe, K. (in press). Chimpanzee alarm call production fulfills three key criteria for intentionality. PLoS One.

Schel, A. M., Tranquilli, S. and Zuberbühler, K. (2009). The alarm call system of two species of black-and-white colobus monkeys (*Colobus polykomos* and *Colobus guereza*). *Journal of Comparative Psychology*, **123**(2), 136–150.

Schenk, T. and McIntosh, R. D. (2010). Do we have independent visual streams for perception and action? *Cognitive Neuroscience*, **1**(1), 52–62.

Schilling, A. (1974). Olfactory communication in prosimians. In G. A. Doyle and R. D. Martin (eds), *The Study of Prosimian Behavior*. New York: Academic Press, pp. 461–542.

Schilling, A. (1979). A study of the marking behavior in *Lemur catta*. In R. D. Martin, G. A. Doyle and A. C. Walker (eds.), *Prosimian Biology*. Pittsburgh: University of Pittsburgh Press, pp. 347–362.

Schilling, A., Perret, M. and Predine, J. (1984). Sexual inhibition in a prosimian primate: a pheromone-like effect. *Journal of Endocrinology*, **102**(2), 143–151.

Schneider, C., Call, J. and Liebal, K. (2012a). Onset and early use of gestural communication in nonhuman great apes. *American Journal of Primatology*, **74**(2), 102–113.

Schneider, C., Call, J. and Liebal, K. (2012b). What role do mothers play in the gestural acquisition of bonobos (*Pan paniscus*) and chimpanzees (*Pan troglodytes*)? *International Journal of Primatology*, **33**(1), 246–262.

Schultz, A. H. (1969). *The Life of Primates*. London: Weidenfeld and Nicholson.

Scott-Phillips, T. C. (2010). Animal communication: insights from linguistic pragmatics. *Animal Behaviour*, **79**(1), e1–e4.

Scott-Phillips, T. C., Blythe, R. A., Gardner, A. and West, S. A. (2012). How do communication systems emerge? *Proceedings of the Royal Society B: Biological Sciences*, **279**(1735), 1943–1949.

Scott-Phillips, T. C., Dickins, T. E. and West, S. A. (2011). Evolutionary theory and the ultimate-proximate distinction in the human behavioral sciences. *Perspectives on Psychological Science*, **6**(1), 38–47.

Sebeok, T. A. and Sebeok, J. U. (eds) (1991). *Speaking of Apes: A Critical Anthology of Two-way Communication with Man*. New York: Plenum.

Semendeferi, K., Lu, A., Schenker, N. and Damásio, H. (2002). Humans and great apes share a large frontal cortex. *Nature neuroscience*, **5**(3), 272–276.

Semple, S. (2001). Individuality and male discrimination of female copulation calls in the yellow baboon. *Animal Behaviour*, **61**(5), 1023–1028.

Semple, S. and McComb, K. (2000). Perception of female reproductive state from vocal cues in a mammal species. *Proceedings of the Royal Society of London. Series B: Biological Sciences*, **267**(1444), 707–712.

Semple, S., McComb, K., Alberts, S. and Altmann, J. (2002). Information content of female copulation calls in yellow baboons. *American Journal of Primatology*, **56**(1), 43–56.

Sereno, M. I. and Allman, J. M. (1991). Cortical visual areas in mammals. In A. G. Leventhal (ed.) *The Neural Basis of Visual Function*. London: Macmillan, pp. 160–172.

Sereno, M. I., Dale, A. M., Reppas, J. B. *et al.* (1995). Borders of multiple visual areas in humans revealed by functional magnetic resonance imaging. *Science*, **268**(5212), 889–893.

Setchell, J. M., Vaglio, S., Moggi-Cecchi, J. *et al.* (2010). Chemical composition of scent-gland secretions in an Old World monkey (*Mandrillus sphinx*): influence of sex, male status, and individual identity. *Chemical Senses*, **35**(3), 205–220.

Seyama, J. and Nagayama, R. S. (2002). Perceived eye size is larger in happy faces than in surprised faces. *Perception*, **31**(9), 1153–1155.

Seyfarth, R. M. and Cheney, D. L. (1986). Vocal development in vervet monkeys. *Animal Behaviour*, **34**(6), 1640–1658.

Seyfarth, R. M. and Cheney, D. L. (1990). The assessment by vervet monkeys of their own and another species' alarm calls. *Animal Behaviour*, **40**(4), 754–764.

Seyfarth, R. M. and Cheney, D. L. (1993). Meaning, reference, and intentionality in the natural vocalizations of monkeys. In H. H. Roitblat, L. M. Herman and P. E. Nachtigall (eds), *Language and Communication: Comparative Perspectives*. New York: Lawrence Erlbaum Associates, pp. 195–220.

Seyfarth, R. M., Cheney, D. L. and Bergman, T. J. (2005). Primate social cognition and the origins of language. *Trends in Cognitive Sciences*, **9**(6), 264–266.

Seyfarth, R. M., Cheney, D. L., Bergman, T., Fischer, J., Zuberbühler, K. and Hammerschmidt, K. (2010). The central importance of information in studies of animal communication. *Animal Behaviour*, **80**(1), 3–8.

Seyfarth, R. M., Cheney, D. L. and Marler, P. (1980a). Monkey responses to three different alarm calls: evidence of predator classification and semantic communication. *Science*, **210**(4471) 801–803.

Seyfarth, R. M., Cheney, D. L. and Marler, P. (1980b). Vervet monkey alarm calls: semantic communication in a free-ranging primate. *Animal Behaviour*, **28**(4), 1070–1094.

Shannon, C. E. (1948). A mathematical theory of communication. *Bell System Technical Journal*, **27**(3), 379–423.

Sherman, R. W., Jarvis, J. U. M. and Alexander, R. D. (1991). *The Biology of the Naked Mole-rat*. Princeton, NJ: Princeton University Press.

Sherwood, C. C., Hof, P. R., Holloway, R. L. *et al.* (2005). Evolution of the brainstem orofacial motor system in primates: a comparative study of trigeminal, facial and hypoglossal nuclei. *Journal of Human Evolution*, **48**(1), 45–84.

Sherwood, C. C., Holloway, R. L., Erwin, J. M. and Hof, P. R. (2004a). Cortical orofacial motor representation in Old World monkeys, great apes, and humans: II. Stereologic analysis of chemoarchitecture. *Brain, Behavior and Evolution*, **63**(2), 82–106.

Sherwood, C. C., Holloway, R. L., Erwin, J. M. *et al.* (2004b). Cortical orofacial motor representation in Old World monkeys, great apes, and humans: I. Quantitative analysis of cytoarchitecture. *Brain, Behavior and Evolution*, **63**(2), 61–81.

Shultz, S. and Dunbar, R. I. M. (2007). The evolution of the social brain: Anthropoid primates contrast with other vertebrates. *Proceedings of the Royal Society B: Biological Sciences*, **274** (1624), 2429–2436.

Shultz, S., Opie, C. and Atkinson, Q. D. (2011). Stepwise evolution of stable sociality in primates. *Nature*, **479**(7372), 219–222.

Simpson, G. G. (1961). *Principles of Animal Taxonomy*. New York: Columbia University Press.

Simpson, J. A. and Weiner, E. S. C. (eds). (1998). *The Oxford English Dictionary* (2nd edn), **Vol. VI**. Oxford: Oxford University Press.

Slocombe, K. E. and Zuberbühler, K. (2005a). Functionally referential communication in a chimpanzee. *Current Biology*, **15**(19), 1779–1784.

Slocombe, K. E. and Zuberbühler, K. (2005b). Agonistic screams in wild chimpanzees (*Pan troglodytes schweinfurthii*) vary as a function of social role. *Journal of Comparative Psychology*, **119**(1), 67–77.

Slocombe, K. E. and Zuberbühler, K. (2006). Food-associated calls in chimpanzees: responses to food types or food preferences? *Animal Behaviour*, **72**(5), 989–999.

Slocombe, K. E. and Zuberbühler, K. (2007). Chimpanzees modify recruitment screams as a function of audience composition. *Proceedings of the National Academy of Sciences*, **104**(43), 17228–17233.

Slocombe, K. E. and Zuberbühler, K. (2010). Vocal communication in chimpanzees. In E. Lonsdorf, S. Ross and T. Matsuzawa (eds), *The Mind of the Chimpanzee Ecological and Empirical Perspectives*. Chicago, IL: University of Chicago Press, pp. 192–207.

Slocombe, K. E, Kaller, T., Call, J. and Zuberbühler, K. (in prep). Chimpanzee food-calls refer to specific food items in captivity: Evidence from a playback study.

Slocombe, K. E., Kaller, T., Call, J. and Zuberbühler, K. (2010a). Chimpanzees extract social information from agonistic screams. *PLoS One*, **5**(7), e11473.

Slocombe, K. E., Kaller, T., Turman, L., Townsend, S. W., Papworth, S., Squibbs, P. and Zuberbühler, K. (2010b). Production of food-associated calls in wild male chimpanzees is dependent on the composition of the audience. *Behavioral Ecology and Sociobiology*, **64**(12), 1959–1966.

Slocombe, K. E., Townsend, S. W. and Zuberbühler, K. (2009). Wild chimpanzees (*Pan troglodytes schweinfurthii*) distinguish between different scream types: evidence from a playback study. *Animal Cognition*, **12**(3), 441–449.

Slocombe, K. E., Waller, B. M. and Liebal, K. (2011). The language void: the need for multimodality in primate communication research. *Animal Behaviour*, **81**(5), 919–924.

Smear, M., Shusterman, R., O'Connor, R., Bozza, T. and Rinberg, D. (2011). Perception of sniff phase in mouse olfaction. *Nature*, **479**(7373), 397–400.

Smith, C. L. and Evans, C. S. (2008). Multimodal signaling in fowl, *Gallus gallus*. *Journal of Experimental Biology*, **211**(13), 2052–2057.

Smith, C. L. and Evans, C. S. (2009). Silent tidbitting in male fowl, *Gallus gallus*: A referential visual signal with multiple functions. *Journal of Experimental Biology*, **212**(6), 835–842.

Smith, C. L., Taylor, A. and Evans, C. S. (2011). Tactical multimodal signalling in birds: facultative variation in signal modality reveals sensitivity to social costs. *Animal Behaviour*, **82**(3), 521–527.

Smith, T. (2006). Individual olfactory signatures in common marmosets (*Callithrix jacchus*). *American Journal of Primatology*, **68**(6), 585–604.

Smith, T. D. and Bhatnagar, K. P. (2004). Microsmatic primates: Reconsidering how and when size matters. *The Anatomical Record Part B: The New Anatomist*, **279B**(1), 24–31.

Smith, T. D. and Rossie, J. B. (2006). Primate olfaction: anatomy and evolution. In W. J. Brewer, D. Castle and C.Pantelis (eds), *Olfaction and the Brain*. Cambridge: Cambridge University Press, pp. 135–166.

Smith, T. D., Dennis, J. C., Bhatnagar, K. P. *et al.* (2011). Olfactory marker protein expression in the vomeronasal neuroepithelium of tamarins (*Saguinus* spp.). *Brain Research*, **1375**, 7–18.

Smith, T. D., Rossie, J. B. and Bhatnagar, K. P. (2007). Evolution of the nose and nasal skeleton in primates. *Evolutionary Anthropology: Issues, News, and Reviews*, **16**(4), 132–146.

Smith, T. D., Siegel, M. I. and Bhatnagar, K. P. (2001). Reappraisal of the vomeronasal system of catarrhine primates: ontogeny, morphology, functionality, and persisting questions. *The Anatomical Record*, **265**(4), 176–192.

Smith, W. J. (1977). *The Behavior of Communicating: An Ethological Approach*. Cambridge, MA: Harvard University Press.

Smith, W. J. (1981). Referents of animal communication. *Animal Behaviour*, **29**(4), 1273–1275.

Snowdon, C. T. and Elowson, A. M. (1999). Pygmy marmosets modify call structure when paired. *Ethology*, **105**(10), 893–908.

Spocter, M. A., Hopkins, W. D., Garrison, A. R. *et al.* (2010). Wernicke's area homologue in chimpanzees (*Pan troglodytes*) and its relation to the appearance of modern human language. *Proceedings of the Royal Society B: Biological Sciences*, **277**(1691), 2165–2174.

Stammbach, E. (1987). Desert, forest and montane baboons: Multi-level societies. In B. B. Smuts, D. L. Cheney, R. M. Seyfarth, R. W. Wrangham and T. T. Struhsaker (eds), *Primate Societies*. Chicago, IL: University of Chicago Press, pp. 112–120.

Stein, B. E. and Meredith, M. A. (1993). *The Merging of the Senses*. Cambridge: MIT Press.

Sterck, E. H. M., Watts, D. P. and van Schaik, C. P. (1997). The evolution of female social relationships in nonhuman primates. *Behavioral Ecology and Sociobiology*, **41**(5), 291–309.

Sterling, E. J. and Richard, A. F. (1995). Social organization in the aye-aye (*Daubentonia madagascarensis*) and the perceived distinctiveness of nocturnal primates. In L. Alterman, G. A. Doyle and M. K. Izard (eds), *Creatures of the Dark: The Nocturnal Prosimians*. New York: Plenum Press, pp. 439–451.

Stevens, J. M. G., Vervaecke, H., de Vries, H. and van Elsacker, L. (2005). Peering is not a formal indicator of subordination in bonobos (*Pan paniscus*). *American Journal of Primatology*, **65**(3), 255–267.

Stoddart, D. M. (1980). *The Ecology of Vertebrate Olfaction*. London: Chapman and Hall.

Strier, K. B. (1992). Atelinae adaptations: behavioral strategies and ecological constraints. *American Journal of Physical Anthropology*, **88**(4), 515–524.

Struhsaker, T. T. (1967). Auditory communication among vervet monkeys (*Cercopithecus aethiops*). In S. A. Altmann (ed.), *Social Communication Among Primates*. Chicago, IL: University of Chicago Press, pp. 281–324.

Struhsaker, T. T. (1975). *The Red Colobus Monkeys*. Chicago, IL: University of Chicago Press.

Struhsaker, T. T. and Leland, L. (1987). Colobines: infanticide by adult males. In B. B. Smuts, D. L. Cheney, R. M. Seyfarth, R. W. Wrangham and T. T. Struhsaker (eds), *Primate Societies*. Chicago, IL: University of Chicago Press, pp. 83–97.

Struhsaker, T. T. and Leland, L. (1988). Group fission in redtail monkeys (*Cercopithecus ascanius*) in the Kibale Forest, Uganda. In A. Gautier-Hion, F. Bourlière, J. P. Gautier and J. Kingdon (eds.), *A Primate Radiation: Evolutionary Biology of the African Guenons*. Cambridge: Cambridge University Press, pp. 364–388.

Stumpf, R. (2007). Chimpanzees and bonobos: diversity within and between species. In C. J. Campbell, A. Fuentes, K. C. MacKinnon, M. Panger and S. K. Bearder (eds), *Primates in Perspective*. New York: Oxford University Press, pp. 321–344.

Sugihara, T., Diltz, M. D., Averbeck, B. B. and Romanski, L. M. (2006). Integration of auditory and visual communication information in the primate ventrolateral prefrontal cortex. *The Journal of Neuroscience*, **26**(43), 11138–11147.

Surridge, A. K., Osorio, D. and Mundy, N. I. (2003). Evolution and selection of trichromatic vision in primates. *Trends in Ecology and Evolution*, **18**(4), 198–205.

Susman, R. L. (ed.). (1984). *The Pygmy Chimpanzee: Evolutionary Biology and Behavior*. New York: Plenum Press.

Sussman, R. W. (1999). *Primate Ecology and Social Structure. Volume 1: Lorises, Lemurs and Tarsiers*. Needham Heights, MA: Pearson Custom Publishing.

Sussman, R. W. and Kinzey, W. G. (1984). The ecological role of the Callitrichidae: a review. *American Journal of Physical Anthropology*, **64**(4), 419–449.

Sutton, D., Larson, C., Taylor, E. M. and Lindeman, R. C. (1973). Vocalization in rhesus monkeys: conditionability. *Brain Research*, **52**, 225–231.

Swaney, W. T. and Keverne, E. B. (2009). The evolution of pheromonal communication. *Behavioural Brain Research*, **200**(2), 239–247.

Taglialatela, J. P. (2007). Functional and structural asymmetries for auditory perception and vocal production in nonhuman primates. *Special Topics in Primatology*, **5**, 120–145.

Taglialatela, J. P., Cantalupo, C. and Hopkins, W. D. (2006). Gesture handedness predicts asymmetry in the chimpanzee inferior frontal gyrus. *Neuroreport*, **17**(9), 923–927.

Taglialatela, J. P., Reamer, L., Schapiro, S. J. and Hopkins, W. D. (2012). Social learning of a communicative signal in captive chimpanzees. *Biology Letters*, **8**(4), 498–501.

Taglialatela, J. P., Russell, J. L., Schaeffer, J. A. and Hopkins, W. D. (2008). Communicative signaling activates 'Broca's' homologue in chimpanzees. *Current Biology*, **18**(5), 343–348.

Taglialatela, J. P., Russell, J. L., Schaeffer, J. A. and Hopkins, W. D. (2009). Visualizing vocal perception in the chimpanzee brain. *Cerebral Cortex*, **19**(5), 1151–1157.

Taglialatela, J. P., Russell, J. L., Schaeffer, J. A. and Hopkins, W. D. (2011). Chimpanzee vocal signaling points to a multimodal origin of human language. *PLoS One*, **6**(4), e18852.

Taglialatela, J. P., Savage-Rumbaugh, S. and Baker, L. A. (2003). Vocal production by a language-competent *Pan paniscus*. *International Journal of Primatology*, **24**(1), 1–17.

Tanner, J. E. and Byrne, R. W. (1993). Concealing facial evidence of mood: perspective-taking in a captive gorilla. *Primates*, **34**(4), 451 457.

Tanner, J. E. (1998). *Gestural Communication in a Group of Zoo-living Lowland Gorillas*. Unpublished Dissertation, University of St Andrews.

Tanner, J. E. (2004). Gestural phrases and gestural exchanges by a pair of zoo-living lowland gorillas. *Gesture*, **4**(1), 1–24.

Tardif, S. D., Harrison, M. L. and Simek, M. A. (1993). Communal infant care in marmosets and tamarins: relation to energetics, ecology, and social organization. In A. B. Rylands (ed.), *Marmosets and Tamarins: Systematics, Behaviour, and Ecology.* Oxford: Oxford University Press, pp. 220–234.

Tardif, S. D., Santos, C. V., Baker, A. J., Van Elsacker, L., Feistner, A. T. C., Kleiman, D. G., . . . and De Vleeschouwer, K. (2002). Infant care in lion tamarins. In D. G. Kleiman and A. B. Rylands (eds), *Lion Tamarins: Biology and Conservation.* Washington DC: Smithsonian Institution Press, pp. 213–232.

Tardif, S. D., Smucny, D. A., Abbott, D. H., Mansfield, K., Schultz-Darken, N. and Yamamoto, M. E. (2003). Reproduction in captive common marmosets (*Callithrix jacchus*). *Comparative Medicine*, **53**(4), 364–368.

Taylor, R. C., Klein, B. A., Stein, J. and Ryan, M. J. (2008). Faux frogs: multimodal signalling and the value of robotics in animal behaviour. *Animal Behaviour*, **76**(3), 1089–1097.

Taylor, R. C., Klein, B. A., Stein, J. and Ryan, M. J. (2011). Multimodal signal variation in space and time: how important is matching a signal with its signaler? *Journal of Experimental Biology*, **214**(5), 815–820.

Tempelmann, S. and Liebal, K. (2012). Spontaneous use of gesture sequences in orangutans: A case for strategy? In S. Pika and K. Liebal (eds), *Recent Developments in Primate Gesture Research.* Herndon VA: John Benjamins Publishing Company, pp. 73–91.

Tempelmann, S., Kaminski, J. and Liebal, K. (2011). Focus on the essential: all great apes know when others are being attentive. *Animal Cognition*, **14**(3), 433–439.

Tempelmann, S., Kaminski, J. and Liebal, K. (2013). When apes point the finger – Three great apes species fail to use a conspecific's imperative pointing gesture. *Interaction Studies*, **14**(1), 7–23.

Terborgh, J. and Goldizen, A. W. (1985). On the mating system of the cooperatively breeding saddle-backed tamarin (*Saguinus fuscicollis*). *Behavioral Ecology and Sociobiology*, **16**(4), 293–299.

Terrace, H. S. (1979). *Nim: A Chimpanzee Who Learned Sign Language.* New York: Columbia Press.

Terrace, H. S., Petitto, L. A., Sanders, R. J. and Bever, T. G. (1979). Can an ape create a sentence. *Science*, **206**(4421), 891–902.

Teufel, C., Gutmann, A., Pirow, R. and Fischer, J. (2010). Facial expressions modulate the ontogenetic trajectory of gaze-following among monkeys. *Developmental Science*, **13**(6), 913–992.

Theall, L. A. and Povinelli, D. J. (1999). Do chimpanzees tailor their gestural signals to fit the attentional states of others? *Animal Cognition*, **2**(4), 207–214.

Thierry, B. (2007). Unity in diversity: lessons from macaque societies. *Evolutionary Anthropology*, **16**(6), 224–238.

Thierry, B. (2008). Primate socioecology, the lost dream of ecological determinism. *Evolutionary Anthropology*, **17**(2), 93–96.

Thierry, B., Demaria, C., Preuschoft, S. and Desportes, C. (1989). Structural convergence between silent bared-teeth display and relaxed open-mouth display in the Tonkean macaque (*Macaca tonkeana*). *Folia Primatologica*, **52**(3–4), 178–184.

Thierry, B., Iwaniuk, A. N. and Pellis, S. M. (2000). The influence of phylogeny on the social behaviour of macaques (Primates: Cercopithecidae, genus *Macaca*). *Ethology*, **106**(8), 713–728.

Tinbergen, N. (1952). 'Derived' activities; their causation, biological significance, origin, and emancipation during evolution. *The Quarterly Review of Biology*, **27**(1), 1–32.

Tinbergen, N. (1963). On aims and methods of ethology. *Zeitschrift für Tierpsychologie*, **20**(4), 410–433.

Tirindelli, R., Dibattista, M., Pifferi, S. and Menini, A. (2009). From pheromones to behavior. *Physiological Reviews*, **89**(3), 921–956.

Tomasello, M. (1996). Do apes ape? In C. Heyes and B. Galef (eds), *Social Learning in Animals: The Roots of Culture*. San Diego, CA: Academic Press, pp. 319–346.

Tomasello, M. (2008). *Origins of Human Communication*. Cambridge, MA: MIT Press.

Tomasello, M. and Call, J. (2004). The role of humans in the cognitive development of apes revisited. *Animal Cognition*, **7**(4), 213–215.

Tomasello, M. and Carpenter, M. (2007). Shared intentionality. *Developmental Science*, **10**(1), 121–125.

Tomasello, M. and Zuberbühler, K. (2002). Primate vocal and gestural communication. In M. Bekoff, C. S. Allen and G. Burghardt (eds), *The Cognitive Animal: Empirical and Theoretical Perspectives on Animal Cognition*. Cambridge: MIT Press, pp. 293–299.

Tomasello, M., Call, J., Nagell, K., Olguin, R. and Carpenter, M. (1994). The learning and use of gestural signals by young chimpanzees: a trans-generational study. *Primates*, **35**(2), 137–154.

Tomasello, M., Call, J., Warren, J. *et al.* (1997). The ontogeny of chimpanzee gestural signals: A comparison across groups and generations. *Evolution of Communication*, **1**(2), 223–259.

Tomasello, M., Carpenter, M., Call, J., Behne, T. and Moll, H. (2005). Understanding and sharing intentions: The origins of cultural cognition. *Behavioral and Brain Sciences*, **28**(5), 675–735.

Tomasello, M., Carpenter, M. and Liszkowski, U. (2007). A new look at infant pointing. *Child Development*, **78**(3), 705–722.

Tomasello, M., George, B. L., Kruger, A. C., Farrar, M. J. and Evans, E. (1985). The development of gestural communication in young chimpanzees. *Journal of Human Evolution*, **14**(2), 175–186.

Tomasello, M., Gust, D. and Frost, G. T. (1989). A longitudinal investigation of gestural communication in young chimpanzees. *Primates*, **30**(1), 35–50.

Tomonaga, M. (2010). Do the chimpanzee eyes have it? Social cognition on the basis of gaze and attention from the comparative–cognitive–developmental perspective. In E. V. Lornsdorf, S. R. Ross and T. Matsuzawa (eds), *The Mind of the Chimpanzee: Ecological and Empirical Perspectives*. Chicago, IL: University of Chicago Press, pp. 42–53.

Townsend, S. W. and Manser, M. B. (2013). Functionally referential communication in mammals: the past, present and the future. *Ethology*, **119**(1), 1–11.

Townsend, S. W. and Zuberbühler, K. (2009). Audience effects in chimpanzee copulation calls. *Communicative and Integrative Biology*, **2**(3), 282–284.

Townsend, S. W., Deschner, T. and Zuberbühler, K. (2008). Female chimpanzees use copulation calls flexibly to prevent social competition. *PLoS One*, **3**(6), e2431.

Townsend, S. W., Deschner, T. and Zuberbühler, K. (2011). Copulation calls in female chimpanzees (*Pan troglodytes schweinfurthii*) convey identity but do not accurately reflect fertility. *International Journal of Primatology*, **32**(4), 914–923.

Tsao, D. Y., Freiwald, W. A., Tootell, R. B. H. and Livingstone, M. S. (2006). A cortical region consisting entirely of face-selective cells. *Science*, **311**(5761), 670–674.

Tsao, D. Y., Moeller, S. and Freiwald, W. A. (2008). Comparing face patch systems in macaques and humans. *Proceedings of the National Academy of Sciences*, **105**(49), 19514–19519.

Uetz, G. W., Roberts, J. A. and Taylor, P. W. (2009). Multimodal communication and mate choice in wolf spiders: female response to multimodal versus unimodal signals. *Animal Behaviour*, **78**(2), 299–305.

Uhlenbroek, C. (1995). *The structure and function of the long-distance calls given by male chimpanzees in Gombe National Park*. Unpublished Ph.D. Thesis, University of Bristol.

Ungerleider, G. and Mishkin, M. (1982). Two cortical visual systems. In D. J. Ingle, M. A. Goodale and R. J. W. Mansfield (eds), *Analysis of Visual Behavior*. Cambridge, MA: MIT Press, pp. 549–586.

van Hooff, J. A. R. A. M. (1967). The facial displays of the catarrhine monkeys and apes. In D. Morris (ed.), *Primate Ethology*. London: Weidenfeld and Nicholson, pp. 7–68.

van Hooff, J. A. R. A. M. (1972). A comparative approach to the phylogeny of laughter and smiling. In R. A. Hinde (ed.), *Non-verbal Communication*. Cambridge: Cambridge University Press, pp. 209–237.

van Hooff, J. A. R. A. M. (1973). A structural analysis of the social behaviour of a semi-captive group of chimpanzees. In M. von Cranach and I. Vine (eds), *Social Communication and Movement: Studies of Interaction and Expression in Man and Chimpanzee*. London and New York: Academic Press, pp. 75–162.

van Lawick-Goodall, J. (1968). The behaviour of free-living chimpanzees in the Gombe Stream Reserve. *Animal Behaviour Monographs*, **1**(3), 163–311.

van Leeuwen, E. J. C., Cronin, K. A., Haun, D. B. M., Mundry, R. and Bodamer, M. D. (2012). Neighbouring chimpanzee communities show different preferences in social grooming behaviour. *Proceedings of the Royal Society B: Biological Sciences*, **279**(1746), 4362–4367.

van Roosmalen, M. G. M. and Klein, L. L. (1988). The spider monkeys, genus *Ateles*. In R. A. Mittermeier, A. B. Rylands, A. F. Coimbra-Filho and G. A. B. de Fonseca (eds), *Ecology and Behavior of Neotropical Primates*, Vol. **2**. Washington DC: World Wildlife Fund, pp. 455–537.

van Schaik, C. P. (1983). Why are diurnal primates living in groups? *Behaviour*, **87**(1/2), 120–144.

van Schaik, C. P. (1989). The ecology of social relationships amongst female primates. In V. Standen and R. A. Foley (eds), *Comparative Socioecology The Behavioral Ecology of Humans and Other Animals*. Oxford: Blackwell Scientific Publications, pp. 195–218.

van Schaik, C. P. (1999). The socioecology of fission–fusion sociality in orangutans. *Primates*, **40**(1), 69–86.

van Schaik, C. P. and van Hooff, J. (1983). On the ultimate causes of primate social systems. *Behaviour*, **85**(1–2), 91–117.

Veà, J. J. and Sabater-Pi, J. (1998). Spontaneous pointing behaviour in the wild pygmy chimpanzee (*Pan paniscus*). *Folia Primatologica*, **69**(5), 289–290.

Vick, S. J., Waller, B. M., Parr, L. A., Smith Pasqualini, M. C. and Bard, K. A. (2007). A cross-species comparison of facial morphology and movement in humans and chimpanzees using the Facial Action Coding System (FACS). *Journal of Nonverbal Behavior*, **31**(1), 1–20.

Visalberghi, E. and Fragaszy, D. M. (2002). 'Do monkeys ape?' Ten years after. In K. Dautenhahn and C. Nehaniv (eds), *Imitation in Animals and Artefacts*. Cambridge, MA: MIT Press, pp. 471–499.

Visalberghi, E., Valenzano, D. R. and Preuschoft, S. (2006). Facial displays in *Cebus apella*. *International Journal of Primatology*, **27**(6), 1689–1707.

von Frisch, K. (1974). Decoding the language of the bee. *Science*, **185**(4152), 663–668.

Waller, B. M. and Cherry, L. (2012). Facilitating play through communication: significance of teeth exposure in the gorilla play face. *American Journal of Primatology*, **74**(2), 157–164.

Waller, B. M. and Dunbar, R. I. M. (2005). Differential behavioural effects of silent bared teeth display and relaxed open mouth display in chimpanzees (*Pan troglodytes*). *Ethology*, **111**(2), 129–142.

Waller, B. M. and Micheletta, J. (2013). Facial expressions in nonhuman animals. *Emotion Review*, **5**(1), 54–59.

Waller, B. M., Bard, K. A., Vick, S.-J. and Smith Pasqualini, M. C. (2007). Perceived differences between chimpanzee (*Pan troglodytes*) and human (*Homo sapiens*) facial expressions are related to emotional interpretation. *Journal of Comparative Psychology*, **121**(4), 398–404.

Waller, B. M., Cray, J. and Burrows, A. M. (2008). Selection for universal emotion. *Emotion*, **8**(3), 435–439.

Waller, B. M., Lembeck, M., Kuchenbuch, P., Burrows, A. M. and Liebal, K. (2012). Gibbon-FACS: A muscle-based facial movement coding system for hylobatids. *International Journal of Primatology*, **33**(4), 809–821.

Waller, B. M., Liebal, K., Burrows, A. M. and Slocombe, K. E. (2013a). How can a multimodal approach to primate communication helps us to understand the evolution of communication? *Evolutionary Psychology*, **11**(3), 538–549.

Waller, B. M., Parr, L., Gothard, K., Burrows, A. M. and Fuglevand, A. (2008). Mapping the contribution of single muscles to facial movements in the rhesus macaque. *Physiology and Behavior*, **95**(1–2), 93–100.

Waller, B. M., Vick, S.-J., Parr, L. A. *et al.* (2006). Intramuscular electrical stimulation of facial muscles in humans and chimpanzees: Duchenne revisited and extended. *Emotion*, **6**(3), 367–382.

Waller, B. M., Warmelink, L., Liebal, K., Micheletta, J. and Slocombe, K. E. (2013b). Pseudoreplication: a widespread problem in primate communication research. *Animal Behaviour*, **86**(2), 483–488.

Wallez, C., Schaeffer, J., Meguerditchian, A. *et al.* (2012). Contrast of hemispheric lateralization for oro-facial movements between learned attention-getting sounds and species-typical vocalizations in chimpanzees: extension in a second colony. *Brain and Language*, **123**(1), 75–79.

Walls, G. L. (1942). *The Vertebrate Eye and its Adaptive Radiation*. Oxford: Cranbrook Institute of Science.

Wang, X. (2000). On cortical coding of vocal communication sounds in primates. *Proceedings of the National Academy of Sciences*, **97**(22), 11843–11849.

Waser, P. M. and Brown, C. H. (1986). Habitat acoustics and primate communication. *American Journal of Primatology*, **10**(2), 135–154.

Watzlawick, P., Bavelas, J. B. and Jackson, D. D. (1967). *Pragmatics of Human Communication: A Study of Interactional Patterns, Pathologies, and Paradoxes*: New York: Norton.

Wedekind, C. and Furi, S. (1997). Body odour preferences in men and women: do they aim for specific MHC combinations or simply heterozygosity? *Proceedings of the Royal Society B: Biological Sciences*, **264**(1387), 1471–1479.

Wedekind, C., Seebeck, T., Bettens, F. and Paepke, A. J. (1995). MHC-dependent mate preferences in humans. *Proceedings of the Royal Society B: Biological Sciences*, **260**(1359), 245–249.

Wheeler, B. C. (2010). Production and perception of situationally variable alarm calls in wild tufted capuchin monkeys (*Cebus apella nigritus*). *Behavioral Ecology and Sociobiology*, **64**(6), 989–1000.

Wheeler, B. C. and Fischer, J. (2012). Functionally referential signals: A promising paradigm whose time has passed. *Evolutionary Anthropology*, **21**(5), 195–205.

Whinnett, A. and Mundy, N. I. (2003). Isolation of novel olfactory receptor genes in marmosets (*Callithrix*): insights into pseudogene formation and evidence for functional degeneracy in non-human primates. *Gene*, **304**, 87–96.

Whiten, A., Goodall, J., McGrew, W. C. *et al.* (1999). Cultures in chimpanzees. *Nature*, **399**(6737), 682–685.

Whitham, J. C., Gerald, M. S. and Maestripieri, D. (2007). Intended receivers and functional significance of grunt and girney vocalizations in free-ranging female rhesus macaques. *Ethology*, **113**(9), 862–874.

Wich, S. A. and de Vries, H. (2006). Male monkeys remember which group members have given alarm calls. *Proceedings of the Royal Society of London. Series B: Biological Sciences*, **273**(1587), 735–740.

Wich, S. A. and Sterck, E. H. (2003). Possible audience effect in Thomas langurs (Primates; *Presbytis thomasi*): an experimental study on male loud calls in response to a tiger model. *American Journal of Primatology*, **60**(4), 155–159.

Wich, S. A., Krützen, M., Lameira, A. R. *et al.* (2012). Call cultures in orang-utans? *PLoS One*, **7** (5), e36180.

Wich, S. A., Swartz, K. B., Hardus, M. E. *et al.* (2009). A case of spontaneous acquisition of a human sound by an orangutan. *Primates*, **50**(1), 56–64.

Wilcox, S. (1999). The invention and ritualization of language. In B. J. King (ed.), *The Origins of Language: What Nonhuman Primates Can Tell Us*, Vol. **351**. Santa Fe, NM: School of American Research Press, pp. 351 384.

Wilson, M. L. and Wrangham, R. W. (2003). Intergroup relations in chimpanzees. *Annual Review of Anthropology*, **32**, 363–392.

Winter, P., Handley, P., Ploog, D. and Schott, D. (1973). Ontogeny of squirrel monkey calls under normal conditions and under acoustic isolation. *Behaviour*, **47**(3/4), 230–239.

Wollberg, Z. and Newman, J. D. (1972). Auditory cortex of squirrel monkey: response patterns of single cells to species-specific vocalizations. *Science*, **175**(4018), 212–214.

Woo, K. L. and Rieucau, G. (2011). From dummies to animations: a review of computer-animated stimuli used in animal behavior studies. *Behavioral Ecology and Sociobiology*, **65**(9), 1671–1685.

Woodruff, G. and Premack, D. (1979). Intentional communication in the chimpanzee: The development of deception. *Cognition*, **7**(4), 333–362.

Wrangham, R. W. (1980). An ecological model of female-bonded primate groups. *Behaviour*, **75**(3/4), 262–300.

Wrangham, R. W. (1987). Evolution of social structure. In B. B. Smuts, D. L. Cheney, R. M. Seyfarth, R. W. Wrangham and T. T. Struhsaker (eds), *Primate Societies*. Chicago, IL: University of Chicago Press, pp. 282–296.

Yahr, P. and Stephens, D. R. (1987). Hormonal control of sexual and scent marking behaviors of male gerbils in relation to the sexually dimorphic area of the hypothalamus. *Hormones and Behavior*, **21**(3), 331–346.

Yarger, R. G., Smith, A. B., Preti, G. and Epple, G. (1977). The major volatile constituents of the scent mark of a South American primate *Saguinus fuscicollis*, Callitrichidae. *Journal of Chemical Ecology*, **3**(1), 45–56.

Young, A. W., Hellawell, D. and Hay, D. C. (1987). Configurational information in face perception. *Perception*, **16**(6), 747–759.

Young, J. Z. (1957). *The Life of Mammals*. Oxford: Oxford University Press.

Young, J. Z. (1962). *The Life of Vertebrates*. New York: Oxford University Press.

Young, M. P. and Yamane, S. (1992). Sparse population coding of faces in the inferotemporal cortex. *Science*, **256**(5061), 1327–1331.

Zald, D. H. and Pardo, J. V. (1997). Emotion, olfaction, and the human amygdala: amygdala activation during aversive olfactory stimulation. *Proceedings of the National Academy of Sciences*, **94**(8), 4119–4124.

Zatorre, R. J. and Belin, P. (2001). Spectral and temporal processing in human auditory cortex. *Cerebral Cortex*, **11**(10), 946–953.

Zatorre, R. J., Belin, P. and Penhune, V. B. (2002). Structure and function of auditory cortex: Music and speech. *Trends in Cognitive Sciences*, **6**(1), 37–46.

Ziegler, T. E., Snowdon, C. T. and Uno, H. (1990). Social interactions and determinants of ovulation on tamarins (Saguinus). In T. Ziegler and F. Berkovitch (eds), *Socioendocrinology of Primate Reproduction*. New York: Wiley Liss, pp. 113–133.

Zimmermann, F., Zemke, F., Call, J. and Gómez, J. C. (2009). Orangutans (*Pongo pygmaeus*) and bonobos (*Pan paniscus*) point to inform a human about the location of a tool. *Animal Cognition*, **12**(2), 347–358.

Zuberbühler, K. (2000a). Interspecies semantic communication in two forest primates. *Proceedings of the Royal Society of London. Series B: Biological Sciences*, **267**(1444), 713–718.

Zuberbühler, K. (2000b). Referential labelling in Diana monkeys. *Animal Behaviour*, **59**(5), 917–927.

Zuberbühler, K. (2001). Predator-specific alarm calls in Campbell's monkeys, *Cercopithecus campbelli*. *Behavioral Ecology and Sociobiology*, **50**(5), 414–422.

Zuberbühler, K. (2002). A syntactic rule in forest monkey communication. *Animal Behaviour*, **63**(2), 293–299.

Zuberbühler, K. (2005). The phylogenetic roots of language evidence from primate communication and cognition. *Current Directions in Psychological Science*, **14**(3), 126–130.

Zuberbühler, K., Cheney, D. L. and Seyfarth, R. M. (1999). Conceptual semantics in a nonhuman primate. *Journal of Comparative Psychology*, **113**(1), 33–42.

Zuberbühler, K., Noë, R. and Seyfarth, R. M. (1997). Diana monkey long-distance calls: messages for conspecifics and predators. *Animal Behaviour*, **53**(3), 589–604.

Species index

Allen's swamp monkey (*Allenopithecus*) **11**
angwantibo (*Arctocebus*) **11**
aye-aye (*Daubentonia*) **11**, 27, 40, 42

baboon (*Papio*) **11**, 17, 19, 86
    chacma baboon (*Papio ursinus*) **22**, 75, 94, 139,
      161
    hamadryas baboon (*Papio hamadryas*) 79–80,
      82–3
    olive baboon (*Papio anubis*) 81, 137, 151
    yellow baboon, *or* savanna baboon (*Papio*
      cynocephalus) 82, 209–10
bamboo lemur (*Hapalemur*) **11**, 15
Barbary macaque, *see* macaque
bearded saki (*Chiropotes*) **11**
black-and-white colobus (monkey), *see* colobus
    (monkey)
black-fronted titi monkey, *see* titi monkey
blue monkey, *see* guenon
bonobo (*Pan paniscus*) **11**, 17–19, 28, 79–80, 123,
    133, 136, **137**, 146–7, 155, 157, 163, 165,
    196–7, 199, 208, 210, 226, 228
Bornean orangutan, *see* orangutan
brown-headed cowbird (*Molothrus ater*) 117
bushbaby, *see* galago

Californian ground squirrel (*Spermophilus beecheyi*)
    121
Campbell's monkey, *see* guenon
capuchin (monkey) (*Cebus*) **11**, 17, 198
    tufted capuchin (monkey) (*Cebus apella*) 81, 91,
      133, 175, 178, 180, 205, 207
    white-faced capuchin (monkey) (*Cebus*
      capucinus) 206–7
chacma baboon, *see* baboon
chimpanzee (*Pan troglodytes*) **10**, **11**, 17–19, 28, **29**,
    33, 40, 58–61, **62**, 64–6, 68–9, 75, 79–82, 85,
    87–8, **88**, 90–1, 94, 96, 99, 123–5, 133, 135–7,
    137, 139–41, 143–51, 155–9, 162–5, 168, 170,
    172, 174–81, 184–6, 197–9, 201, 208–13, 219,
    226
colobus (monkey) (*Colobus*) 11, 17
    black-and-white colobus (monkey), *or* Guereza
      (*Colobus guereza* 17, 165, 205

western black-and-white colobus (monkey), *or*
    king colobus (*Colobus polykomos*) 205
common squirrel monkey, *see* squirrel monkey
Coqui frog (*Eleutherodactylus coqui*) 116
cotton-top tamarin, *see* tamarin
crab-eating macaque, *see* macaque
crested (black) macaque, *see* macaque
crested gibbon (*Nomascus*) **11**
crested mangabey (*Lophocebus*) **11**
croaking gourami (*Trichopsis vittata*) 132

De Brazza's monkey, *see* guenon
Diana monkey, *see* guenon
domestic dog (*Canis lupus familiaris*) 222
douc langur (*Pygathrix*) **11**
drill (*Mandrillus leucophaeus*) 137
dusky titi monkey, *see* titi monkey
dwarf gibbon (*Hylobates*) **11**, *or* gibbons
    (Hylobatidae)
    agile gibbon (*Hylobates agilis*) 164
    white-handed gibbon, *see also* lar gibbon
      (*Hylobates lar*) 79–80, 165
dwarf lemur (*Cheirogaleus*) **11**, 42

fork-crowned lemur (*Phaner*) **11**
fowl (*Gallus gallus*) 117, 120

galago, *or* bushbaby (*Galago*) **11**, 15, 40, 42, 52,
    73
    lesser galago, *see also* lesser bushbaby (*Galago*
      senegalensis) 38
    mohol bushbaby (*Galago moholi*) 27
gelada (*Theropithecus gelada*) **11**, 17, 137, 147
giant mouse lemur (*Mirza*) **11**
gibbons, *or* hylobatids (Hylobatidae) 15, 27, 39, 88,
    93, 159, 164–5, 180, 193
Goeldi's monkey (*Callimico goeldii*) **11**
golden lion tamarin (*Leontopithecus rosalia*) **11**, 207
gorilla (*Gorilla*) **11**
    mountain gorilla (*Gorilla beringei beringei*)
      17, 79
    western gorilla (*Gorilla gorilla*) **22**, **36**, 79–82,
      133, 137, **137**, 142–3, 149, 155, 163, 174, 178,
      181, 184, 186, 198, 227

greater galago (*Otolemur*) **11**
grey langur (*Semnopithecus*) **11**, 17
grey squirrel (*Sciurus carolinensis*) 121, 202
guenon (*Cercopithecus*) **11**, 17
  blue monkey (*Cercopithecus mitis*) 17
  Campbell's monkey (*Cercopithecus campbelli*)
    161, 166–7, 205
  De Brazza's monkey (*Cercopithecus neglectus*)
    138
  Diana monkey (*Cercopithecus diana*) 148, 161,
    166–7, 203–5
  putty-nosed monkey (*Cercopithecus nictitans*)
    166, 202, 206
  red tail monkey (*Cercopithecus ascanius*) 17
guereza, *see* colobus (monkey)

hamadryas baboon, *see* baboon
honeybee (*Apis mellifera*) 120
hoolock gibbon (*Hoolock*) **11**
howler monkey (*Alouatta*) **11**, 17, 33, 35, 93
  mantled howler monkey (*Alouatta palliata*) 123
human (*Homo sapiens*) 3–6, **11**, 33–4, **37**, 37–41,
    43–5, **44**, 47–50, **49**, 53–5, **57**, 61, 64–5, 67, 77,
    84, 87, 93, 99, 110–11, 113, 125–6, 143–4, 149,
    154, 162, 170, 174, 178, 181, 194–9, 223–5

indri (*Indri*) **11**, 15, 27, 93

Japanese macaque, *see* macaque
jumping spider (*Phidippus clarus*) 117

leaf monkey, *or* surili (*Presbytis*) **11**, 19
  Thomas langur (*Presbytis thomasi*) 174, 184
long-nosed monkey, *see* proboscis monkey
long-tailed macaque, *see* macaque

macaque (*Macaca*) **11**, 17, 19, 21, 25, 27, 52, 58, 86,
    100, 159, 167, 177
  Barbary macaque (*Macaca sylvanus*) **22**, **28**, **44**,
    79, 144, 210
  crested (black) macaque (*Macaca nigra*) 5, 7, **22**,
    122, 122
  Japanese macaque (*Macaca fuscata*) 97, 134–5
  long-tailed macaque, *or* crab-eating macaque
    (*Macaca fascicularis*) 134, 161, 210
  pigtail macaque (*Macaca nemestrina*) **44**, 57
  rhesus macaque (*Macaca mulatta*) 25, **51**, 54–5,
    57, 65, 67, 69, 73, 75, 88, 91–2, 97, 99, 116,
    122, 125, 134–6, 138, **141**, 141–2, 148, 161,
    175, 178, 198, 204, 207–8, 211–12
  stump-tailed macaque (*Macaca arctoides*) 25, 176
  toque macaque (*Macaca sinica*) 207
mandrill (*Mandrillus sphinx*) 5, **11**, 137, **145**, 145
marmoset (*Callithrix*) 10, **11**, 15, 33, 36, **44**
  pygmy marmoset (*Callithrix (Cebuella) pygmaea*)
    11, 146
  white-headed marmoset (*Callithrix geoffroyi*) 207

meerkat (*Suricata suricatta*) 222
Mohol bushbaby, *see* galago
mouse lemur (*Microcebus*) **11**, 15, 35, 42, 52
  brown mouse lemur (*Microcebus rufus*) 27
  lesser mouse lemur (*Microcebus murinus*) 52
moustached tamarin, *see* tamarin
muriqui, *or* woolly spider monkey (*Brachyteles*)
    11

needle-clawed bushbaby (*Euoticus*) **11**

olive baboon, *see* baboon
olive colobus (*Procolobus*) **11**
orangutan (*Pongo*) **11**, 88, 133, 137, 141, 144–5,
    149, 159, 174, 180–1, 186, 197, 206, 227
  Bornean orangutan (*Pongo pygmaeus*) 15, 81,
    135, 148
  Sumatran orangutan (*Pongo abelii*) 15, 80–1, **137**,
    155, 163, 184, 196–9
owl monkey (*Aotus*) 11, 15, 42, 52

patas monkey (*Erythrocebus patas*) **11**, 17
pigtail macaque, *see* macaque
potto (*Perodicticus*) **11**
proboscis monkey, *or* long-nosed monkey (*Nasalis
    larvatus*) **11**, 39
putty-nosed monkey, *see* guenon
pygmy marmoset, *see* marmoset

red colobus (monkey) (*Piliocolobus badius*) **11**, 17,
    21
red tail monkey, *see* guenon
red-bellied tamarin, *see* tamarin
red-capped mangabey, *see* white-eyelid mangabey
red-fronted lemur (*Eulemur fulvus rufus*) 205
red-winged blackbird (*Agelaius phoeniceus*) 116
rhesus macaque, *see* macaque
ring-tailed lemur (*Lemur catta*) 5, **11**, 17, 19, **35**, 35,
    40, 75, **76**, 77, 121, 205
ruffed lemur (*Varecia*) **11**, 42, 206

saddle-back tamarin, *see* tamarin
saki monkey (*Pithecia*) **11**
satin bowerbird (*Ptilonorhynchus violaceus*) 116
savanna baboon, *see* baboon
siamang (*Symphalangus syndactylus*) **11**, 27, **38**, 39,
    79–80, **88**, 136, 155, 163
sifaka (*Propithecus*) 11, 17
  white sifaka (*Propithecus verreauxi*) 205
slender loris (*Loris*) **11**, 42
slow loris (*Nycticebus*) **11**
snub-nosed monkey (*Rhinopithecus*) **11**
spider monkey (*Ateles*) **11**, 17, 19, 27–8, 178
sportive lemur (*Lepilemur*) **11**
squirrel monkey (*Saimiri*) **11**, 65, 100, 134, 175
  common squirrel monkey (*Saimiri sciureus*) 35,
    69, 81, 180

stump-tailed macaque, *see* macaque
Sumatran orangutan, *see* orangutan
surili, *see* leaf monkey

talapoin (*Miopithecus*) **11**
tamarin (*Saguinus*) 10, **11**, 15, 33, 36, **44**
　cotton-top tamarin (*Saguinus oedipus*) **25**, 165,
　　174, 207
　moustached tamarin (*Saguinus mystax*) 205
　red-bellied tamarin (*Saguinus labiatus*) 207
　saddle-back tamarin (*Saguinus fuscicollis*)
　　205
tarsier (*Tarsius*) **11**, 42
telencephalon 46–8
Thomas langur, *see* leaf-monkey
titi monkey (*Callicebus*) **11**, 15
　black-fronted titi monkey (*Callicebus nigrifrons*)
　　205
　dusky titi monkey (*Callicebus moloch*) 165
toque macaque, *see* macaque
tree squirrel, *see* grey squirrel

true lemur (*Eulemur*) **11**
Túngara frog (*Engystomops pustulosus*) 119

uakari (*Cacajao*) **11**

vervet monkey (*Chlorocebus aethiops*) **11**, 98,
　132, 134–5, 161, 173, 175, 191, 201, 203–4

western black-and-white colobus (monkey),
　or king colobus, *see* colobus (monkey)
white Carneaux pigeon (*Columba livia*) 120
white sifaka, *see* sifaka
white-eyelid mangabey (*Cercocebus*) **11**, 180
　red-capped mangabey (*Cercocebus torquatus*)
　　147, 157
wolf spider (*Schizocosa ocreata*) 117, 119
woolly lemur (*Avahi*) **11**
woolly monkey (*Lagothrix*) **11**
woolly spider monkey, *see* muriqui

yellow baboon, *see* baboon

Subject index

accessory olfactory system (AOS) 32, 34, 50–2, 74–5, 77
acoustic
    analysis 93–7, 108, 151, 158, 210, 214
    structure 62, 92, 94–7, 134, 139, 146, 148–51, 159, 161, 167, 176, 204, 208–12
    variability 95–6, 138, 148, 157, 208–12
Acoustic Adaptation Hypothesis 151
acquisition 66, 80, 89, 97, **111**, **113**, 131–53, 226
action unit (AU) 87, **88**, 224–5
acuity
    olfactory acuity 77
    visual acuity 41, 43
adaptation 5, 8–9, **10**, 13, **41**, **86**, 87, 117, 151
    adaptive behaviour 85, 117
    adaptive function **7**, 8, 86, 212, 219, 225
    adaptive response 93, 109, 157, 160–1, 202–5
    preadaptation 38
Adobe Audition, *see software*
affective, *see emotional*
affiliation 148
    affiliative contact 27, 122–3, 185, 223
    affiliative context, *see context*
    affiliative interactions 13, 158, 198
    signal of affiliation **28**, 122, 137, 157
aggression **20**, **22**, 27, 32, 52, 99, 119–20, 176, 211, 224
    aggressive context, *see also* context – agonistic context
    aggressive defence 21
    aggressive inter-group encounters 18, 27, 205
    aggressive within-group encounters 17–18, 27, 176, 185
agonistic context, *see context*
air sac 39, 119–20
airborne odours, *see* odours, volatile odours
alarm barks, *see* vocalizations, types of
alarm calls, *see vocalizations*
alloparenting, 15, 225
American Sign Language 147, **226**
amplitude 92–3, 119
amygdala 47, 51–5, 92
anogenital glands, *see* glands

anterior cingulate gyrus 64–5
anticipatory feeding stance 207
AOS, *see* accessory olfactory system
ape 'language' projects 147, 226
    Chantek (orangutan) 197, **227**
    Gua (chimpanzee) **226**
    Kanzi (bonobo) 147, 226, **228**
    Koko (gorilla) **227**
    Loulis (chimpanzee) 226, **227**
    Nim Chimpsky (chimpanzee) **227**
    Sarah (chimpanzee) **226**
    Vicky (chimpanzee) **226**
    Washoe (chimpanzee) **226**
apomorphic 8
appeasement 27, 54, 74, 137, 157
approach face, *see* gestures, types of, muzzle to muzzle
arboreal 18, 24, 28, **41**, 108
arm-raise gesture, *see* gestures, types of
arousal 50, 163, **171**, 175–6, 185, 187–92, 205–6
articulators 63–4
attention 48, 50, 87, 98, 101, 161, 212, *see also* joint attention
    attention getter 44, **60**, 66, 68, 81, 119, 123, 140–1, 144, **173**, 180–3, **182**, **189**, 191, 193, **226**
    attentional state 78, 81–2, 123, 147, **173**, 179–81, 184, **189**, 191
    directing attention (of others) 178, 194–6, 213
    visual attention 42, 78, 121, 123, 176, 179, 181
AU, *see* action unit
audience
    audience effects 78, 173–7, 179, 188, 192, 197
    audience monitoring 177
    composition of audience 175–7, 191, 193
    target audience 92–3, 121, 178, 184, 192
audio-visual
    neuron **57**
    signal 67, 116, 122–3
audition 31–2, 36, 40, 46, 123, *see also* auditory, vocalization, sound *and* non-vocal sound
auditory, *see also* audition, sound, vocalizations *and* calls
    cortex 47–8, 62–3, 67–8, 161

gestures, *see* gestures
information 24, 37, 47, 53, 62–3, 67
nerve 40, 62
primary auditory cortex **49**
signals 24, 37, 45, 62, 67, 75, 119–23, 125, 162, 174, 202
thalamus 62
aversive stimulus 91, 134
Avisoft, *see* software
axon 51–2, 56

barks, *see* vocalizations, types of
basal ganglia **47**, 47
basilar membrane **39**, 40, 62
begging gestures, *see* requesting
behavioural ecology 5
belt (brain area) 62, 67
bimodal neurons 67
binocular vision, *see* vision
blended display 137, **164**, 164, 225
blind spot 43
blowing a kiss, *see* gestures, types of
boom calls, *see* vocalizations, types of
brachial (wrist) glands, *see* glands
brain imaging 58, 92, 100, 125–6
    functional magnetic resonance imaging (fMRI) 92, 100
    magnetic resonance imaging (MRI) 58, 65, 85
    positron emission tomography (PET) 58, 65, 68, 85, 100
brain size 13, 91–2, 218
brain h shaking gesture, *see* gestures, types of
Broca's area **49**, **57**, 61, 68, 125
Brodmann areas (22 39, 40, 44, 45) **49**

cackle calls, *see* vocalizations, types of
call structure, *see* acoustic structure
calling rate 176–7, 204
calls, *see* vocalizations
captive setting, *see* setting
cells
    glia cells 56
    hair cells 40, 62
    mitral cells 51–2
    photoreceptor cells 42–3, 52
    receptor cells 34, **50**, 50
central nervous system, *see* nervous system
central visual area 43
cerebellum **47**, 47
cerebral cortex **9**, 47–8
Chantek, *see* ape 'language' projects
chemical senses, *see* olfaction *and* taste
chemical signals 26, *see also* olfactory signals, odours *and* pheromones
chest beat gesture, *see* gestures, types of
chest-rubbing, *see* gestures, types of
chewing, *see* mouth actions

children 77, **113**, 172–3, 186, 196–8
chirps, *see* vocalizations, types of
chirrups, *see* vocalizations, types of
choroid 43
chorus, *see* song
chutters, *see* vocalizations, types of
ciliary muscles 43
cochlea **39**, 40, 62
coevolution **60**, 223, 225
cognition 47–50, 57, 77, 91, 99, 126, 131, 147, 154, 194
    cognitive mechanisms 73, 82, 109, 126, 152, 154, 199, 202, 214
    cognitive processes 3, **7**, 89, 131, 154, 170, **171**, 175, 179, 199, 213, 222
    cognitive skills x, 108, **110**, 111–12, 138, 147, 152, 160, 168, 188, 191, 223
colour vision 43, *see also* vision
    dichromatic 43
    monochromatic 43
    trichromatic 43
combination of signals 148–9, *see also* sequential signals, gesture sequence
    flexibility in, *see* flexibility in signal combination
    into multimodal signals 33, 66–8, 115, 117–18, 120, 143, 162
    into sequences 69, **113**, **115**, 155, 167–8
competition 18–19, 20–1, 35, 93
composite signals, *see* signals
conditioning 91, 97, 134, **226**, *see also* cooperative-conditioning paradigm
cones, *see* cells, photoreceptor cells
contact barks, *see* vocalizations, types of
context
    affiliative context 28, 93
    agonistic context 13, 28, 93, 99, 108, 132, 149, 157–9, 202, 209, 211–12, 214, 218
    evolutionarily urgent context 108, 118, 157–8, 202
    food context 93, 96, 135, 146, 159, 173, 202, 214
    functional context xi, 22, **111**, 133, 181, 187
    of predator encounters 108, 157, 202, 214
    of sexual behaviour 22, 28, 33, 75, 93, 123, 176, 202, 209, 214
    play context 28, 93, 108, 136, 149, 157, 163, 218
    submissive context **20**, **28**, 81, 83, 122, 133, 149, 176
context-independent signals 200, 214
context-specific signals 155–9, 167, 187, 200, 213–14
    facial expressions 157
    gestures 181
    multimodal signals 167, 213, 218
    vocalizations 158–9, 165–6, 168, 212
convergence
    of call structure 146
    of signals 151, 225
convergent evolution 222

coo calls, *see* vocalizations, types of
cooperative 223
　communication 5, 84
　contexts 225
　species 10, 15, 199, 223
cooperative-conditioning paradigm 134
copulation calls, *see* vocalizations, types of
core (brain area) 53, 62, 67
cornea 43
corpus callosum 47
courtship displays 117, 119–20
cranial nerve nuclei 46
cranial nerves 55
crook tail 136
cross-fostering 82, 97, 134–5
cross-modal integration, *see* multisensory integration
cross-sectional design 80, 83
crying 64–5, 93, 133
crypsis 27, 202
culture 141, 149
　call cultures 141, 144, 146
　cultural transmission 142–3, 145–6, 150, 226,
　　227, **228**
　human culture 146–7, 196
cytoskeleton 56

decision-making 52, 116
declarative 201, **228**, *see also* pointing, declarative
　pointing
defence 223
　against predators 21, 108, 157, 202, 214
　of food 18–19, 22, 27, 32, 35–6
　of mates 15, 17, 27, 35
　of social status 32, 211
　of territory 15, **16**, 27, 35–6
development 8–9, **10**, 32, 77, 80, 131–44, 152–3,
　163, 191, 194–5, 197, 219–20
　developmental mechanisms 76, 82–3, 89, 97
dichromatic, *see* colour vision
diencephalon 46–7, **47**
directed scratch, *see* gestures, types of
directed use (of signals) 78, 83, 116, 161–2, 172,
　176–7, 179, *see also* intentionality
discreteness 90, **110**, **114**, 144, 200
discriminant function analysis 96
displacement **110**, **226**, 226, **228**
diurnal species 19, 21, 32, 34, 36, 40, 42–4,
　see also nocturnal species
divergence of calls 148
domestication 199
dominance display 23, 136
dominance structure 17–18, 20–1, 25, 27, 99, 151,
　159, 211–12, 224–5
drumming, *see* gestures, types of
duality of patterning **110**, 162
duetting, *see* song
dyadic communication 163, 176–7, 195

ear
　external ear **39**, 40
　inner ear **39**, 40, 62
　middle ear **39**, 40, *see also* middle ear bones
　oval window 40
　tympanic membrane 40
eardrum, *see* ear, tympanic membrane
eavesdropping **4**, 93, 117, 120, 202
effort grunts, *see* vocalizations, types of
elaboration 78, **173**, 183, 185–7, 191–3
ELAN, *see* software
electrical stimulation 64, 87–8, 100
electrophysiological recording 77
emotional
　component **60**, **171**, 213
　emotionally driven **171**, 187, 225
　experience 7, 89
　information 213–14, 225
　process 7, **171**, 201
　signals 53, 89, 91, 109, 170, 201, 229
　state 55, 64, 86, 89, 157, 168, **171**, 185, 187, 191,
　　201, **226**, **227**
　system 229
empathy 86
encephalization 48
encephalon, *see* brain
enculturated apes 147, 197–8, *see also* ape
　'language' projects
entorhinal cortex 51
epiglottis 37, 45
ethogram 80, 86–7
evolution, of
　binocular vision **41**, 41
　human communication 3, 5, 26, 126, 219, 229
　language xi, 38, 45, **57**, 63, 68, 104, 109–15,
　　126–7, 172, 222, 229
　multimodal communication 116, 126, 217–18,
　　223–9
　primate communication 9, 23–30, 92–3, 139, 151,
　　158, 181, 202, 206, 217–29
　primate societies 9, 18–21
　the brain 48–9
evolutionarily urgent context, *see* context
exaggeration 148, 176–7
Excel, *see* software
experimental method 64, 77, 79, 81–3, 86, 90–1,
　98–9, 105–7, 118, 127, 132, 134, 174,
　178, 183
experimenter
　competitive 84, 177
　cooperative (helpful) 177, 198–9, *see also* setting,
　　cooperative setting
extant species 8, 13, 34, **41**, 111, 220
extended arm gesture, *see* gestures, types of
extended food grunt, *see* vocalizations, types of
extended grunt, *see* vocalizations, types of
external ear, *see* ear

eye 5, **41**, 42–3
  iris 43
  lens 43
  pupil 154
  retina 43–4, 50, 52–3
  sclera 5, **44**, 44–5, 178
eye contact 44–5, 84, 148, 177–8, *see also* gaze
eye covering gesture, *see* gestures, types of
eyetracking technique 178

F5 area **57**, 58
face representation area 55
face-selective neurons 54–5
facial (motor) nucleus 55, 64, 91–2
Facial Action Coding System (FACS) 78, 87, **88**, 159
facial area of the motor cortex, *see* motor cortex
facial display, *see* facial expressions
facial expressions 6, **7**, 27–8, 42, 47–8, 78, 85–92, 143
  perception of 53–5, **57**, 89–91
  processing of 67, 91–2
  production of 55–8, 64, 69, 86–7, 89
  structure of 87–9
facial expressions, types of
  horizontal bared teeth display 149
  lip protrusion **57**, 92
  lipsmacking **28**, 54, **57**, 86, 92, 122, 133, 141, 148, 177
  open mouth bared teeth display 149, **160**
  open-mouth silent bared-teeth display 133
  open mouth threat face 54, 133
  playface **7**, **29**, **61**, 85–6, 133, 137–8, **138**, 147, **160**, 174, 218, 225
  scalp-lifting 133
  silent bared-teeth display **61**, 133, 149, 157, **160**
  smile 87, 133, 223–5
  stretch pout whimper **164**, 164
  teeth chatter 57
  tense face 157
  tongue protrusion 92
  vertical bared teeth display 149
facial gestures, *see* facial expressions
facial landmarks 85, **88**
facial movements 78, 85, 87–9, 122, 125, 141, 143, 146
facial repertoire 25, 86, **113**, 218
FACS, *see* Facial Action Coding System
fear 87, 135, 157, 185, 201
feeding, *see also* context - food context
  feeding behaviour 15, 21, 80, 207–8
  feeding competition 19
feeling state, *see* emotional state
fission–fusion, *see* social organization
fitness **4**, 31, 158, 202, 222
flavour 52
flexibility, in
  combination of signals 168

receiver 160
  usage 63, 109, **111**, 143, **144**, 155–9, 167–8, 170, 175, 187, **189**
fMRI, *see* brain imaging
focal animal sampling 101
food, *see also* context - food context
  distribution 19, **20**
  quality 18, 31, 34, 68, 81, 123, **183**, 196, 206–7
  type 19, 146, 208
food sharing 101, 157, 159
food-associated calls, *see* vocalizations
food-begging paradigm 179, **183**, 186
forebrain **47**, 47
form (of signals) 75–6, 81–2, 87–9, 94–6
fovea 43, 53
frequency (of sound) 37, 40, 62, 92, **95**, 134, 148, *see also* amplitude, temporal pattern
  frequency range of hearing 40
  fundamental frequency 95, 148
  peak frequency 95
frontal lobe 48, 51
functional magnetic resonance imaging, *see* brain imaging
functional reference 109, 120, 194–214, *see also* referentiality, referent *and* pointing
  functionally referential vocalization 112, 158, 167, 195, 202–3, 206–12, 214, 225
funny faces 146
fussing 133

gas chromatography-mass spectrography (GC-MS) 75, **76**
gaze 148, 177–9, *see also* eye contact
  alternation **173**, 178–9, 191–2, 197
  aversion 177
  direction 178, 193, 202, 206
  gaze-following 5, 161, 195
GC-MS, *see* gas chromatography–mass spectrography
Generalized Linear Mixed Models 103
gestation **10**
gestural repertoire 80, 82–3, **113**, **136**, 142–3, 218
gesture sequence 162–3, 168, 183–5
gestures 77–85
  auditory gestures **36**, 78, 81, 133, 181–3
  manual gestures 42, 58, **60**, 66, 68, 73, 82, 84, **113**, 142, 151, 213
  perception of 83–4
  processing of 84–5
  production of 79–83
  speech-accompanying gestures 78, 82, **113**, **171**
  structural properties of 81–2
  tactile gestures 28, 78, 155, 182–3
  visual gestures 28, 53, 58–61, 78, 115, 123, 133, 151, 155, 180, 183, 191, *see also* visual signals

gestures, types of
  arm-raise 140
  blowing a kiss 79
  branch shaking 168
  chest beat **36**, 37, 78, 82
  chest-rubbing 28
  directed scratch 209
  drumming 165, 213
  extended arm 83, 157, 159
  eye covering gesture **145**
  ground slap 181
  hand-clasp 145, 150
  leaf-clipping 150, 181
  muzzle to muzzle **137**, 137
  offer food with arm 150
  poke at 181
  slap gesture 149, 218
  throw stuff 181
  throwback head gesture 136
  thumbs up 79
  waving goodbye 79
girney vocalization, *see* vocalizations, types of
glands
  anogenital glands 35, 74
  brachial (wrist) glands 35, 74
  pectoral glands 27, 74
  pituitary gland 47, 52
  scent glands 33–5, 74, 76, 121
glia cells, *see* cells
GLMM, *see* General Linear Mixed Models
glomerulus 51
goal-directed 135, 169–70, 179, 184–5,
    see also intentionality *and* directed use
    (of signals)
grasping 8, **57**, 58, 85, 141
greeting 28, 133, 139
gregarious species 13
grimace 122
groaning 64–5
grooming 5, **22**, 28, 122, 145, 150, 209, 223
ground slap, *see* gestures, types of
group identity, *see* identity
group size 13, 15–19, **20**, **25**, 25, 56, 152, 218, 223
group-living species, *see* social organization
gruffs, *see* vocalizations, types of
grunts, *see* vocalizations, types of
Gua, *see* ape "language" projects
gyrification 50

habituation–dishabituation 99, 161, 208
hacks, *see* vocalizations, types of
hair cells, *see* cells
hand-clasp, *see* gestures, types of
harmonic arches, *see* vocalizations, types of
hearing, *see* audition
hemimouth area **61**
hemisphere **47**, 47, **49**, 55, 58, 84–5

left hemisphere **48**, **49**, **58**, 61, 84
right hemisphere **49**, **60**, 61
hindbrain **47**
hippocampus 47
hok call, *see* vocalizations, types of
hok-oo call, *see* vocalizations, types of
homologues 55, 58, 69, 87, 112–13, 220, 223–5
  Broca's homologue **57**, 57, 68, 125
horizontal bared teeth display, *see* facial expressions,
    types of
hormones 15, 47, 52, 76
human infants, *see* children
huu alarm call, *see* vocalizations, types of
hypoglossal nucleus 46, 64

identity
  facial identity 53–4
  group identity 149
  individual identity 32, 61, 74, 87, 93, 148, 177,
    200, 210
  kin identity 32, 74
imitation **111**, 136, 140–2, 153, **226**
  neonatal imitation **141**, 142
imperative 196, 198, 201, 206, *see also* pointing -
    imperative pointing
incus, *see* middle ear bones
individual identity, *see* identity
inferior colliculus 47
inferior frontal cortex 68
inferior frontal gyrus **49**, **60**, 61, 66, 68
inferior temporal cortex 53
infinite productivity, *see* productivity
inflexible signals, *see* involuntary
innate **7**, 7, 64–5, 97, 111, 142, 150, 152–3,
    219, 222
inner ear, *see* ear
instrumental actions 78
insular cortex 51
intensity (of signals) **171**, 183, 186
intention 140, 177–80, 182–3, 194, 199, 201, 214
  intention movements 139, 142,
    see also ritualization
  intentional communication 78, 169–73, 197,
    see also signals - intentional signals
  marker of intentionality 179–80, 185, 187–91
intentionality 108, 169–93, 220, *see also* Theory of
    Mind
  first order intentionality 170
  second order intentionality 169–70
  shared intentionality **110**, *see also* joint attention
Interact, *see* software
interaction
  inter-group interaction 27, 205–6
  intra-group interaction 27
inter-group calls, *see* vocalizations
internal state, *see* emotional state
intra-group calls, *see* vocalizations

involuntary 43, 93, 112, **113**, 154, 159–61, 170, **171**, 181, 204
iris, *see* eye
isolation experiment 134, 142

jaws, *see* facial landmarks
joint attention **110**, 194–5, 197

Kanzi, *see* ape 'language' projects
kin identity, *see* identity
kin recognition 45, 74
kiss (squeak), *see* non-vocal sounds
Koko, *see* ape 'language' projects
krak call, *see* vocalizations, types of
krak-oo call, *see* vocalizations, types of

language 3, **4**, 97, **110**, 166, 169, 194, 197, 199, 220–2, 225–9
    artificial language systems 104, 222
    comprehension **49**, 84
    processing **49**, 58, **60**
    production **49**, **57**, 61, 84, **113**
    sign language 82, **113**, 147, **226**
language-trained apes, *see* ape 'language' projects
larynx 37–8, 45–6, **49**, 64–5, 86, 92, 145
lateral fusiform gyrus 53
lateral geniculate nucleus 52
lateral sulcus **49**, **63**
laterality **60**, 84
    lateralized processing **49**, 58, 84
    lateralized signal **58**, **60**, 84
laughter 64–5, 93, 133, 158, 220, 225
leaf-clipping, *see* gestures, types of
learning 47, **113**, 131, 140, 152, 191–2, 226–8
    motor learning 65
    social learning **10**, 111, 136, 140
    vocal learning 97, **114**, 144, 150
lens, *see* eye
lesioning 64, 92, 99–100
lexigrams 147, **228**
limbic system **47**, 47
linguistic abilities xi, **4**, 112, **113**, 220, 226
lips 28, 39, 63–4, 85, 196, **226**
lipsmacking, *see* facial expressions, types of
long-distance vocalizations, *see* vocalizations, types of, loud calls
long-term memory, *see* memory
longitudinal design 80, 83, 133
loud calls, *see* vocalizations, types of
Loulis, *see* ape 'language' projects

magnetic resonance imaging, *see* brain imaging
main olfactory system (MOS) 34–5, **50**, 50–2, 74–5, 77
major histocompatibility complex (MHC) **33**, 76
malleus, *see* middle ear bones
manual gestures, *see* gestures

marking, *see* scent marking
match-to-sample paradigm **90**, 90, 123, **124**, 125, 162
mating, *see* sexual behaviour
mating system 13–18, **16**, 24–30
    monogamy 13, 15, **16**, 27, 223
    polyandry **10**, 15, **16**
    polygamy 13, 15
    polygyny **16**, 17
    promiscuity 15, **16**
matriline 27, 211
McGurk effect 125
means-ends dissociation 83, 155, 187, 191,
        see also flexibility
mechanistic aspects of behaviour, *see* proximate level
medial geniculate complex 62
mediofrontal cortex 64
medulla oblongata 46, 64
memory 50–1
    long-term memory 47
    working memory 48
mental representation 160, 201, 204, 222
mental state 111, 169, *see also* Theory of Mind
mesencephalon 46–7
metencephalon 46, **47**
MHC, *see* major histocompatibility complex
microsmats 31, 45
midbrain **47**, 64
middle ear, *see* ear
middle ear bones 40
middle ear ossicles, *see* middle ear bones
mirror neurons 57, 58, 65, 92, **113**, 141–2
mitral cells, *see* cells
modality 23, **24**, 31, 40, 46, 50, 73, 78, 104–9, 112, 116, 125, 146, 148, 151, 158, 172–3, 176, 186, 192–4, 200–1, 209, 212–14, 217
model
    predator model 94, 165, 174–5, 177, 179, 185, 201, 205
    robotic model 118–21, 213
modification of signals 144, 146–53, 158, 166
monochromatic, *see* colour vision
monogamy, *see* mating system
monophyletic group 8
morphemes **110**
morphology 5, 8–14, 31–45, 73
MOS, *see* main olfactory system
motivational state 157, 200, 209, *see also* emotional state
motor cortex **49**, 64
    facial area of the motor cortex 55
    primary motor cortex 56, 64, 91
motor nuclei 64
mouth actions
    communicative mouth actions **57**, 92
    ingestive mouth actions **28**, 54–5, 137
mouth movements, *see* mouth actions

MRI *see* brain imaging, magnetic resonance imaging
multicomponent signals, *see* signals
multi-level societies **16**, 17
multimodality 6, **66**, 66, **67**, 104–27, **122**, 132, 163, 167–8, 186, 192–3, 202, 212–14, 217–18, 225–9
multiple message hypothesis 115
multisensory integration 57, 66–8, *see also* McGurk effect
muzzle to muzzle, *see* gestures, types of
myelencephalon 46, **47**

nasal epithelium, *see* olfactory neuroepithelium
nasal septum 34
neocortex **47**, 47–8, 52
neonatal imitation, *see* imitation
nervous system 46, **47**
    central nervous system 46, **47**, 50, 61
    peripheral nervous system 46, **47**
neuroendocrine system 32
neuroepithelium, *see* olfactory neuroepithelium
neurofilament protein, *see* cytoskeleton
neuroimaging, *see* brain imaging
neuronal ensembles 63
neuropil 56
Nim Chimpsky, *see* ape 'language' projects
nocturnal species 13, 15, 21, 24, 27, 32, 34, 36, 40–3, *see also* diurnal species
non-speech sounds 93
non-vocal sounds 66, 78, *see also* sound, vocalizations, calls
    kiss (squeak) 37, 66, 68
    lipsmacking 37, 92
    raspberries 66, 68, 92, 141, 144–6
    whistling 92, 145
novel signals **113**, 144–6, 166, 183
    novel vocalizations 144–5
noyau, *see* social organization
nucleus ambiguus 46, 64, **65**

object-choice 84, 147, 198
observational method 74–5, 81, 84, 87, 89–90, 93–4, 97–8, 106–7, 127, 132, 183, 198
Observer, *see* software
occipital lobe 48, 52
occipitotemporal cortex 53
odours 26, **33**, 34, 52, 77
    liquid-based odours, *see* pheromones
    non-volatile odours 52, *see also* pheromones
    odour profile, *see* signature
    volatile odours 32, 34, 51–2
oestrogen 74
oestrus **10**, 17, 75
offer food with arm, *see* gestures, types of
olfaction 6, 31–6, 47, 50–2, 73–7, 105, 121–2, 217, *see also* olfactory signals

olfactory
    bulb 34, 47–8, **50**, 51–2
    cortex 51
    nerves 33–4, **50**, 51, 77
    neuroepithelium 33–4, 50, 77
    repertoire 218
olfactory signals 10, 26–7, 31–3, 121–2, 160, 173, *see also* olfaction, odours *and* pheromones
    perception of 34–5, **50**, 77
    processing of 50–2, **51**, 77
    production of 35–6, 74–5, 77
    structure of 75–6
    on vocalization, *see* vocalizations, types of
one-to-one correspondence 154, 157, 167–8, 194, 208
ontogenetic ritualization, *see* ritualization
ontogeny, *see* development
open mouth bared teeth display, *see* facial expressions, types of
open-mouth silent bared-teeth display, *see* facial expressions, types of
open-mouth threat face, *see* facial expressions, types of
optic nerve 43, 52
optical convergence, *see* vision binocular vision
orbitofrontal cortex 51
orientation, *see also* attentional state
    of the body 180, 182
    of the face 180
orienting responses 98, **173**, 177–9, 212
origin of language, *see* evolution of language
orofacial
    movements 55–6, **57**, **60**, 64, 78
    representation 55–6, 64, 91
oval window, *see* ear
ovulation 9, **10**, 15, 33, 75

pair-living species, *see* social organization
palmtop computers (PDAs) 100–1
pant hoot, *see* vocalizations, types of
pant-grunt, *see* vocalizations, types of
pant-threat, *see* vocalizations, types of
parabelt 62
parasympathetic division **47**
parietal lobe 48, **49**
PDAs, *see* palmtop computers
pectoral glands, *see* glands
peep calls, *see* vocalizations, types of
peering, *see* gestures, types of, muzzle to muzzle
periaqueductal grey 64
peripheral nervous system, *see* nervous system
persistence 78, **173**, 183–7, **189**, 191, 193, 197
PET, *see* brain imaging
pharynx 37–8
pheromones **10**, 31–2, 52, 74–5, 77
    in urine 34–5, 45, 74
    primer pheromones 32
    releaser pheromones 32
    secretion of pheromones 32, 34, 74–5

philopatric 18, **20**
phonation 38, 64, 100
phoneme **110**, 144, 162
phonology 165
photoreceptor cells, *see* cells
phylogenetic inertia 222
phylogenetic ritualization, *see* ritualization
phylogeny 9–13, *see also* Tinbergen's four questions
piriform cortex 51
pituitary gland, *see* glands
play **29**, 83, 86, 137, 140, 158, 172, 181,
    see also context, play context
    feeling playful **7**, 86
playback experiments 94, 98–9, 107–9, 116–21,
    123, 132, 135, 150, 159, 164, 166–7, 176,
    203–14
playface, *see* facial expressions, types of
play-hit 140
pointing
    comprehension of pointing 198–9
    declarative pointing 195–8
    imperative pointing 195–8, 213
poke at, *see* gestures, types of
polyandry, *see* mating system
polygamy, *see* mating system
polygyny, *see* mating system
pons 47, 55, 64
pooling fallacy, *see* pseudoreplication
Positron Emission Tomography, *see* brain imaging
posture 27–8, 42, 78–9, 167, 213, **227**
Power Asymmetry Hypothesis of Motivational
    Emancipation 25, 159, 224
Praat, *see* software
predator detection 34, 117, 173, 202,
    see also defence, predator defence, alarm call,
    crypsis
preferential sniffing paradigm 77
prefrontal cortex 48, 50, 61
pre-linguistic communication 172–3
premotor cortex 53, 58, 64, 92
    ventral premotor cortex **57**
pre-post-event histograms 89
pre-supplementary motor area 64
primary motor cortex, *see* motor cortex
primary sensory cortex 62
primer pheromones, *see* pheromones
productivity 110, 226
    infinite productivity 111
promiscuity, *see* mating system
prosody **49**, **171**
proximate level 6, 8, 22, 24, 29, 170, 219, 222–3,
    see also Tinbergen's four questions
proximate mechanisms 76, 82, 89, 97, 193, 225
pseudoreplication 99, 101–3
pupil, *see* eye
pyow-hack sequence 167
pyows, *see* vocalizations, types of

radioactive tracer isotope 85
rapid-fire gesturing, *see* gesture sequences
raspberries, *see* non-vocal sounds
rate coding 63
Raven, *see* software
receiver psychology 24, 86, 93
receptor cells, *see* cells
referent 187, 194, 197, **226**, *see also* referential,
    referentiality *and* functional reference
referential
    gestures 83, 195, 206, 209, 214
    multimodal signals 120, 202, 212
    pointing 196–8
    vocalization, *see* functional reference
referentiality **110**, **113**, 140, 179, 194–214,
    see also functional reference, referent *and*
    pointing
relaxed open mouth face, *see* facial expressions,
    types of, playface
releaser pheromones, *see* pheromones
reliability 102
repertoire
    closed repertoire **113**, 144, 167–8
    communicative repertoire **25**, 25–6, 101, 136, 143,
        146, 172, 195, 218
    repertoire size **25**, 136, 140
    species-typical repertoire 143, 147, 152
requesting 58, 68, 81, 140, 143, 145–6, 157, 174,
    180–1, 183, 186, 195, 197, 226, *see also* food-
    begging paradigm
research environment, *see* setting
respiratory activity 64
response
    neuronal response 63, 67
    receiver response 77, 91, 98, 107, 118–19, 121–5,
        159–61, 187, 204, 211
response-waiting 163, 187, 191–2
reticular formation 64
retina, *see* eye
retinotopical organization 53
ritualization
    ontogenetic ritualization 139–40, 142–3, 153,
        see also intention movements
    phylogenetic ritualization 79, 139, 187
roars, *see* vocalizations, types of
rods, *see* cells, photoreceptor cells
rough grunts, *see* vocalizations, types of

saliva 35, 74
sampling error, *see* pseudoreplication
Sarah, *see* ape 'language' projects
scalp-lifting, *see* facial expressions, types of
scent glands, *see* glands
scent marking 32, **35**, 45, 52, 74
sclera, *see* eye
scream, *see* vocalizations, types of
secondary visual areas 52

secretion 15, 32, 34, 47, 74–5
selection pressure 8, 92, 157, 170, 217, 219,
    222, 229
self-control 48
semanticity 110, **147**, 165, 204, **226**, 226, **227**
semicircular canals 40
semi-gregarious species 24
sensory information 46–8, 50, **57**, 66–9
sequential structures **110**, **113**, **155**, 162,
    see also simultanious structures
setting
    captive setting 80, 86, 105–8, 127, 132, 136, 142,
        144–5, 152, 172, 184, 196–9, 205, 208, 211,
        214, 220
    competitive setting 84, 177, 199
    cooperative setting 84, 177
    laboratory setting 73, 75, 97, 133, 135–6, 147
    nursery setting 133, 141, 147
    wild setting 28, 79–80, 86, 105–8, 121, 127, 132,
        135–6, 142, 144, 152, 172, 184, 196, 212
severe screams, *see* vocalizations, types of
sexual behaviour 10, 22, 31–3, 36, 74–5, 101, 123,
    136, 181, 209–11, *see also* context of sexual
    behaviour
sexual maturity 10, 17–18, 132
Shannon's Theory of Information **4**
shrieks, *see* vocalizations, types of
signalling play 7, 86, 137
signals, *see also* context-specific *and* context-
    independent
    composite signals 117–18, **118**, 167, 213, 218
    conventionalized signals 79, 149
    idiosyncratic signals 66, 140, 146
    intentional signals 170, 172, 179, **189**
    multicomponent signals 116, **171**
    redundant signals 116, **118**, 118, 120, 163
signature 32, 37, 74–7, **76**, 150
silent bared-teeth display, *see* facial expressions,
    types of
simultaneous structures 82, 115, 126, 162–3,
    see also sequential structures
single-cell recording 85, 92
slap gesture, *see* gestures, types of
smile, *see* facial expressions, types of
sniff 28, 33, 77
social bonding 148–9, **171**, 223, 225
social cohesion 13, 19, 93, 223
social facilitation 175, 177, 191
social organization 13–15, 17, 21, 24, 30,
    133, 211
    fission–fusion **16**, 17, 19, 27, 212
    group-living species 13, 149
    noyau 15, **16**, 26
    pair-living species 13, 15, 26–7
    solitary species 13–15, 26
social referencing 135
social structure 13, 25–6, 159, 218

despotic **20**, 25, **160**
    egalitarian **20**, 25, **160**
social system 3, 10, 13–21, 24–30, 45, 219
Socioecological Model 18–21
soft grunt, *see* vocalizations, types of
soft palate 37, 45
software
    Adobe Audition 94
    Avisoft 94
    ELAN 101
    Excel 101
    Interact 101
    Observer 101
    Praat 94
    Raven 94
    SPSS 101
solitary species, *see* social organization
somatic (skeletal) nerves **47**
somatosensation 31, 47, 73
song 27, 93, 116–17, 148, 164–6
sound, 37, **49**, **57**, **60**, 66–7, **110**, 125, 161,
    see also non-vocal sounds, vocalizations, calls
    attention-getting sounds, *see* attention-getter
    novel sounds, *see* novel signals
    propagation of 92–3, 150–1
    sound recording, *see* acoustic analysis
speech-accompanying gestures, *see* gestures
spinal cord 46, **47**
spreadsheets 100–1
SPSS, *see* software
stapes, *see* middle ear bones
staring 122, 178
stereotypical displays 79, 86
sternal (chest) glands, *see* glands, pectoral glands
steroids 74
stimulus enhancement 209
stimulus-response behaviour 154–5
stink fights 35
stretch pout whimper, *see* facial expressions, types of
structural properties of gestures 78, 143, 159, 174,
    186, 193
subcortical areas 47, 61
submissive context, *see* context
suffixation 166
superior colliculus 47
superior temporal gyrus **49**
superior temporal plane 62, **63**
superior temporal sulcus 53–5, 58, 67, 161
supplementary motor area 64
supra-additivity 67, 125
swelling 5, **10**, **32**, 33, 75, 210–11
symbol **4**, **110**, 226
sympathetic division **47**
sympatric species 93, 166, 205
syntax **49**, **110**, 165, **228**
    lexical syntax 165
    phonological syntax 165

tactile **9**, 31, 33, 47, 66, 73, 104, 123, 133, 140, 173, 182, 191–3
tactile gestures, *see* gestures
tail flagging signals 121, 202
tantrum scream, *see* vocalizations, types of
tapetum lucidum 44
taste 31, 51, 73
teeth chatter, *see* facial expressions, types of
telencephalon **47**
temporal coding 63
temporal lobe 48, 53–4, 62
   anterior temporal lobe 53
   medial temporal lobe 51
   posterior temporal lobe 49
temporal pattern 92, 138
tense face, *see* facial expressions, types of
territorial display 27, 119
territorial marking, *see* scent marking
territorial vocalizations, *see* vocalizations
testosterone 74
thalamus 47
Theory of Mind **111**, 131, 169
threat call, *see* vocalizations, types of
threat face, *see* facial expressions, types of
throw stuff, *see* gestures, types of
throwback head gesture, *see* gestures, types of
thumbs up, *see* gestures, types of
tid-bitting display 117, 120
time sampling methods 101
time-frequency spectrogram **95**, 95
Tinbergen's four questions 6–7, *7*, 218–23
ToM, *see* Theory of Mind
tongue 38, 46, **49**, 63–4, *see also* facial expression, tongue protrusion
   tongue protrusion **141**, 141
tool
   tool making 225
   tool use 148, 197
touch screen **90**, 90, 107, 214
travelling 151, 159
triadic interaction, *see* referentiality
trichromatic, *see* color vision
trigeminal motor nucleus 64
trill calls, *see* vocalizations, types of
tympanic membrane, *see* ear

ultimate level 6–7, **7**, 21, 24, 29, 219–22, *see also* Tinbergen's four questions
unimodality 66, 105, 112–23, 126–7, 139, 143, 152–3, 163–4, 167–8, 172, 186, 192, 212, 217–18
urine-marking 33, 35, 45, 121
urine-washing 33, 74
uterus
   bicornuate uterus **10**
   simplex uterus **10**

vasopressin 52
ventrolateral prefrontal cortex 67
vertical bared teeth display, *see* facial expressions, types of
vestigial vomeronasal organ 34, 52, 77
Vicky, *see* ape 'language' projects
victim screams, *see* vocalizations, types of
violation of expectancy paradigm 99
vision 40–2, *see also* visual signals *and* colour vision
   binocular vision **9**, 41, 41–3, 45
   photopic vision 42
   scotopic vision 42
   stereoscopic vision **41**, 41, 45
visual cortex 53, 92
   extrastriate visual cortex 52
   primary visual cortex 52
visual field **41**, 41, 45, 53, 183
visual perspective taking 179, 191
visual signals **10**, 27–8, 33, 42, 108, 119, 121–6, 133, 159, 176, 180, **182**, 182, 191–3, 200, 202, 212–14, *see also* gestures *and* facial expressions
   perception of 43–4, 53 5, 57–8
   processing of 52–61
   production of 55–61
VNO, *see* vestigial vomeronasal organ
vocal cords, *see* vocal folds
vocal folds 37, 64, 66, 78, 86
vocal plasticity, *see* novel signals – novel vocalizations
vocal repertoire 24 **5, 26**, 96, 144, 140, 157, 218, 223
vocal tract **37**, 38, 45, 97
   supralaryngeal vocal tract 64
vocalizations 26, 66–9, 86, 92–100, 107–9, 112, 119, 121–5, 132–6, 160–1, 164–7, 173–6, 179, 181–2, 184–5, 200–14, *see also* auditory signals, sound, non-vocal sound
   perception of 40, 61–3, 97–9, 116, 139
   processing of 61–6, 99–100
   production of 37–40, 63–6, 93–4, 97, 138–9
   structure of 94–7, *see also* modification of signals *and* acoustic structure
vocalizations, types of
   alarm barks 139, *see also* vocalizations, alarm calls
   alarm calls 19, 27, 93–4, 98, 135, 148, 157, 161, 166–7, 173, 175, 179, 184–6, 191, 201–6, 213, 222
   barks 121–2, 158, 165, 208, 213
   boom calls 166
   cackle calls 134
   chirps 165, 207
   chirrups 165
   chutters 161

vocalizations, types of (cont.)
    contact barks 94, 139, *see also* vocalizations, contact calls
    coo calls 67, 125, 135, 138, 207
    copulation calls 176, 209–11
    effort grunts 133
    extended food grunt 66
    extended grunt 66, 68, 141, 145
    food-associated calls 68, 94, 96, 120, 175, 204, 206–9
    girney vocalization 122, 149
    gruffs 135
    grunts 151, 207–8
    hacks 167, *see under* calls - pyow, pyow-hack sequence
    harmonic arches 207
    hok call 166, *see also* hok-oo call, krak call, krak-oo call
    hok-oo call 166, *see also* hok call, krak call, krak-oo call
    huu alarm call 177
    inter-group calls 93
    intra-group calls 93
    krak call 166, *see also* krak-oo call, hok call, hok-oo call
    krak-oo call 166, *see also* krak-call, hok call, hok-oo call
    loud calls 27, 93, 108, 148, 150, 166, 178, 205
    on vocalization 138
    pant hoot 94, **124**, 144, 148, 150, 159, 162
    pant-grunt 139
    pant-threat 67
    peep calls 134, 147, 208
    pyows 167, *see also* pyow-hack sequence
    roars 93, 205
    rough grunts 146, 208
    scream 99, 122, 125, 162, 176, 211–12
    severe screams 212
    shrieks 205
    soft grunt **122**, 122
    tantrum screams 212
    territorial vocalizations 27, 36, 93, 164
    threat call 125
    trill calls 146
    victim screams 212
    waa bark 177
    warbles 207
    whistles 165
    wɪɪ vocalization 132, 161
    yaps 205
    yelps 208
voice region 62
voluntary control, *see* intentionality
vomeronasal organ, *see* vestigial vomeronasal organ

waa bark, *see* vocalizations, types of
waggle dance 120
warbles, *see* vocalizations, types of
Washoe, *see* ape 'language' projects
waveform (sound visualization) **95**, 95
waving goodbye, *see* gestures, types of
Wernicke's area **49**
whistles, *see* vocalizations, types of
whistling, *see* non-vocal sounds
wild setting, *see* setting
words 77, **110**, **113**, 162, 194–5, 199, **226**, **227**, **228**
working memory, *see* memory
wrr vocalization, *see* vocalizations, types of

yaps, *see* vocalizations, types of
yawning 86
yelps, *see* vocalizations, types of